The Endangered Earth

READINGS FOR WRITERS

Sarah Morgan
Park College

Dennis Okerstrom
Park College

ALLYN AND BACON
Boston   London   Toronto   Sydney   Tokyo   Singapore

Executive Editor: Joseph Opiela
Senior Editorial Assistant: Amy Capute
Editorial-Production Administrator: Rowena Dores
Editorial-Production Service: Editorial Inc.
Text Designer: Pat Torelli
Cover Administrator: Linda Dickinson
Composition Buyer: Linda Cox
Manufacturing Buyer: Megan Cochran

Copyright © 1992 by Allyn and Bacon
A Division of Simon & Schuster, Inc.
160 Gould Street
Needham Heights, Massachusetts 02194

All rights reserved. No part of the material protected by this copyright notice may be reproduced or utilized in any form or by any means, electronic or mechanical, including photocopying, recording, or by any information storage and retrieval system, without written permission from the copyright owner.

Library of Congress Cataloging-in-Publication Data

Morgan, Sarah, (date)
    The endangered earth : readings for wrtiers / Sarah Morgan, Dennis Okerstrom.
    p. cm.
    ISBN 0-205-13218-9
    1. Environmental protection. 2. Human ecology. 3. Ecology.
I. Okerstrom, Dennis. II. Title.
TD170.3.MB7 1992
363.7—dc20   91-25648
                                                                                                 CIP

This book is printed on recycled acid-free paper.

Acknowledgments
Chapter 1
Page 8. "Soil." From *The Living Landscape*, by Paul B. Sears. © 1962, 1966 by Paul B. Sears. Reprinted by permission of Basic Books, Inc., Publishers, New York.
Page 17. "Vast Green Seas Shrivel to Desert," by Fiona Sunquist. Copyright 1990 by the National Wildlife Federation. Reprinted from the March/April 1990 issue of *International Wildlife*.

Credits continued on page 474, which constitutes an extension of the copyright page.

Printed in the United States of America

10 9 8 7 6 5 4 3 2 1   96 95 94 93 92 91

Table of Contents

Preface xv

Introduction 1

1 Using the Land 5

PAUL B. SEARS, **Soil** 8

"*The generations of plants and animals that form the soil and are sustained by it during their brief existence pass on. What remains, as evidence of their activity in catching and transforming the sun's energy, is the organized soil. It is vulnerable to our blundering, immensely rewarding to good husbandry.*"

FIONA SUNQUIST, **Vast Green Seas Shrivel to Desert** 17

"*Fortunately, desertification need not be irreversible.*"

SANDRA POSTEL, **Halting Land Degradation: Abuse It and Lose It** 22

"*The four principal causes of land degradation—overgrazing on rangelands, overcultivation of croplands, waterlogging and salinization of irrigated lands, and deforestation—all stem from excessive human pressures or poor management of the land.*"

JOHN MADSON, **A Wilderness of Light** 29

"*Today the long-grass prairie is nearly gone, although some scraps of it linger on a few farms, along old railroads, and in neglected country cemeteries.*"

ANNE W. SIMON, **The Thin Edge** 33

"*The coast keeps us from drowning, maintaining the present global balance of one-third land, two-thirds water.*"

THOMAS A. LEWIS, *The Frontier Dream We Call Alaska*   41
"How much of Alaska's wealth can we go after without destroying Alaska?"

GRETEL EHRLICH, *The Solace of Open Spaces*   48
"There is no wildnerness left; wildness, yes, but true wilderness has been gone on this continent since the time of Lewis and Clark's overland journey."

2   Species and Endangerment   60

JARED DIAMOND, *Playing Dice with Megadeath*   63
"Everywhere humans have gone, they have wiped out whole species of birds, mammals, reptiles, fish, and other forms of life."

JAKE PAGE, *The Owls of Night*   73
"Must we be responsible for everything? Can't we distinguish?"

RUDY ABRAMSON, *Sharks Under Attack*   82
"But as humans have added shark's fin soup and shark filets to more and more menus, the predators increasingly have become the prey."

MICHAEL MILSTEIN, *Unlikely Harbingers*   86
"In what could be a potent omen of human-wrought environmental damage, lakes, ponds, and streams once melodic with frogs' deep, intimate mating calls have now fallen deadly quiet."

MARGARET KNOX, *Africa Daze Montana Knights*   94
"As the globe shrinks and human populations explode, conflicts over the last unpolluted, unpaved, and untilled patches of soil are becoming ever more desperate and violent."

LEWIS THOMAS, *Natural Man*   105
"Perhaps we are the invaded ones, the subjugated, used."

3   Controlling the Rivers   111

LOREN EISELEY, *The Flow of the River*   114
"If there is magic on this planet, it is contained in water."

CONGER BEASLEY, JR., **The Return of Beaver to the Missouri River** 122

"*Gradually, the Corps has exerted more and more control over the river, reducing it in size to a tawny ribbon whose least impulse can be carefully monitored.*"

HENRY DAVID THOREAU, **Concord River** 130

"*I had often stood on the banks of the Concord, watching the lapse of the current, an emblem of all progress, following the same law with system, with time, and all that is made. . . .*"

JOHN MUIR, **The River Floods** 136

"*Storms are fine speakers, and tell all they know, but their voices of lightning, torrent, and rushing wind are much less numerous than the nameless still, small voices too low for human ears; and because we are poor listeners we fail to catch much that is fairly within reach.*"

JOHN MCPHEE, **Los Angeles Against the Mountains** 144

"*Debris flows amass in stream valleys and more or less resemble fresh concrete. They consist of water mixed with a good deal of solid material, most of which is above sand size. Some of it is Chevrolet size.*"

ANNIE DILLARD, **The Present** 155

"*Live water heals memories. I look up the creek and here it comes, the future, being borne aloft as on a winding succession of laden trays.*"

CONSTANCE ELIZABETH HUNT, **Creating an Endangered Ecosystems Act** 161

"*If rare habitats were protected by federal statute, the number of species requiring protection under the Endangered Species Act in order to survive would be much smaller.*"

4 Developing the Deserts 168

MARY AUSTIN, **Land of Little Rain** 171

"*The sculpture of the hills here is more wind than water work, though the quick storms do sometimes scar them past many a year's redeeming.*"

JOHN ALCOCK, *A Treasure of Complexities* 178
"This is beautiful country and much of its beauty and value lies in the variety of species that the land sustains."

GARY PAUL NABHAN, *An Overture* 184
"The rains came that night—they changed the world."

CHARLIE HAAS, *Desert Sojourn* 189
"Respect is the only choice, because nature is in charge here."

JOHN NICHOLS, *Meeting on the Mesa* 196
"At heart, most of the men are fatalistic. None carry an illusion that their way of life will survive much beyond themselves. When they die, most likely it will be over."

PETER STEINHART, *The Water Profiteers* 201
"Environmentalists and city water users have long pitted themselves against farmers in the West's water battles."

KIM HEACOX, *A Poet, a Painter, and the Lonesome Triangle* 214
"The East Mojave isn't Wisconsin; it takes a lot of desert to feed one cow."

5 Oceans, Lakes, and Contamination   228

PAUL COLINVAUX, *The Ocean System* 231
"All the chemicals in the sea are cycled slowly through it. They come from the rivers, they spend a time in the oceans, diluted and in suspense, then they are taken in by the mud, held for a brief few million years, and then thrust back onto the land in a prison of rock."

RACHEL CARSON, *Wealth from the Salt Seas* 237
"It is a curious fact that there is little similarity between the chemical composition of river water and that of sea water."

PETER SEARS, *Oil Spill* 249
"The ocean is leather in slow motion."

ART DAVIDSON, *In the Wake of the* Exxon Valdez: *Marine Birds* 250

"But on this cold April evening when they tied up in Snug Harbor, oil was 18 inches deep on the surface of the water."

PETER NULTY, **What We Should Do to Stop Spills** 257

"The time has come to dam the spillage of oil. On this much we all agree."

JODI L. JACOBSON, **Holding Back the Sea** 263

"For most of recorded history, sea level has changed slowly enough to allow the development of a social order based on its relative constancy. Global warming will radically alter this."

WILLIAM ASHWORTH, **Sludge** 274

"Like an animal accumulating pollutants in its body fat, the lake accumulates them in its sediments—a process that is every bit as dangerous for the lake as it is for the animal."

JOSEPH S. LEVINE, **The Tainted Cup** 286

"Though knowledge of the sea and its ecosystems is growing rapidly, the ability of human beings to alter those ecosystems permanently is growing even faster."

6 Wilderness and Intrusion   296

ANNA QUINDLEN, **Our Animal Rites** 299

"But out where the darkness has depth, where there are no street lights and the stars leap out of the sky, condescension, a feeling of supremacy, what the animal-rights types call speciesism, is impossible."

BARRY LOPEZ, **Yukon-Charley: The Shape of Wilderness** 302

"The insistence of government and industry, that wildnerness values be rendered solely in economic terms, has led to an insidious presumption, that the recreational potential of wild land, not its biological integrity, should be the principal criterion of its worth."

EDWARD ABBEY, *The Great American Desert* 314

"To save what wilderness is left in the American Southwest—and in the American Southwest only the wilderness is worth saving—we are going to need all the recruits we can get."

SUE HUBBELL, *Summer* 323

"Most of the snakes around here are harmless or, like the black rat snakes, which eat rodents, beneficial, and I have no sympathy with the local habit of killing every snake in sight."

CHRISTIAN KALLEN, *Eco-Tourism: The Light at the End of the Terminal* 327

"Through treks, river rafting, visits to natural history reserves and countless other offerings, adventure travel promises close and meaningful encounters with the natural world, in contrast to more traditional vacations."

LORI NELSON, *The Dolphins of Monkey Mia* 335

"After extensively testing the pollution levels of the shore water, the EPA determined that high levels of sewage contamination had occurred just before the dolphins disappeared."

BROOKS ATKINSON, *The Great Swamp* 342

"Although planners and engineers can destroy Great Swamp, no organization of human beings and inhuman machines could have built it. The structure is too vast in scale."

7 Destroying the Forests — 349

JOHN HAY, *Cove and Forest* 352

"We have relatively narrow means by which to approach a tree. It may be in the way, or it may have ornamental value. For those who deal in lumber it will have another; and most people do not know its name."

LAURA TANGLEY, *The Last Stand?* 358

"Why have the dry forests fared even worse than rain forests?"

DOUG STEWART, **Green Giants** 363

"Although the giant sequoia and the coast redwood have both succeeded in adapting to major changes in their habitat since their earliest days on Earth, researchers worry that pollution-linked changes could be altering the habitats so quickly that the trees will not be able to adjust."

MICHAEL FROME, **Forestry: Only God Can Make a Tree, But . . .** 369

"Clearcutting has been subject to so many challenges and criticisms, and may do such serious long-range damage to soils and streams of the nation, that it needs to be curbed at once and restricted to experimental uses only, until answers are fully known."

JOHN NIELSEN, **Expanses of Trees Fall Sick and Die** 378

"By all accounts, centuries of relentless logging have irrevocably changed the nature of temperate forests around the world."

DON HINRICHSEN, **Acid Rain and Forest Decline** 383

"Acid rain spares nothing."

JULIE SLOAN DENSLOW, **The Tropical Rain Forest Setting** 395

"The rain forest and the people who make their living from it are inextricably interwoven."

8 Attitudes and Practices 406

PAUL R. EHRLICH AND ANNE H. EHRLICH, **Making the Population Connection** 409

"Global warming, acid rain, depletion of the ozone layer, vulnerability to epidemics, and exhaustion of soils and groundwater are all, as we shall see, related to population size."

WENDELL BERRY, **Think Little** 417

"Nearly every one of us, nearly every day of his life, is contributing directly to the ruin of this planet."

VINE DELORIA, JR., **The Artificial Universe** 427

"In recent years we have come to understand what progress is. It is the total replacement of nature by an artificial technology."

REVEREND JESSE L. JACKSON, SR., **Making Lions Lay Down With Lambs** 439

"*Don't wait for an environmental organization to give you back your clean air—go out and fight for it yourselves.*"

ELLEN GOODMAN, **The Killer Bee Syndrome** 444

"*Anybody could make a mistake.*"

ALBERT L. HUEBNER, **The Medfly Wars** 447

"*After the first aerial spraying of her community, her son developed swollen glands, a sore throat, and a fever.*"

ALDO LEOPOLD, **Conservation** 453

"*Harmony with land is like harmony with a friend: you cannot cherish his right hand and chop off his left.*"

DAN GROSSMAN AND SETH SHULMAN, **Down in the Dumps** 462

"*To the workers of the Garbage Project the excellent condition of the trash they find has thrown into question the entire notion of biodegradability.*"

Environmental Organizations 473

Index 477

Rhetorical Table of Contents

Argument/Persuasion

Jared Diamond, *Playing Dice with Megadeath*	63
Constance Elizabeth Hunt, *Creating an Endangered Ecosystems Act*	161
Paul Colinvaux, *The Ocean System*	231
Thomas A. Lewis, *The Frontier Dream We Call Alaska*	41
Peter Nulty, *What We Should Do To Stop Spills*	257
Michael Frome, *Forestry: Only God Can Make a Tree, But . . .*	369
Paul R. Ehrlich and Anne H. Ehrlich, *Making the Population Connection*	409
Reverend Jesse L. Jackson, Sr., *Making Lions Lay Down With Lambs*	439
Edward Abbey, *The Great American Desert*	314
Aldo Leopold, *Conservation*	453
Lewis Thomas, *Natural Man*	105
Barry Lopez, *Yukon-Charley: The Shape of Wilderness*	302
Wendell Berry, *Think Little*	417
Vine Deloria, Jr., *The Artificial Universe*	427

Cause/Effect

Joseph S. Levine, *The Tainted Cup*	286
William Ashworth, *Sludge*	274
Mary Austin, *Land of Little Rain*	171
John Muir, *The River Floods*	136
Sandra Postel, *Halting Land Degradation: Abuse It and Lose It*	22
Loren Eiseley, *The Flow of the River*	114
Fiona Sunquist, *Vast Green Seas Shrivel to Desert*	17
Anne W. Simon, *The Thin Edge*	33
Michael Milstein, *Unlikely Harbingers*	86
Conger Beasley, Jr., *The Return of Beaver to the Missouri River*	122
John McPhee, *Los Angeles Against the Mountains*	144
Constance Elizabeth Hunt, *Creating an Endangered Ecosystems Act*	161
John Alcock, *A Treasure of Complexities*	178
Peter Steinhart, *The Water Profiteers*	201
Art Davidson, *In the Wake of the* Exxon Valdez: *Marine Birds*	250

Peter Nulty, *What We Should Do To Stop Spills* 257
Jodi L. Jacobson, *Holding Back the Sea* 263
Lori Nelson, *The Dolphins of Monkey Mia* 335
John Nielsen, *Expanses of Trees Fall Sick and Die* 378
Julie Sloan Denslow, *The Tropical Rain Forest Setting* 395
Don Hinrichsen, *Acid Rain and Forest Decline* 383
Paul R. Ehrlich and Anne H. Ehrlich, *Making the Population Connection* 409
Wendell Berry, *Think Little* 417
Albert L. Huebner, *The Medfly Wars* 447
John Madson, *A Wilderness of Light* 29
Michael Frome, *Forestry: Only God Can Make a Tree, But . . .* 369
Ellen Goodman, *The Killer Bee Syndrome* 444

Classification and Division

Peter Steinhart, *The Water Profiteers* 201
Paul B. Sears, *Soil* 8
Sandra Postel, *Halting Land Degradation: Abuse It and Lose It* 22
Lewis Thomas, *Natural Man* 105
John Alcock, *A Treasure of Complexities* 178
Rachel Carson, *Wealth from the Salt Seas* 237
Sue Hubbell, *Summer* 323
Christian Kallen, *Eco-tourism: The Light at the End of the Terminal* 327
Brooks Atkinson, *The Great Swamp* 342
John Hay, *Cove and Forest* 352
Doug Stewart, *Green Giants* 363
Don Hinrichsen, *Acid Rain and Forest Decline* 383
Aldo Leopold, *Conservation* 453
Edward Abbey, *The Great American Desert* 314
Thomas A. Lewis, *The Frontier Dream We Call Alaska* 41

Comparison & Contrast

John Alcock, *A Treasure of Complexities* 178
Thomas A. Lewis, *The Frontier Dream We Call Alaska* 41
John Madson, *A Wilderness of Light* 29
Anne W. Simon, *The Thin Edge* 33
Jared Diamond, *Playing Dice with Megadeath* 63
Margaret Knox, *Africa Daze Montana Knights* 94
Conger Beasley, Jr., *The Return of Beaver to the Missouri River* 122
Henry David Thoreau, *Concord River* 130
Gary Paul Nabhan, *An Overture* 184

Charlie Haas, *Desert Sojourn* — 189
Kim Heacox, *A Poet, a Painter, and the Lonesome Triangle* — 214
Sue Hubbell, *Summer* — 323
Laura Tangley, *The Last Stand?* — 358
Vine Deloria, Jr., *The Artificial Universe* — 427
Lewis Thomas, *Natural Man* — 105

Definition

Paul B. Sears, *Soil* — 8
John McPhee, *Los Angeles Against the Mountains* — 144
Constance Elizabeth Hunt, *Creating an Endangered Ecosystems Act* — 161
John Alcock, *A Treasure of Complexities* — 178
Christian Kallen, *Eco-tourism: The Light at the End of the Terminal* — 327
Brooks Atkinson, *The Great Swamp* — 342
Reverend Jesse L. Jackson, Sr., *Making Lions Lay Down with Lambs* — 439
Michael Frome, *Forestry: Only God Can Make a Tree, But . . .* — 369
John Hay, *Cove and Forest* — 352
Gretel Ehrlich, *The Solace of Open Spaces* — 48
Anna Quindlen, *Our Animal Rites* — 299
Vine Deloria, Jr., *The Artificial Universe* — 427
Aldo Leopold, *Conservation* — 453

Description

Paul B. Sears, *Soil* — 8
John Madson, *A Wilderness of Light* — 29
Anne W. Simon, *The Thin Edge* — 33
Lewis Thomas, *Natural Man* — 105
Loren Eiseley, *The Flow of the River* — 114
Conger Beasley, Jr., *The Return of Beaver to the Missouri River* — 122
Henry David Thoreau, *Concord River* — 130
John Muir, *The River Floods* — 136
John McPhee, *Los Angeles Against the Mountains* — 144
Annie Dillard, *The Present* — 155
Mary Austin, *Land of Little Rain* — 171
Gary Paul Nabhan, *An Overture* — 184
Charlie Haas, *Desert Sojourn* — 189
Kim Heacox, *A Poet, a Painter, and the Lonesome Triangle* — 214
Brooks Atkinson, *The Great Swamp* — 342
John Hay, *Cove and Forest* — 352
Dan Grossman and Seth Shulman, *Down in the Dumps* — 462

Gretel Ehrlich, *The Solace of Open Spaces* — 48
Joseph S. Levine, *The Tainted Cup* — 286

Exemplification

Jake Page, *The Owls of Night* — 73
Ruby Abramson, *Sharks Under Attack* — 82
Margaret Knox, *Africa Daze Montana Knights* — 94
Peter Steinhart, *The Water Profiteers* — 201
Don Hinrichsen, *Acid Rain and Forest Decline* — 383
Albert L. Huebner, *The Medfly Wars* — 447
Aldo Leopold, *Conservation* — 453
Ellen Goodman, *The Killer Bee Syndrome* — 444

Narration

Mary Austin, *Land of Little Rain* — 171
Jake Page, *The Owls of Night* — 73
Henry David Thoreau, *Concord River* — 130
John Muir, *The River Floods* — 136
John McPhee, *Los Angeles Against the Mountains* — 144
Anne Dillard, *The Present* — 155
Gary Paul Nabhan, *An Overture* — 184
Charlie Haas, *Desert Sojourn* — 189
John Nichols, *Meeting on the Mesa* — 196
Gretel Ehrlich, *The Solace of Open Spaces* — 48
Kim Heacox, *A Poet, a Painter, and the Lonesome Triangle* — 214
Art Davidson, *In the Wake of the* Exxon Valdez: *Marine Birds* — 250
William Ashworth, *Sludge* — 274
Anna Quindlen, *Our Animal Rites* — 299
Barry Lopez, *Yukon Charlie: The Shape of Wilderness* — 302
Edward Abbey, *The Great American Desert* — 314
Lori Nelson, *The Dolphins of Monkey Mia* — 335
John Nielsen, *Expanses of Trees Fall Sick and Die* — 378
Dan Grossman and Seth Shulman, *Down in the Dumps* — 462
Joseph S. Levine, *The Tainted Cup* — 286

Process

Jared Diamond, *Playing Dice with Megadeath* — 63
John McPhee, *Los Angeles Against the Mountains* — 144
Edward Abbey, *The Great American Desert* — 314
Doug Stewart, *Green Giants* — 363
Michael Frome, *Forestry: Only God Can Make a Tree, But . . .* — 369

Preface

THE ENVIRONMENTAL MOVEMENT has become, in many ways, the victim of its own success. But it seems to us that much of what passes today for environmental concern is trivial, often self-serving, and rarely helps us see the interrelatedness of the various issues. What's worse, many of the truly crucial issues facing the Earth and those of us who inhabit it are virtually ignored, or treated superficially. And what to us is the most unsettling result of this newly achieved success of the environmental movement is the lack of understanding on the part of some college students. Many seem to believe that by recycling aluminum cans or not using aerosol hair spray, they have done everything needed to solve all of the serious environmental problems that promise to become more critical with each passing year.

This book grew from our frustration at not finding integrated sources on the environment for our writing students and from our recognition of the importance of this issue in years to come. But it was shaped by our teaching experiences and beliefs. As we put this book together, we had in mind the "research" component of most first-year composition classes. The problems in such classes are multiple: Students often have too little background to guide their research and not enough time within the semester to gain it. Students' interest, the starting point for all research, varies widely or perhaps not enough; the instructor, then, tries to help each student focus and direct attention and research. Often, students cannot find adequate material. Sometimes students' disparate searches (or is that desperate?) block the building of a sense of community, a sense that can do much to hasten the writer's progress.

The writing course is also a thinking course. Focusing on the wide topic of environment offers endless possibilities for critical thinking and writing. Students may forge a sense of how interrelated the issues within one topic truly are. They may examine issues deeply and widely, analyze thoroughly, connect causes and effects as they begin to own knowledge of environmental matters. They are, then, thinking within a discipline and within a context; critical thinking demands context as well as knowledge.

Yet the topic of environment is large; within it one may pursue information on acid rain or America's first conservationists, rising sea levels or natural history writers. The "greening" of America; the politics and the greed; the social consequences as well as the health-related concerns; the class-related decisions regarding landfills or toxic discharges; these and countless other possibilities present themselves for study and research. One may focus on the world or the country, the state or city, the backyard or the basement. Thus environment as a topic focuses but does not constrain student interest; the topic is as large as Carl Sagan's universe or as small as Robert Frost's mite.

The Endangered Earth: Readings for Writers contains eight thematic chapters, each of which presents six to eight readings. The authors of the readings are scientists, essayists, columnists, researchers, reporters, naturalists, teachers. (One of the first of our discoveries in putting this book together was that a job description neither indicates nor limits environmental concern or the ability to articulate it.) Thus, the readings were selected from various magazines, newspapers, and books; they present their variety of perspectives in a variety of writing styles.

Each chapter opens with an introduction to help students build a context for the chapter's selections; preceding each selection is a headnote that provides background and prereading suggestions to help students find an old hook on which to hang new knowledge. Questions for discussion and writing as well as writing ideas—individual and collaborative—follow the selections. Each chapter closes with writing possibilities that ask students to forge links among the chapter's readings; research ideas provide a starting point for students who are shaping new questions to answer.

We appreciate the suggestions and advice from those at other colleges and universities who responded to the evolving versions of *The Endangered Earth: Readings for Writers* and thus thank them for their expertise: Stephen Goldman, University of Kansas; William Lutz, Rutgers University—Camden; Linda Palmer, California State University at Sacramento; Sally Barr Reagan, University of Missouri at St. Louis; Louise Smith, University of Massachusetts at Boston; Richard Zbracki, Iowa State University.

We thank also editor Joe Opiela, whose aim we found unwavering; Amy Capute and Rowena Dores of Allyn and Bacon for their assistance and guidance; Marsha Finley for her sharp eye and pencil; our colleagues at Park College and elsewhere who have offered advice and encouragement. And finally, we wish to acknowledge our families, for they received less acknowledgment than they deserved while we worked on *The Endangered Earth*.

Introduction

THE REVEREND JESSE JACKSON, SR., generally known for his civic leadership, summed up better than many professional environmentalists the need for all of us to examine the global, national, and local issues of pollution, waste, and destruction. The quotidian concerns of eating, sleeping, and staying warm, of making grades, of paying bills, of keeping the car running, of friendships and families and lovers—all these pale when considered alongside the possible consequences of unchecked destruction of the environment.

"Everyone who breathes is an environmentalist," Jackson said.

Whether we want to be or not.

The issue becomes, then, not whether we care about the problems of global warming, ozone depletion, rain forest destruction, ocean pollution, acid rain, trash accumulation, land desertification, groundwater contamination, or species extinction—a list mind-numbing in its complexity and length. These problems will not go away no matter how hard we ignore them, so the issue becomes, what are we going to do about them?

The essays and articles here may help you formulate or refine your own position, your own response.

To solve environmental problems, we first must understand them. As we started research for this collection of essays, we learned very quickly the interrelatedness of the issues and the complexity of the entire web of life on planet Earth. Ours is the only known planet that sustains life, a fragile and intricate system of creatures and plants, food chains, water, soil, and atmosphere. Everywhere, it seems, the system is under attack by industrial and transportation emissions, by greed and ignorance, by too many humans and too few resources, by our choices to live without regard for consequences.

At first the problems may seem overwhelming, unsolvable, depressing. Daily we are hammered by new and dire forecasts: at the present rate of extinction, half of all the species on Earth today will be gone in 30 years; the population of the Earth will reach 14 billion in the next century; global warming will raise sea levels by one meter in the next 50 years, possibly devastating large regions of the Earth; pollution is killing the oceans, source of all life.

These predictions are realistic, but they are not yet reality. Our purpose is to acquaint readers with some of the problems, their connections, and consequences, and with the perspectives of situations and solutions offered by various experts.

We have attempted to simplify the complexity somewhat by grouping the issues: some chapters deal with the geographical features of deserts, forests, rivers, oceans, and lakes; other chapters focus on wilderness and species. The final chapter suggests how our attitudes have defined our practices, practices that have created problems; it also suggests changes that lead toward solutions. Throughout this volume you will be asked to do more than read the essays. Questions at the end of each selection are designed to help you understand issues and actively engage problems. Our belief is that by reading about these issues, by discussing them with classmates and friends, by exploring your own experiences with and beliefs about certain problems, and by writing in a variety of ways about the issues presented here, you will gain a deeper understanding not only of problems but of solutions and of how issues connect.

Undoubtedly, you will care more about some issues than others. But each issue connects to the others. The Earth is much like a human body; unchecked cancer in the lungs will not simply kill the lungs—it will kill the entire body.

The writing here represents a broad spectrum of views and styles. You will recognize some of the names: Henry David Thoreau, Annie Dillard, Barry Lopez, Edward Abbey, Gretel Ehrlich, John McPhee, and Wendell Berry. These people have articulated their concerns through metaphoric, image-filled language that sings. Many of the writers you may not have heard of; they may be the stars of environmental writing in the future. Others we chose for the depth of their knowledge, the clarity of their explanation of complex problems. We believe the mix of artistic expression and reportorial writing is necessary not only for breadth of exposure to a variety of problems but for the experience of examining a variety of rhetorical styles.

What you will not find here is straight nature writing. You will find instead writers of the present and from the past, scientists who write well and writers who love science. The mix is important, we believe, because both the aesthetic and the scientific aspects of life are necessary. Life without art would be empty; life without science might be impossible. Together, the two perspectives form a holistic view of ecology, of the biosphere and its ecosystems. We all know the world both aesthetically and scientifically; the collection that follows reflects the ways we make that knowledge.

1

Using the Land

IN 1852 DR. THOMAS WHITE AND HIS FAMILY left their home in La Grange County, Indiana, and traveled by covered wagon to Oregon. In a letter to a friend back in Indiana, White described the Willamette Valley:

> . . . the greater portion of the Lower part of it is timbered land, a great share of it the most beautiful (entirely) as the Irishman would say, standing almost as thick as it is possible for it to stand, & as streigh[t] as hemp, from 1 to 2 ft. in diameter, & 2 hundred ft high. . . . You can find 30, 40 & up to an 100 acres or more, on which the trees would not average over from 1 to 10 trees to the acre, & generally these are fir, & by boring 2, 2 inch auger holes near the root, & puting in them some lightwood, you may burn down the largest of them in a short time. . . . I think it is the best land in the valley. . . .

For many Americans from 1607 on, as for Dr. White, the siren call of the New World was not gold or fur or even freedom in all its guises. Rather, it was the land—an entire continent, sparsely populated, rich in resources and ripe for taming. America was a new Eden, and the predominently Christian settlers took literally the biblical injunction to subdue the earth, to transform it from what it had been to what they wanted or needed it to be.

Today we are only beginning to realize the consequences of nearly four centuries of subduing the land. Dr. White, like most others seeking livelihood or profit, saw land as space to clear for farming—but the land isn't always suited to the transformation humans effect. Few of us understand the extent of the problem, or the complexity of interrelated systems and individual organisms that cling to the crust of the Earth.

John Lovelock, a British scientist, has offered the theory that the planet is a living thing, a being he calls Gaia. In his view the planet approaches the ancient goddess-like persona of Mother Earth, a dynamic organism of living soil and living seas. Lovelock's theory has not gained wide acceptance, but his fundamental principle has: if the seas and soil die, so do we all.

In this chapter on land and how we use it, writers look at land from a variety of perspectives. Sandra Postel explains abusive practices and their wide consequences. Paul Sears reminds us that land is first of all soil and that its composition determines what grows and what doesn't. Further, soil that is intensively used in agriculture must be renewed, either naturally—a slow process—or through chemicals, which in themselves can cause serious problems.

Anne Simon takes a look at coasts, the ephemeral point at which sea meets soil—continuously shifting, an infinitesimal line of demarcation that is in danger of dying from overuse. Concrete and asphalt replace sand and tidal lowlands, pollution and oil and sewage kill marine plant and animal life, tourism and development doom coral reefs thousands of years old. In the same way, the bright, clear wind-buffeted prairie that greeted early pioneers in the Midwest has been replaced with farmland, a transformation that saddens John Madson. The prairie as a unique ecosystem has been overlooked in the rush to subdue the land, to make it a vital part of the growth of America, and in our eagerness to plant crops, we have lost something important, Madson writes.

Alaska is for many, perhaps, the embodiment of frontier and wilderness as we close in on the twenty-first century. But it too, despite its vastness, is in danger of being exploited for short-term gain. Thomas A. Lewis describes the process and enumerates the problems in trying to keep the dream of wilderness alive in a world that favors technology and profit.

As you read the selections in this chapter, ponder the interrelatedness of soil, of space, of wilderness and coast and prairie. Consider the wonder of Dr. White as he viewed a new land and the consequences of our efforts to subdue it.

Paul B. Sears

Soil

In a chapter of The Living Landscape *author Paul B. Sears describes soil as "a thin carpet that has developed upon the geosphere." The geosphere is, of course, the Earth, and the essay that follows examines that thin carpet.*

When we speak of land use, we speak of using both the space and the surface of the land. An acre of land-space in Georgia equals an acre of land-space in Wisconsin, but the surface of these two acres differs. The characteristics of each may determine how each is used or how it will have to be transformed before it can be put to the use humans desire. Thus, the location of the land's space and the nature of its surface have much to do with how a piece of land fares in both its own and human ecological systems.

In the following pages Sears discusses how soil is formed, how it connects to other systems, and how it can be depleted or enriched. One of the principles he espouses is that if land is to be productive, its "materials must either be recycled or restored." Write a quick explanation of what you think that principle means.

> So cream that was richest
>   And meat that was rare
> Were free to them always
>   As water and air.
>   —Palmer Cox

So went the final lines of a rhyme for children, written less than a lifetime ago. The plot, which involved animals, need not concern us here, but the comparison does. Water is no longer free, while anyone who has gasped and choked in the atmosphere of some of our great urban and industrial centres should know that it will cost money, and a great deal of it, to give us all pure fresh air to breathe.

There is another old expression 'cheap as dirt' which is beginning to sound quaint and out of fashion. The word 'dirt' does not here refer to filth, of which we have more than ever, some of the worst of it invisible. Instead it means the soil that covers the ground and which the farmer tills. It has become a costly commodity. True, the highest prices are not for the soil itself, but for the space it occupies. But when the space alone is purchased, it is for such structures as highways and buildings that end its usefulness as soil. In the United States not less than a million acres a year are lost in this way to the production of crops. In Holland the amount of land so painfully reclaimed from the shallow sea margin is barely enough to equal that which is required for purposes other than farming.

One cannot have a sound perspective on human history if he ignores the role that soil has played. Nor can he properly interpret today's problems, domestic and international, without taking soil into consideration. Or better, we should speak of soils, for soil exists in great variety, and it is this variety which has been so significant for plants, animals and man. Yet the relationship is a curiously intricate one, for without living organisms there would be no soils. The stuff that covers the earth becomes soil only by being lived in and upon. It comes about by the interaction of air, water, and life upon rock-stuff, the energy for this complex operation being furnished by the sun.

Rock materials which form the matrix of soils differ greatly from place to place, both in texture and chemical make-up. Climates likewise vary, and the resulting patterns of moisture, temperature and light influence changes in the exposed rock materials as well as the activities of living organisms. The several sciences, physics, chemistry, geology and biology, have all been needed to help us find out what is in the various soils and what is going on in them. The result has been an elaborate catalogue of types. Rock particles range in size from coarse gravels through sands, silts and to impalpably fine clays, mixed in various proportions. The kinds and amounts of organisms and dead organic material likewise differ, as does the content of air and water. Furthermore—and this is most important—soils differ in the amount and kind of materials that can be dissolved in soil water and so be available for plant nutrition.

A huge catalogue of this kind of information about soils was built up during the nineteenth century, yet something was needed to bring order out of this knowledge. For one thing, the soils of Western Europe, then the centre of scientific activity, had been greatly changed from their original condition by centuries of heavy agricultural pressure. This fortunately was less true in the vast expanses of Russia and the United States, where natural or virgin soils were still to be found.

In Russia, since the days of Peter the Great, it had been customary to sponsor men of high talent in the arts and sciences. Though few in number, they were of exceptional ability, and among them was one—later others—who turned his attention to a study of undisturbed natural soils. And as original scientific activity increased in the United States, such work got under way here.

Many great scientific advances seem surprisingly simple after they have been made. Soil science is no exception. Progress came by studying not just what soils are like but how they have been formed. Trenches or cuts extending down into the parent material give what is called a profile. This enables one to trace the changes that have transformed raw rock into fertile soil. If then, as has happened, profiles be examined in different regions whose climate is expressed by vegetation—forest grassland or arid steppe and desert—a significant pattern begins to emerge.

Let us take a few examples. Under forest cover, for example, one finds a dark rich layer seldom more than eight or ten inches thick. Below this is normally a light-coloured mineral layer poor in such plant nutrients as lime and potash. Then comes a zone of transition to the parent rock. Under clearing and agriculture such soils are usually very productive until the thin black top layer disappears. Unless unusual care is taken, as it has been in Denmark, Britain and France, one is likely to find abandoned farms such as those of New England where once there was forest.

In a forest most of the organic material falls to the surface as leaves, twigs and dead trunks. There are few animals that burrow beneath the surface. Droppings and remains of deer, squirrels and birds accumulate on the forest floor, while insects, worms and organisms of decay are confined to the surface litter. This litter is moist and compact, favouring the production of organic acids by fermentation, a process similar to the making of sauerkraut and ensilage. These acids in turn are carried downward by percolating rain, dissolving out the alkaline nutrients in the layer just below. That is why this leached mineral layer is relatively infertile when at length it is laid bare by the destruction, through practices which fail to renew it, of the rich surface layer of leaf mould.

Roughly speaking, leafy forests are found where rainfall exceeds evaporation. Where the reverse is true, trees will not grow unless there is an extra supply of soil moisture, as from irrigation. As one moves from forest to successively drier conditions, he will first encounter the tall-grass prairie of subhumid climates; passing further towards the continental interior he will find the grasses growing shorter and the vegetation sparser; assuming that he does not encounter mountains on his way, he will next meet with thorny, wide-spaced woody scrub such as one finds in driving through northern Mexico. Beyond this he will pass by degrees into true desert, where evaporation far exceeds rainfall. (Some find this statement confusing. It simply means that the capacity of the air to remove moisture is much greater than the amount of water available.)

So much for the vegetation and climate. What of the soils passed over on such a journey? Those of the tall-grass prairie or subhumid steppe are in dramatic contrast with that formed under the adjoining forest. Instead of less than a foot of dark humus, they exhibit three to five feet or more of rich black topmost layer. Below this instead of a leached zone, there will be a transition to the parent material, often containing rather than lacking nutrient minerals.

The deep organic layer of the prairie is densely occupied by the fine fibrous roots of grasses and owes its character to their activity while alive, the products of their decay after they die. In this zone are also the storage organs of other plants, the burrows and tunnels of countless animals, large and small, and the abode of a rich variety of micro-organisms. Both in prairie and forest food is manufactured in the sunlit leaves. But in the forest much of the surplus is stored in great trunks and branches or eaten by browsers and tree dwellers. In the prairie much of this surplus is moved below ground and there stored or eaten and ultimately converted back for reuse by green plants. That which remains available above ground as tender leaves or nutrient seeds serves as food to the herds of grazing animals and flocks of birds which in turn support a population of meat eaters. But ultimately the wastes and remains of animal life are returned to the soil and used again.

In addition there are hosts of micro-organisms, some of which, sustained either on organic wastes or on the roots of the many legumes (clovers, beans and the like), draw nitrogen from the air and make it available for the building of plant proteins. Small wonder then that the prairie or moist steppe regions of the world have become its most fertile source of human food when placed under cultivation. The orderly cycles of nature with their beautiful economy of accumulation and reuse through the ages that native vegetation and wildlife have been at work have made this possible. If the United States were suddenly deprived of its prairie soils, our civilization would be profoundly altered, if indeed it did not collapse.

This is equally true of other continents, the sub-humid steppes of Russia, for example, and those of the Argentine, as the ruinous meddling of Perón demonstrated recently. The true wealth of his country lies in its grasslands with their rich black soil. But mesmerized by the thought that the secret of prosperity lay in urbanization and industry, Perón diverted capital and energy away from the land. The result was, as we know, catastrophic.

Because prairie soils are so productive, remnants of their original cover of plant and animal life are much scarcer today in our Mid-west than are forest preserves farther east. Such as remain should be protected jealously, both as living museums and for their aesthetic beauty. They are not, as we might suppose, monotonous stretches of grass, but carpets of infinite variety whose colours change throughout the seasons with the growth, flowering and fruiting of many kinds of herbaceous plants.

In this rich variety lies the secret of that resilience which has enabled the prairie to flourish through the millennia in an environment too harsh to permit the growth of trees. This environment is marked by sudden changes of temperature, daily and seasonal, by strong winds and hot sun. Drought years tend to come in groups, interspersed between groups of moist years. Genial spring weather may be followed by biting frost. To all of this the great grasslands have become adjusted, as they have to the moving herds of grazing animals, and even to fire. For fires were set by lightning long before the invasion of man.

Curiously, grassland completely protected from fire and grazing deteriorates. Dead stems and leaves become matted on the surface, locking up needed nutrients and choking the new growth needed to maintain vigour and production. These remnants above ground do not decay rapidly as do fallen leaves and twigs in the more humid air of the forest. In short, the natural prairie is a beautiful example of a system organized for permanence in the face of recurrent hazards, even fire, and a model to man for intelligent use of the landscape.

In the drier, short-grass steppes which we call the High Plains in North America growth is far less luxuriant. The mat of roots penetrates less deeply than in the prairie. Food production is less, yet remarkably dependable, sustaining animal life both above and below ground. The organic layer is shallow, grey or brown rather than black since there is no great surplus of humus. Whenever the infrequent rains moisten the soil, the subsequent evaporation serves to bring up soluble plant nutrients, making the soil potentially fertile. Only the scarcity of water limits production. If these soils happen to be ploughed up and planted during a time of better than average rainfall, they produce good crops of wheat at low cost. But our cereal crops do not withstand the inevitable groups of dry years. These bring crop failure and destruction of the thin soil by wind, with resultant distress to man. The continued attempts to exploit the short-grass, semi-arid, lands have brought several periods of such distress, while their heavy and economical production of wheat during normal years is a major source of our present costly surplus. Yet this land is admirably adapted to extensive grazing and even permits the growing of forage in selected places. Some wheat growers who have returned to livestock production are doing very well. Our agriculture and general economy might benefit if more did so, but so long as every state has two senators and profits are high during good years, a miracle would be required to bring about such a change through political means.

In desert soils there is practically no production of humus by decay. Plant and animal remains dry out on the surface. Yet here, to an even greater degree than in the semi-arid grassland, nutrient salts are concentrated at the surface by evaporation. Except in low alkaline spots where these salts are too concentrated, desert soil can be highly productive if supplied with water. But irrigation has to be conducted with great skill, lest it result in too great an increase of mineral salts.

This is not the only hazard. When, as often happens, erosion from upstream loads the irrigation channels with silt, fields may be converted into marsh and swamp. This has taken place in the region around Albuquerque, where engineers have become occupied with flushing and draining the soil to keep it reasonably productive. And as one travels north from that city he sees a panorama of arroyos that have been cut into the foothills draining into the upper Rio Grande and contributing their silt. Even if the silt does not create marshy conditions, it may vastly increase the labour and cost of production by clogging the ditches that carry water to the fields. In Mesopotamia this task had become oppressive long before the irrigation works were destroyed by invaders after some thousands of years of use.

The reckless expansion of irrigation at great public expense in a time of food surplus makes heavy demands upon water in regions where it is scarce, increases competition for farmers with heavy costs elsewhere, and has at times resulted in failure from soil deterioration. In discussing a large appropriation to extend irrigation, Senator Frank J. Lausche from Ohio noted that we are now paying farmers as much as $50 an acre to limit production, while at the same time we propose to expend $500 to $2,000 an acre on land priced at $7 to bring it into production. His objection was overruled.

Each of the great climatic types of soil owes its character to the orderly economy of nature in which energy is channelled and materials recycled—a process which affords the model for any permanent human activity, a model too often ignored. Within each climatic type there are many variations due to such factors as slope, exposure, drainage and nature of parent material. This last is important, as we have seen, for if necessary chemical materials are scarce or absent, they will limit production.

Doubtless in nature, before the day of fences and other manmade barriers, these local differences were offset to some degree by the horizontal movement of animals. For animals have a notable ability to make up mineral deficiencies by visiting salt-licks and mineral springs, or migrating to favourable feeding grounds. In this way their droppings and remains have tended to equalize soil differences to some extent.

There are places where this process has gone on through thousands of years. Often around and about them may be found the bones of animals long extinct, as in the bone-licks of Kentucky. Here the rich accumulation of organic matter was exploited for fertilizer before its scientific record could be properly catalogued and interpreted.

Recurring floods and their residues dropped in valley floors have likewise helped to distribute nutrient minerals and organic matter. The Nile is the best known but by no means the only example of this function and it has sustained an intensive human economy for thousands of years. Our present policy of using flood-plains for urban and industrial expansion, safeguarded by the building of dikes and the pre-emption of productive valley land for reservoirs might well be re-examined in the light of natural and human history.

Man is sometimes referred to as a Pleistocene mammal, which means that his existence has been coextensive with the glacial, or Pleistocene Age—approximately one million years. This fact has influenced his activity, even his modern economy, in more ways than we realize. As to soil specifically, the moving glaciers have presented him with mineral material from remote areas. In New England, for example, the ice masses have brought in the weathered products of granites and sandstones containing little of the precious calcium so necessary to a prosperous agriculture. Once the natural soils had been exhausted or washed away as a result of cultivation, New Englanders have had little choice but to give way or invest heavily in fertilizer. Two-thirds of this area was in farms in the early 1800's. Today an equal amount is back in inferior second- or third-growth forest, since the best seed-trees have long since been harvested.

Further west much of the mineral material brought in by the glaciers was better provided with plant nutrients. When the late Louis Bromfield established his farm in Ohio on glacial soil, he took over 'worn-out' land whose fertility had apparently been exhausted. He was obliged to restore some of the missing materials by fertilizing. But he made use of deep-rooting crops. Presently these began to bring up the needed mineral elements from the rich storehouse of glacial till below. Then instead of selling off everything he grew, he returned as much of it to the land as possible, patterning his operations after the recycling economy of nature which we have described. The sound practices which he demonstrated have had a wide influence in Ohio and elsewhere. Invidious tongues have complained that his farm was not a money-maker, ignoring the fact that he was an artist, not an accountant, and that throughout his life he lavished baronial hospitality. He was a master farmer who understood the model that nature has given us which we so generally choose to ignore.

Much has been said about using the tropics to feed the world's expanding population. A short answer is that had it been possible, it would have been done by this time. Here again, the answer lies in the soil. In the moist tropics there is ample water, while temperatures are suitable to luxuriant plant growth. But the tropical forest is about as different as can be from a grain field. It is maintained by an extremely rapid and efficient turnover of a minimum amount of mineral nutrients, under a protecting canopy of many species which shields the ground from driving rain and searing sunshine.

Once a tropical soil is exposed to open cultivation, it is doomed. The combination of heat and heavy rain leaches and oxidizes the nutrient materials. Only dense growths of jungle or useless grasses can take over, until by a long process the forest is restored. In Indonesia, for example, former forestland has been taken over by a dense growth of coarse grass, useful only to furnish fuel. Where the tropics are being used effectively it is by growing tree-crops, or other crops with sheltering trees, as in the coffee plantations of El Salvador, shaded by the well-named Madre de Cacao.

To repeat, the great climatic zones have stamped their character on the soils formed within them. Occasionally, by luck, we find bits of fossil soils preserved and so are able to deduce climatic conditions during times long past. And since climates have shifted measurably, even within the past ten thousand years, we occasionally find soils that are changing their character under the impact of climatic change. This is notably true in the wavering borderline between forest and prairie in the middle states.

Yet most farmers know that even on their limited acres the soils may vary from place to place and that these differences are important to proper use and management. Slight differences in elevation or greater differences in parent material may serve to modify the effects of climate, or even to mask it. So does the history of the soil itself. The chalk downs of England, though formed by the weathering of lime rock, may become deficient in lime through generations of rainfall and pasturing. In time, even the rich phosphate soils of the Blue Grass region or the dairy lands of Wisconsin may have to be restored with phosphates from Florida, to replace that which has gone to market in the form of animals and animal products. There is no escape from the relentless principle that, to keep land productive, materials must either be recycled or restored.

Coal, gas and oil are the legacy to us of past sunshine, stored below the surface of the earth. Soil, the dark carpet upon the earth, is a similar legacy and a more precious one because it is more important that we eat than delegate our work to machines. The generations of plants and animals that form the soil and are sustained by it during their brief existence pass on. What remains, as evidence of their activity in catching and transforming the sun's energy, is the organized soil. It is vulnerable to our blundering, immensely rewarding to good husbandry.

Content Considerations

1. What ingredients compose soil? What variations affect soil, and how?
2. How does nature provide a model for human use of the land?
3. What does the author say about intelligent use of the land?

The Writer's Strategies

1. Sears's essay begins with a quotation from a children's rhyme. How does the quotation relate to the essay's content?
2. The author details the composition of various kinds of soil: forests, prairies, high plains. How does knowing the chemical composition of these soils contribute to your understanding of land use?
3. This piece uses a cause-and-effect organization to detail possibilities that may arise from misuse or ignorance of the soil. How does this organization direct your reading?

Writing Possibilities

1. Working with a classmate, speculate on the ways in which soil (not dirt) matters to various people. How might a city dweller view soil? How might a farmer? An urban planner? A developer? Who should take precedence? Write a report of your collaborative speculation.
2. Describe in an essay an area that you have visited where soil has been used up, worn out, abused. What went wrong? According to Sears, what should have been done?
3. Land is valuable in several ways. Two mentioned by the author are land as space and land as soil. Write an essay in which you enumerate and discuss other ways in which land has value.

FIONA SUNQUIST

Vast Green Seas Shrivel to Desert

To many urban and suburban Americans, grass is the green stuff full of chiggers that has to be mowed every week or so in the summertime, the stuff that has to be trimmed off the sidewalks, fertilized, raked, treated for grubworms and other voracious pests. The reward of such attention, of course, is a lush expanse of greenness surrounding the house.

For millions of other people grasslands have become or are potential croplands, places where various kinds of grain can be raised for food or profit. Or the grasslands have become grazing fields for sheep, goats, and cattle, the original wild grasses having been replaced by domestic varieties. Most wild grasslands have been converted for crops or grazing—and the consequences of such conversion have been devastating, by many accounts.

Alternatives to conversion do exist. This article by writer Fiona Sunquist, published in International Wildlife *in 1990, talks about some of them and discusses the practical value and benefits of natural grasslands. Before you read her article, write for a few minutes about what those benefits could be.*

As THE SETTING SUN lights up the towering Himalayan peaks in the distance, Ram Bhadur lays down his sickle and bundles up the grass he has cut. He struggles to his feet under a 60-pound load and joins the rest of his relatives for a 20-mile walk home to his Nepalese village. It has been a good day's work. His family has cut enough grass and cane to build a new house.

Fifty years ago, when the grasslands seemed inexhaustible, Bhadur's ancestors harvested the green bounty within sight of their village. Times are different now, and except in a handful of parks and preserves, there is no tall grass left in Nepal's lowlands. So Bhadur is thankful for the preserves—and for a farsighted Nepalese government program that allows local people to harvest grass there. The tall grass quickly grows back, the people get their harvest, and everyone learns the importance of conservation.

Grass is one of the most successful of plant families. Nearly 6,000 species circle the globe, covering nearly a quarter of the world's land surface. Almost every continent has a major grassland region, such as the pampas of South America, the prairies of North America, the Asian steppe and Africa's Serengeti. Today, however, much of the original

1

2

3

natural grassland has been plowed or fenced, turned into croplands or converted to pasture for livestock. In the U.S. state of Missouri, for instance, less than half of one percent of the original tallgrass prairie remains.

The conversion of wild grassland to farms and ranches has long been regarded as the price we pay to feed the world, but now, many scientists and governments are beginning to realize that the conversion has gone too far. Each year in semi-arid parts of the world, "desertification"—a disastrous combination of soil erosion, overgrazing and overcultivation—turns millions of acres into wasteland. What's more, many of the disappearing wild grasses are themselves valuable as potential sources of new crops or hardier genes that could improve existing crops.

To many experts, the proper course of action is obvious—restore and preserve wild grasslands wherever possible. "Conservation of these plants will offer important benefits for future generations," says biologist Samuel McNaughton of Syracuse University.

Such interest in nurturing wild grasslands comes at a time when scientists are beginning to understand the intricacies of how this unique ecosystem works. The basic principle is simple enough. Unlike the leaves of trees, a blade of grass is like a never-ending ribbon. Cut it and the blade responds by putting out new growth from the protected base. This allows grass to be cropped continuously by everything from termites to gazelles to elephants.

The grazers in turn provide food for a host of predators and scavengers. Aardvarks and anteaters dine on tiny grass-eating insects. Medium-sized carnivores like long-legged serval cats and foxes prey on seed-eaters like mice and birds. And the big predators—like lions, pumas, wolves and wild dogs—eat wildebeests and deer. The whole system is amazingly productive: a single acre of the richest grassland may produce as much as 9,000 pounds of leaves, stalks and seeds in a year, more than six times the average yield of the world's cornfields.

People have found it easy to tap into this productive system. On a grassland, there are no trees to fell and burn. The land is flat and fertile, and, once the steel mold-board was developed, the soil was easy to plow. Moreover, corn and other domesticated species of grass grow well where natural prairie once flourished, and in areas of low rainfall, where grain crops do not thrive, wild grazers can be replaced easily with sheep and cattle.

As a result, the world's people now get more than half their protein and calories from just three species of grass—wheat, rice and maize. Meat and milk from domestic grass-eaters such as sheep, goats and cattle make up a large part of the remainder of their diet. In addition, the human species uses grasses for numerous other purposes, from furniture (made with bamboo) to insect repellent (extracted from citronella).

Unfortunately, the grassland system that produces so much to benefit people is a delicate one. Its maintenance depends on a combination of fire, rainfall and grazing. A few years of too little rain, coupled with overgrazing, can transform a once productive grassland into barren uselessness.

Over the centuries, many groups of people have come to understand this balance. Hundreds of years ago, in what is now Syria, pastoralists developed a system known as "Hema." Each family was granted a license to graze a certain number of sheep in an area, thus preventing overgrazing. Even today, the pastoral nomads of the Turkana region of Kenya adjust everything from the pattern of their travels to the mix of their animals in order to take no more from the environment than their semi-arid grassland can give.

Now, many of the ancient lessons have been forgotten. All over the world, expanding populations and overgrazing by many of the world's three billion cattle, sheep, goats and camels are taking a serious toll on the land that feeds them. "The people know how to live in harmony with the land—they have done so for generations," says Mohamed T. El-Ashry, a vice president of the World Resources Institute in Washington, D.C. "But population pressures and government policies are leaving people with few alternatives to overusing the land."

Land abuse is not the only symptom that worries scientists. As the grass of the steppes or the pampas succumbs to the plow, so too do animals that are part of the ecosystem. Already, we have lost vast herds of bison. Now creatures like African hunting dogs, cheetahs, prairie chickens and black-footed ferrets are in danger.

Even more troubling, because it could affect the future of humankind, the disappearance of natural grasslands means the irrevocable loss of grass species themselves. These may have future value as pasture grasses, new food crops or the genetic material that plant breeders need to improve current crops. A species of wild oat found in Algeria, for example, provided genes that made oat crops resistant to a crown rust fungus that once devastated North American farms.

Fortunately, desertification need not be irreversible. Studies have shown that overgrazed land can recover if given a chance, although the process is usually painfully slow. Some governments and communities are trying to solve parts of the problem. China, Ethiopia and several other countries have developed national plans to combat desertification, while others are successfully restoring or protecting wild grasslands.

In the U.S., biologists are now sowing tallgrass seed across hundreds of acres of Wisconsin prairie, part of a program to convert millions of acres of highly erodible cropland back into grassland. In Syria, communities of sheep herders have revived the ancient "Hema" system of management. By reducing overgrazing, the system has enabled the revegetation of 17 million acres of rangeland, more than a third of Syria's total land area.

This list of successes, short though it is, offers hope for the future. Even so, few conservationists are ready to celebrate. They know that no strategy of preserving grasslands will work over the long term unless the local people can derive some benefit from the venture. 17

That is why Nepal's policy of allowing Ram Bhadur and his fellow villagers to harvest grass from the country's national parks is especially encouraging. In the lowlands of Nepal, grass has always been used for everything from roofs to fish traps. As the population has grown over the past few decades, however, the grass has disappeared. 18

The situation could have led to widespread grass poaching in four parks and preserves where grasslands remain healthy. That did not happen. Instead of keeping local villagers out, Nepal's National Parks and Wildlife Department allows more than 100,000 people into the preserves during a ten-day period each year. Ram Bhadur and his family get the grass that they need—and the plants quickly bounce back. People benefit, and the scheme seems to offer proof that vital wild grasslands can be saved. 19

Content Considerations

1. What are the basic features of the grassland ecosystem? (An ecosystem is a natural community of organisms and their environment.)
2. List reasons why natural prairie grasses have diminished around the world.
3. What consequences follow grassland destruction, according to the writer?

The Writer's Strategies

1. This piece begins and ends with a narrative about one man in Nepal. Why did the author choose this framing device?
2. What is the tone of this piece? Is it shrill? Angry? Calm? Dispassionate? How does the tone affect the believability of the piece?
3. Near the end of this article the author offers a brief list of successful grasslands conservation programs. Why?

Writing Possibilities

1. Sunquist suggests that no preservation attempt can succeed unless "local people can derive some benefit from the venture." Working with a group of classmates, list the kinds of benefits that might persuade others to participate in a preservation program. Then write letters to the editor of your local newspaper urging readers to adopt your suggestions.
2. Focus an essay on the reasons why people who know "how to live in harmony with the land" refuse to do so.

SANDRA POSTEL

Halting Land Degradation: Abuse It and Lose It

When land degradation or desertification occurs, the land is no longer able to protect itself from elements that erode, deplete, and ruin. In the following report Sandra Postel explains direct and indirect causes of desertification; both kinds are related to basic issues regarding how land is used.

The way humans use land—for planting, for grazing, for building—affects the land's ability to sustain life. In the last few centuries much has been learned about how to minimize damage, how to set land into its restorative cycle, how to produce more by abusing less. But this knowledge is often offset by social and economic structures. Problems remain, and some have even intensified as the earth's population increases, making more and varied demands upon the soil and surface of the land.

Before you read the following report, published in Focus *in 1989, list possible causes for desertification. Then try to relate those causes to the societal or economic conditions from which they might stem.*

MAJOR DROUGHTS in Africa, China, India, and North America over the last four years have spotlighted an immutable reality for much of the world: despite a myriad of sophisticated technologies and scientific advances, humanity's welfare remains tightly linked to the land. Millions in these drought-stricken regions have watched their economic futures—or in the worst case, their chances for survival—fade. For the first time in more than a decade, global food security has come into question.

While these headline-making events ignite concern for a few weeks or months, the true tragedy goes unnoticed. Much of the world's food-producing land is being sapped insidiously of its productive potential through overuse, lack of care, or unwise treatment—a process scientists call "desertification." While the term conjures up images of the Sahara spreading beyond its bounds to engulf new territories, its most worrisome aspects are less dramatic. Desertification refers broadly to the impoverishment of the land by human activities. Perhaps a more appropriate term is land degradation, which we will use interchangeably with desertification.

Each year, irreversible desertification claims an estimated 6 million hectares (14.8 million acres) world-wide—a land area nearly twice the size of Belgium, lost beyond practical hope of reclamation. An additional 20 million hectares (49 million acres) annually become so impoverished that they are unprofitable to farm or graze. Most of the affected land, however, lies on the degradation continuum, somewhere between fully productive and hopelessly degraded. Unfortunately, much of it is sliding down the diminishing productivity side of the scale.

Although the technologies to restore resilience and productivity to stressed lands exist, so far the political will does not. The majority of people affected are poor farmers and pastoralists living at society's margins and lacking a political voice. A lasting victory over land degradation will remain a distant dream without social and economic reforms that give rural people the security of tenure and access to resources they need to improve the land. And with degradation rooted in excessive human pressures, slowing population growth lies at the heart of any effective strategy.

Land degradation may be difficult to rally around and adopt as a cause. Yet its consequences—worsened droughts and floods, famine, declining living standards, and swelling numbers of environmental refugees—could not be more real or engender more emotion. A world of 5.1 billion people, growing by nearly 90 million each year, cannot afford to be losing the productivity of its food base. Without good land, humanity quite literally has nothing to grow on.

More than a decade has passed since government representatives from around the world gathered at the United Nations Conference on Desertification. Held during the summer of 1977 in Nairobi, Kenya, the meeting followed on the heels of a devastating drought that struck much of western and north-central Africa from 1968 through 1973. It focused the world's attention for the first time on the problems and prospects of fragile lands.

Out of Nairobi came a Plan of Action to Combat Desertification, which recommended 28 measures that national, regional, and international institutions could take to halt land deterioration around the world. Sadly, the action plan never got off the ground, a victim of inadequate funding and lack of sustained commitment by governments. When severe drought and famine repeated themselves in Africa in 1983 and 1984, again bringing tragedy, Canadian meteorologist F. Kenneth Hare remarked grimly: "It is alarming that ten years later . . . the news stories should be so familiar."

Seven years after the Nairobi conference, the United Nations Environment Programme (UNEP) took a more careful look at the overall status and trends of desertification worldwide. This included sending a questionnaire to 91 countries with lands at risk. These data—incomplete, sketchy, and lacking in geographic detail though they are—remain the best available and are more than sufficient to grasp the severity of the problem.

According to UNEP's 1984 assessment, 4.5 billion hectares (11 billion acres), or 35 percent of the earth's land surface, are threatened by desertification. Of this total—on which a fifth of humanity makes its living—three fourths has already been at least moderately degraded. Fully one third has already lost more than 25 percent of its productive potential.

What lies behind these numbers is a deteriorating relationship between people and the land that supports them, a situation all the more tragic because people themselves are not only degradation's victims but its unwitting agents. The four principal causes of land degradation—overgrazing on rangelands, overcultivation of croplands, waterlogging and salinization of irrigated lands, and deforestation—all stem from excessive human pressures or poor management of the land.

Rangelands and the animals that graze them play an important role in the global food supply. The 3 billion cattle, sheep, goats, and camels that roam the world's pastures can do something humans cannot: They convert lignocellulose—a main product of photosynthesis that is indigestible to humans—into meat and milk that provide the human population with high-quality protein. Shifts to livestock fed on grain or forage have diminished dependence on grazing animals in some regions. But in much of Africa and the Middle East, and in parts of India and Latin America, roaming ruminants still underpin subsistence economies and support millions of pastoralist families.

Degradation on rangelands mainly takes the form of a deterioration in the quality and, eventually, the quantity of vegetation as a result of overgrazing. As the size of the livestock herds surpasses the carrying capacity of perennial grasses on the range, less palatable annual grasses and shrubs move in. If overgrazing and trampling continue, plant cover of all types begins to diminish, leaving the land exposed to the ravages of wind and water. In the severest stages, the soil forms a crust as animal hooves trample nearly bare ground, and erosion accelerates. The formation of large gullies or sand dunes signals that desertification can claim another victory.

Ironically, years of abundant rainfall—seemingly beneficial to pastoral peoples—can often sow the seeds of further degradation and hardship. During wetter periods the area suitable for grazing expands, leading pastoralists to increase the sizes of their herds as insurance against another drought. When the next dry spell hits, the number of livestock exceeds what the reduced area of grass can sustain. The result is overgrazing and accelerated land degradation, a pattern most visible in Africa, where more than half the world's livestock-dependent people live.

Livestock watering holes, a popular feature of international development projects, have contributed to rangeland desertification as well. Cattle cannot go more than three days without water, so digging water holes to sustain herds during dry seasons seems logical. But the concentration of livestock around the watering points leads to severe localized overgrazing, which gradually spreads outward from this central area. When drought strikes again, the animals rarely will die from thirst, but rather from lack of forage.

For more than two decades, farmers in south-central Niger have lamented in Hausa to development workers that *kasar mu, ta gaji* "the land is tired." Peasants in western parts of the country strike a more ominous chord in Zarma with *Laabu, y bu,* "the land is dead." The phrases aptly depict land suffering from overcultivation, which now affects at least 335 million hectares (827 million acres) of rainfed cropland worldwide (excluding the humid regions), more than a third of the global total.

Agricultural land left without vegetative cover or situated on steeply sloping hillsides is subject to the erosive power of wind and rainwater. An inch of soil takes anywhere from 200 to 1,000 years to form; under the most erosive conditions, that same soil can be swept off the land in just a few seasons. Erosion saps the land's productivity because most of the organic matter and nutrients are in the upper layers of soil. According to one estimate, about half the fertilizer applied to U.S. farmland each year is replacing soil nutrients lost through erosion. In addition, erosion degrades the soil's structure and diminishes its water-holding capacity. As a result, crops have less moisture available to them, which, especially in drier regions, is often erosion's most damaging effect.

Only a handful of countries have attempted to estimate their rates of soil loss in any detail, so the magnitude of the problem worldwide is difficult to gauge in other than broad terms. One useful measure is the load of earth materials carried to the sea by rivers and streams. This figure totals at least 20.3 billion tons per year, which includes 15.3 billion tons of suspended sediment, 4 billion of dissolved material, and 1 billion of coarser bed load. Since this accounts only for material reaching the sea—and excludes, for instance, sediment trapped behind dams—it underestimates the total amount of soil lost from the land.

A look at the geographic distribution of these sediment loads gives a quick sense of where severe erosion is taking place. A 1987 expedition of the internationally sponsored Ocean Drilling Program estimated that the Ganges and Brahmaputra rivers on the Indian subcontinent transport 3 billion tons of sediment to the Bay of Bengal each year, far more than any other river system. The sediment fan on the floor of the bay now spans 3 million square kilometers (1.6 million square miles). Much of that sediment originates in the Himalayas, where deforestation and cultivation of steep slopes in recent decades has added to millions of years of massive erosion from natural geologic activity.

In China, the Huang He (Yellow River), with a drainage area half that of the Ganges-Brahmaputra system, carries more than a billion tons of sediment to the sea each year. About half of it comes from the Loess Plateau, in the Huang He's middle reaches, among the most water-eroded areas on earth. Deeply cut gullies and denuded hillsides span 430,000 square kilometers (165,900 square miles), and erosion rates average some 65 tons per hectare annually.

Today roughly one third of the world's food is grown on the 18 percent of cropland that is irrigated. Irrigated fields typically yield two to three times more than those watered only by rain, and, because crops are protected from the ravages of drought, provide a crucial degree of food security.

Unfortunately, poor irrigation practices have degraded much of this valuable cropland. Over time, seepage from canals and overwatering of fields cause the underlying water table to rise. In the absence of adequate drainage, water eventually enters the root zone, damaging crops. Farmers belonging to a large irrigation project in the Indian state of Madhya Pradesh have referred to their once fertile fields as "wet deserts."

In dry regions, salinization usually accompanies waterlogging as moisture near the surface evaporates, leaving behind a layer of salt that is toxic to plants. An air view of severely salinized fields can give the impression they are covered with snow. UNEP's assessment placed the irrigated areas damaged by salinization at 40 million hectares (99 million acres).

About half of the affected area is in India and Pakistan, but other regions suffering from salinization's effects include the Tigris and Euphrates basins in Syria and Iraq, California's San Joaquin Valley, the Colorado River basin, China's North Plain, and Soviet Central Asia. In the Soviet republic of Turkmenistan, the government blamed salinization for a cotton harvest shortfall of one third in 1985. Meanwhile, another salinity threat has struck Soviet Central Asia: Because so much irrigation water is being siphoned off from the two major rivers flowing into the Aral Sea, the sea's surface area has shrunk by 40 percent since 1960 and its volume has dropped by two thirds. Winds picking up dried salt from the basin are now annually dumping some 43 million tons of it on more than 15 million hectares (37 million acres) of cropland and pasture surrounding the shrinking sea.

The last major cause of land degradation—deforestation—cuts across all land use types. By accelerating soil erosion and reducing the soil's water absorbing capacity, deforestation often accentuates the effects of overcultivation and overgrazing. Moreover, though forest clearing in humid regions was not included in UNEP's desertification assessment, in many cases it results in a net decline in the productivity of land. Most of the nutrients supporting moist tropical forests are held in the vegetation, so forest clearing removes them as well. Having lost its inherent fertility, the land cannot long support intensive agriculture. Large areas of pasture and cropland that replaced tropical forest in the Brazilian Amazon, for example, have been abandoned.

The U.N. Food & Agriculture Organization (FAO) estimates that each year 11.3 million hectares (28 million acres) of tropical forest are lost through the combined action of land clearing for crop production, fuel-wood gathering, and cattle ranching.

Recent satellite data from Brazil, however, indicate that 8 million hectares (19.7 million acres) of forest were cleared in 1987 in the Brazilian Amazon alone—strongly suggesting that widely cited FAO figures are far too low. Some portion of deforested land goes into sustainable land uses such as traditional shifting cultivation, which includes a fallow period that restores the land's fertility—but the bulk of it does not. In the tropics today, deforestation usually translates into land degradation.

Desertification's direct causes—overgrazing, overcultivation, salinization, and deforestation—are easy to enumerate, but only by grappling with the complex web of conditions leading to these excessive pressures is there hope of stopping desertification's spread. Though they vary greatly from place to place, these underlying forces generally are rooted in population densities greater than the land can sustain and, more fundamentally, in social and economic inequities that push people into marginal environments and vulnerable livelihoods.

Content Considerations

1. What practices lead to land desertification or degradation?
2. How is drought related to desertification?
3. How does land degradation affect what happens to rainfall?

The Writer's Strategies

1. How is this article organized? How does this organization help your understanding?
2. The author discusses land use in several geographic areas—Africa, China, India, North America, South America. What does she gain by including such diverse areas in her article?

Writing Possibilities

1. Summarize the causes of land degradation and how these causes relate to weather and population.
2. Working with classmates, speculate upon how its human population has affected the land of the area in which you live.
3. How might land degradation in India or Africa affect life in the United States?

JOHN MADSON

A Wilderness of Light

The subject of this essay is the prairie, the open spaces of the middle states that once rolled across America in waves of wild grasses and flowers. The space is still there, of course, but now it is filled with farms, pastures, and cities, for that is how its residents have used it.

Early settlers found the prairie land irresistible. Its soil was rich with thousands of years of growing in undisturbed cycles; few forests had to be cleared to make way for crops. John Madson, author of several books and many articles, gives a brief history of the land and settlers of this region, speaks of soil and things that grow, and evokes questions about what we give up in order to gain.

Before you read Madson's essay, one in a collection entitled Stories From Under the Sky *(1961), quickly describe your impressions of the prairie—how it looks and how it is used. As you read, compare your descriptions to those he gives.*

NOT SO LONG AGO, between the eastern forests and the buffalo plains, there was a sea of grass and flowers.

The midland of the continent was open, rolling, long-grass prairie and settlers emerging from the woods and snug fields of the east were stunned by a blaze of sunlight and an immense sweep of sky.

The old settlers said it was like a sea with long, heavy groundswells. It was neither angular and abrupt nor flat and monotonous, but a vast reach of grassland broken with stately groves and dissected by timbered stream valleys. One man wrote that where the groves crowded into the prairie the effect was like that of a rugged shoreline, with the surrounding forest indented to form bays and headlands in the grass and sometimes—when regarded across leagues of prairie—the distant forest was like "a dim shore beheld at a great distance from the ocean."

It was a unique, unliterary wilderness without pestilential swamps or black walls of forest. This was an open wilderness of birds, flowers, grass and sun. In the groves and timbered valleys were elk, deer, bear and turkey. On the open prairie were grouse beyond number and the eastern fringes of the great bison herds that blanketed the short-grass plains beyond the Platte.

A large land, whose breathtaking immensity of sky disturbed many travelers. It was this pitiless quality of *openness,* a relentless intimidation of wind and distance—even more than the marauding Sioux—that drove some settlers back to the east. It dwarfed them beyond their endurance and they fell back from this new land's largeness and its jarring excesses of light and space.

There was, and is, no other region in North America with such climatic extremes. With no large water bodies to temper the weather and no forests or mountains to check the wind, the prairie summers and winters were elemental entities, neither possessing any of the attributes of the other. The first storms of autumn might leave snow three feet deep on the level prairie and fill some sheltered ravines with drifts that lingered until early June. After the blizzards came weeks of bright, bone-cracking cold with seventy degrees of frost, only to be followed by a summer with blazing afternoons of a hundred and fifteen degrees and incessant winds that desiccated the prairies and their people and drove the great fires that consumed them both. Even if the settler survived winter, wildfire, sick oxen, renegade Wahpekutes, diphtheria and tetanus, there was still the aching loneliness and the prison of empty, illimitable horizons.

But it was worth it, and people crowded into the open land.

For never has there been such a place of incredible richness. This land had not invested its strength in trees, but had renewed and rebuilt itself annually as the lush prairie grasses and forbs had died and enriched the earth. There evolved a soil that was light, black and fluffy with organic matter. Even after the hardest rain and hottest sun, a man could walk over prairie loam and scuff his boots in it, and the soil was soft and flocky. It was much later, after the soil had been mined by decades of cash-grain farming, that it became as heavy and solid as the men who worked it.

Today the long-grass prairie is nearly gone, although some scraps of it linger on a few farms, along old railroads, and in neglected country cemeteries. I know of only a few hundred-acre prairies in Iowa today, and these anachronisms are invariably ringed by farmers who eye them as wistfully and hungrily as orphans regard a jar of peppermints. Such men have fine farms and all the good land that they can comfortably manage, but they know that the keen edge of that land has been blunted by a century of cultivation. They ache to drill their modern seed into an ancient earth that has been storing up richness for ten thousand years.

The primeval soils of these prairies are also coveted by scientists. A few years ago a botanist with the Atomic Energy Commission told me that only virgin prairie areas can give chemists the patterns of our best original northern soils—patterns that will serve as reference indices in the event of atomic contamination of croplands. This could be vital, they say, in assaying the effects of atomic reactors and reactor wastes on soils and plants—and the atomic reactor age is upon us.

I'm too dull to be a farmer and too sharp to be an atomic scientist. But for reasons of my own I also prize the prairie, and am selfish enough to believe that those reasons are as good as any man's. I think we owe something to the life forms that have been hard-put to survive our technology, whether they be whooping cranes, prairies, or schoolteachers who spank.

One of my favorite places is a hundred and eighty acres of Iowa prairie that has never been plowed nor grazed. It has not been invaded by volunteer crop plants nor even exotic weeds, for the native prairie species are well-entrenched in their home sod and doggedly resist invasion. Except for a barn on the horizon and the fences that bar the covetous Holsteins and Herefords, the little patch of prairie is unchanged from the last ice age.

By late August, as I write this, that prairie is a grandmother's quilt of form and color. The ironweed is heavy with purple, flat-topped blooms and here and there are a few arrow-straight, magenta spikes of blazing-star. Most of the cornflowers are gone by this season, but you can see the rattlesnake-master with its spiny leaves and odd, bulbous seed heads and if you look carefully, you can find low shrubs of redroot with their huge, plow-busting, mahogany rootstocks.

There are wild grasses everywhere, the big bluestem waving higher than a man's head in lofty stands that were once cut for wild hay—a grass that makes the finest bedding for hunting dogs that I know of. There are open, sun-washed flats of little bluestem, dropseed grass and buffalo grass with the late summer wind running across them. Being in such a place is being in yesterday, and sometimes I can look across this prairie and feel my great-grandfather coming toward me, with his bullhide boots and linsey-woolsey breeches and walking with a two-furrow stride.

If you are a woman standing in that prairie, you can look at flowers that you have never seen before and may never see again. Think of a sod house with the nearest woman forty miles away; of hearing your children struggle with McGuffey's Reader by the light of bayberry candles.

If you are a man, walk out into this prairie and say: "I have just served in the war with Mexico, I was wounded at Churubusco, and Congress has given me this quarter-section as my land warrant."

If you are a small boy, find a soft place in a patch of yellow stargrass and dream of Injuns.

Content Considerations

1. What difficulties did the prairie present for early settlers?
2. Why do atomic scientists prize untilled prairie soil?
3. Paraphrase Madson's description of what prairie regions were like before settlers arrived.

The Writer's Strategies

1. How does the title of the essay relate to its content and tone?
2. The author of this essay counts heavily on a historical perspective. Outline this perspective, then decide its effectiveness. Is it missing anything?

Writing Possibilities

1. Many ideas exist for "wilderness" that are outside the exotic visions usually evoked by that word. Write an essay in which you explore one such wilderness of your own.
2. Support or argue with the notion that rich land is wasted if it is not being farmed.
3. With classmates, discuss how Native Americans might have viewed the prairies before European settlers arrived. Which of Madson's ideas might they have shared?

Anne W. Simon

The Thin Edge

> The coast that separates land from water serves a variety of natural functions: It protects the earth from the sea; it provides the habitat in which countless fish, birds, and plants grow—organisms whose relation to both sea and land is intricate and complex. Yet the coast is fragile, and the ways in which we use it may cause damage that cannot be seen or, sometimes, even suspected. In most industrialized countries, in fact, natural coastlines do not exist. They have been put to other uses.
>
> Author Anne W. Simon argues in The Thin Edge: Coast and Man in Crisis *(1978)* that transforming natural coastlines to human uses has dangerous ecological consequences. Beaches, oceans, wetlands, and estuaries are part of an ecosystem; development of coastlines threatens the way all parts of the system work. The piece that follows is the first chapter of her book, which received the American Book Award for science. Before you begin to read, make a list of the changes humans have made to the earth's coasts.

THE COAST, that bright thin edge of the continent where you can sit with your back to the crowds and gaze into seemingly infinite space, is now a theatre of discovery. On the seashore where terrestrial life began, we have to use all the wits man has developed to figure out how life can continue, how to design our complex, fast-moving, energy-consuming existence without destroying nature's system of life support in the process. It is a compelling adventure. Wherever it leads, neither man nor coast will be the same again.

Survival on land and in the sea depends on a functioning coast. The coast keeps us from drowning, maintaining the present global balance of one-third land, two-thirds water. It nurtures fish and shellfish, birds and plant life, as it nurtures the ocean, the essential source of a third of the world's oxygen, the largest source of its protein. Its multiple processes are arranged in dozens of natural systems with myriad parts, each neatly slotted into an operation as sophisticated as the latest computer, as intricate as a vast jigsaw puzzle. Its abilities are exquisite in their detail, awesome in their grand accomplishments.

Universally we yearn for the coast with an inexplicable need for its serene horizons, for the endless, timeless rhythm of waves on rough rocks or smooth beaches, for the amplitude—plenty of sand, water, seagulls, seaweed, a harvest of sea-worn pebbles and minute sea animals in every wave. Here where the sea is shallowest, land is lowest, rivers slowest, there is dynamic interchange between water and earth, a phenomenon often believed to make passions run higher, emotions keener, the sense of well-being quickened. We come closer to our primitive selves on the thin edge, at once nurtured and excited by it.

Ever-obliging, its generous compliance has provided poetry, joy, convenience and profit. But now we glimpse its deeper nature, stern, inflexible, firm in principle and in its limits. Innocent use has been cut short in the 1970s by what has been discovered about the coast, most of it in the past half century, much in the last decade, and by what is tantalizingly beyond our present vision. The existing information is unequivocal; the coast is different from any other place on earth and has different requirements. There is no man-made substitute for its manifold natural functions. We do not want to get along without a working coast and we now realize that we literally cannot get along without it.

Around the globe, coast functions falter under the encrustations of twentieth-century civilization. The east coast of the United States is vividly representative of any coast, anywhere, magnifying every coastal dilemma in its 28,000 miles of shoreline—coast, offshore islands, sounds, bays, rivers and creeks—stretching from Maine to Florida, from rigorous to tropical climate, the rocky northern shore testifying to its glacial past, the long stretches of wide southern beach, having escaped the glacier, relatively flat. Thirteen states have a slice of this coast under their domain and separate laws. It is heavily developed, industrialized, crowded, with hardly distinguishable towns wedged into the megalopolis solidifying between Boston and Washington, although there are, almost miraculously, still a few places—Down East Maine, some of Georgia's barrier islands—almost as they were when our ancestors first set eyes on their virgin marvels.

Ever since the last Ice Age gave way to a warming sea, the coast had been a magnificently productive system. Enormous trees, enormous quantities of fish and fertile black soil on the Maine coast amazed early explorers. The few remnants of the towering forests are remarkable today where they have survived, mementoes of a time when they covered the shore: ". . . goodly tall Firre, Spruce, Birch, Beech, Oke very great and good," says James Rosier, clerk on Sir George Waymouth's *Archangel*, sailing on a fair June day in 1605 past Monhegan Island, "a meane high land," to George's River "as it runneth up the maine very nigh forty miles toward the great mountains." Upon the hills, Rosier says, "notable high timber trees, masts for ships of 400 tons."

The adventurous men in their small ships were no less amazed by the waters teeming with huge fish of all varieties, a sight we can only imagine. "While we were at shore," Rosier relates, "our men aboard with a few hooks got about thirty great Cods and Haddocks, which gave us a taste of the great plenty of fish which we found afterward wheresoever we went upon the coast." The *Archangel* found "Whales, Scales, Cod very great, Haddocke great, Plaise, Thornbacke, Rockefish, Lobster great, Crabs, Muscles great with pearls in them, Tortoises, Oisters"—the list is long. Haddock and lobsters were so thick in the waters that some fishermen scooped them out with a bucket, salting them down in the hold for the long voyage home.

Even the much-traveled Captain John Smith was impressed. "Besides the greatness of the timber . . . the greatness of the fish and the moderate temper of the air," he writes, "who can but approve this a most excellent place for health and fertility?"

We can still see something of this excellent Maine coast. So too on a few remaining wilderness islands to the south, lush and semitropical, there are still wide sand beaches which sweep into a protecting line of dunes, while, behind them, gargantuan live oaks, pines and palms combine in a primeval forest, an enchanting world apart. When Giovanni da Verrazzano, still searching for the way to Cathay, explored this coast in 1524, wilderness was everywhere. He saw the beaches, dunes and estuaries that we struggle to keep fragments of. "The shoare," he says, "all covered with small sand, and so ascendath up for the space of 15 foote, rising in form of little hills . . . small rivers and armes of the sea washing the shoare on both sides as the coast lyeth."

It was a land "as pleasant and delectable as is possible to imagine." And on it, Verrazzano reported to his French sponsors, a delectable population, "people of color russet [who] go altogether naked except that they cover their privie parts with certain skins of beastes . . . which they fasten onto a narrow girdle made of grasse very artificially wrought, hanged about with tayls of divers other beastes."

In the millennia that man has inhabited the ancient edge of this continent, he has taken the short-term view of gratifying his desire for pleasure and security, shooting a few birds in the marsh for dinner, trapping furred animals, going fishing. Great piles of shells unearthed by archeologists, some charred, some halved, testify to the enjoyment of clambakes and clams on the half shell as early as 4000 B.C. Along with the shells in the middens, there are bones of otters, seals, whales, all sorts of fish, suggesting that their fur, fat and meat were used and valued. Coast dwellers took what they found when they wanted to, there being no apparent reason not to.

The coast's abundance welcomed colonists with the necessities—the fertile soil, wild game and fish, great timbers—that made settlement of the new world possible. The number of settlers was small, their requirements modest; the shore provided food and shelter and the water their only means of transportation. It was not until 1722 that for the first time a team of horses was driven from Connecticut to Rhode Island on a dirt path, winding through dense woods from one coastal clearing to the next. The coast, apparently constant and indestructible, continued to perform its functions.

It continued, in fact, through centuries of settlement and development, continued valiantly through industrialization. It continued, if somewhat less efficiently, as the east coast population zoomed from 29.8 million people living within fifty miles of this narrow strip in 1940, to 48 million people in 1970, almost a quarter of all Americans, and the proportion increases three times as fast as the national average. All U.S. coasts together (including the Great Lakes) contain the nation's seven largest cities, account for 53 percent of its population and 90 percent of its population growth, and it is anticipated that, by the year 2000, two hundred millions will squeeze themselves into smaller and smaller segments of the thin edge.

Unprecedented numbers of people swarming onto the coast make unprecedented use of it in this technological age. The ravenous growth society devours many parts of the continent for its expansion requirements, but the thin edge, with its special attributes, is most delectable of all. It is a magnet for growth.

It is a magnet for people. On every coast the people business burgeons, lining the shore with stacked condominiums interspersed with mobile-home parks, marinas, dense second-home developments. A roaring tourist industry ricochets off coastal highways with its accompanying eateries, motels, neon-lighted putting greens. Coast recreation—surfing, swimming, snorkeling, sport fishing—escalating to an average ten days per year per American, is a profitable business.

The great bays of the east coast—Chesapeake, Delaware—and other estuaries are a magnet for heavy industry, refineries, power plants using water for cooling processes. Forty percent of the nation's industrial complexes edge its estuaries, 50 percent of its manufacturing facilities, and the east coast has more than its share. On the Delaware River, for example, utility companies plan 42 new power plants by 1986; one of these alone will evaporate 54 cubic feet of water per second, a loss equal to that of a small city. The activities of all these people and industries bring waste in unprecedented quantity to the shore, dumping it into the water.

Offshore waters are a magnet for the oil business. The colonists' lifeline becomes the tankers' trek as they ply and break up alongshore. Superships, with moorings approved for the Gulf coast in 1977, will soon rock in east coast offshore swells, and tracts of its Outer Continental Shelf have been leased to oil companies, now preparing to start full-scale oil fields in the Baltimore Canyon depths and in Georges Bank off Massachusetts, relatively shallow water warmed by the Gulf Stream and long a fabled fishing ground.

In the last ten years the coast's magnetic pull has become stronger than ever—more industry, more oil, more people, hotels, motels, boatels, more sewage, more waste . . . and more pressing evidence that the coast has limits, an idea hardly known and little considered until now. Sometimes quietly, sometimes violently, the coast is informing us that there is a saturation point beyond which its natural functions no longer flourish, often diminish, or simply cease.

The fastest-growing area in the United States is said to be the Florida Keys, a sixty-mile strip of islands and reefs some ten miles wide. At the present rate, the two millions who now crowd this reef will increase to ten millions before the century ends. Under the jammed Keys, reef-building corals, the only such colony in U.S. continental waters, are dead, their massive branches skeletons, covered with white spots where the organisms once grew. If you go snorkeling there, gliding past the dead coral mass, any fish you see, a dwindling population, are likely to be diseased and deformed. Biologists say coral requires warm, well-oxygenated water, that too much sewage and too much silt from dredging and filling for new buildings have suffocated the coral that built the Keys that are attracting humans faster than any other place in the country.

A coral reef, suffocated by the human life it supports, is a signal, quiet enough to go unremarked in the rush to cover the remaining inches of the Keys with concrete. Although such action can't further harm the coral, already smothered, it has other effects, noticeable wherever man transforms the soft sand shore into an inflexible wall.

The coast protects higher land by using wave and wind energy and gravity to build sand barriers that resist storms and pounding surf, as in the surprisingly sturdy barrier beaches that guard much of the east coast. In this era when the sea level is rising throughout the world, water encroaches on the shore and the coastline retreats. These barrier beaches reveal the remarkable ability to move inland along with the shore, rolling over on themselves to migrate with their entire ecosystem—beach dune, marsh—intact. A North Carolina island has just performed such a giant somersault in less than a century. Pace is the key: the shore must move at its own speed, when and where it will. Interfere with its pace and it will neither guard nor turn somersaults. Before this life-saving information was discovered, much of the shore had been covered with mammoth concrete development, preventing free movement of sand and water, a matter of considerable concern. It will be of more concern if the hurricane cycle, which has been in an unusual and seductive lull during the 1970s, years of the most concentrated seaside building, returns as expected, roaring along the concretized and thus dangerously vulnerable Atlantic coast. "The cost in dollars and lives of the next Camille-size hurricane will be staggering," a scientist predicts.

Behind the barriers where rivers empty into protected bays, the coast manufactures food for marine life by a mix of fresh and salt water, wetland grasses, sun and tide, delivers it to coastal species, for many of which these sheltered spaces are a necessary habitat during part or all of their lives. It has recently been found that, acre for acre, wetlands are the most productive land on earth. Without protective barriers they will drown. Already, thousands of such acres along the eastern seaboard have been irretrievably lost; they were filled in, converted to high land, dredged or otherwise stressed, before their value became known, and even after.

The man-coast love-hate relation changes with each discovery of a new facet of coastal character. We begin to see limits beyond which the coast cannot function, where its nurturing nature turns hostile, antagonistic to life, suffocating, drowning, poisoning. The signals are ever stronger, ever deadlier.

The Glorious Fourth weekend of the nation's Bicentennial, when New York was momentarily a festival city, applauding the muster of the tall ships in its rivers and harbors, the skipper of the *Faye Joan,* trawling for whiting off New Jersey, winched in his net, spilled a thousand pounds of fish on the deck. "The contents stank," David Bulloch, an observer, says. "The fish were dead, a few dying, most decaying. The crew worked, barely breathing, to shovel the fish over the side. The urge to vomit was overpowering." Diving to investigate, Bulloch and others found unusual dark brown water and below it, on the cold bottom, piles of dead fish, crabs, lobsters, mussels, "a foot-thick black mass of decay swaying with the surge of the sea." By August the killing sea extended for three thousand square miles. "Everything was dead," a microbiologist on the scene reports. "Nobody can remember anything like this." The level of dissolved oxygen in the water, required to sustain marine life, fell to zero, overpowered by the torrents from the celebrating city's sewers.

A different coast from the natural shore explorers found, different from the settled coast of the start of this century, different, even, from what it was just a decade or two ago when we apparently passed many of its limits without even knowing it! Each day we venture further into the unknown character of a world without a working coast, to date the only generation to experience such terror. But each day we are better equipped to stake out the limits for man on the coast. We begin to decode the signals that issue from the thin edge.

Each change of one part of the coast system affects other parts.

Some connections have been discovered, such as the linkage of barriers and wetlands; some are still unknown. It may be that the shore is so complex that we will never completely quantify the results of a change, that we must always play Russian roulette with our coastal intrusions. Consider the mid-seventies decision to explore for oil in Georges Bank. The results circle out like ripples from a pebble thrown into a quiet pond, with no end in sight. One such ripple catches up coastal flora and fauna in an interconnection never imagined.

Marine biologists, of late particularly interested in the common seaweed kelp because it can be cleanly, cheaply converted to fuel, are surprised to find bald areas in underwater kelp forests off the northeast coast. More than eight times the expected number of sea urchins in great herds are grazing the kelp down to bed rock, John Culliney says in *The Forests of the Sea*. Some suspected increased sewage in the waters, enjoyed by the prickly half-sphere animals, might be responsible; others believed that overharvesting lobsters, the urchins' most avid predators, could account for the multiplying urchin armies and the vanishing kelp.

There are fewer lobsters to eat urchins and to be eaten by man for a reason that these strange succulent creatures have long kept hidden. Only in 1970 was it discovered that some lobsters, primitive, awkward and slow-moving as they may seem, have each fall for thousands of years walked 150 to 200 miles across the rock and sand bottom of the offshore sea to Georges Bank, enjoying the winter in the nonfreezing temperatures there, walked back again in spring to coastal waters to copulate under sheltering rocks where, in a miracle of precise timing, the females shed their hard shells to make it possible for the waiting males to enter their bodies and deposit sperm.

Something new has happened in the underwater world of Georges Bank, a change so fascinating to lobsters that they hang around like hooked junkies, Culliney says, the vernal journey back to shore and its primal purpose forgotten. There is oil in these waters now, oil from exploratory digs, oil from tanker spills, more oil than ever before, and the lobsters, it has just been found out, are mightily attracted to it, will attack and eat kerosene-soaked paper in laboratory tanks, seek it out in their wintering grounds. If Georges Bank oil wells start in earnest, propagation in the wild of the migratory branch of *Homarus americanus* may be over forever.

As a single change of balance it could be inconsequential; sea urchins are unlikely to take over the world, lobsters can, perhaps, be successfully cultivated. But as representative of countless changes, revealed or still unknown, the kelp-urchin-oil-lobster cycle is grave and deeply troubling.

The change is an archetype coast module, the module that appears in hundreds of fragments and forms, in unexpected places with sometimes inconvenient, sometimes punishing, sometimes murderous effect. The more such a module is pieced together, the clearer it becomes, suggesting as it does that the essential coast character is its intricate, indisputable interconnection. Discovery of the coast's amazing systems advances our knowledge of this interlocking nature of the thin edge where we stand, precariously, listening to its silent scream.

Content Considerations

1. Summarize the author's reasons for asserting that coasts are important.
2. What specific examples does Simon give to support her contention that humans are destroying the coastal ecosystem?
3. Why does industry cluster along coasts, estuaries, and rivers?

The Writer's Strategies

1. What images are evoked in Simon's contrast of Maine 350 years ago and the dead waters off New Jersey at the nation's Bicentennial? What does the author achieve through this contrast?
2. Besides historical journals, what other evidence does Simon use in building her case for protection of coasts?
3. How does Simon try to persuade you to share her views about coastal ecosystems?

Writing Possibilities

1. Write about a coastal area with which you are familiar, focusing on how Simon's assertions might hold true for that area.
2. Write an essay speculating on the reasons so many people choose to live on coasts and explaining whether you share those reasons.
3. Brainstorm with other students about the possible effects—biological, economic, political, and other—of restricting development of coasts. Write about your conclusions.

THOMAS A. LEWIS

The Frontier Dream We Call Alaska

In many ways the land that makes up Alaska is a special case. It is rich in natural resources and relatively unpopulated; large portions of it remain wild and nearly inaccessible. Many people treasure its wildness and wish to see it remain; others see in it opportunities for development, ways of gathering resources to bolster both private and public economies. Such gathering necessarily tames the land, domesticates it by putting people in charge.

Alaska is in a unique position. Current laws prevent its being settled and developed the way the rest of America has been, yet the urge to subdue this land is as great as it was for other portions of America as westward expansion pushed the boundary toward the coast. Some people call for restraint, for leaving the land as it is. Others see idle land and think it wasteful if not producing material benefit for human beings.

Before you read the following pages, predict what resources and land conditions could be causing conflict in Alaska. Then write a statement expressing your position on land that remains wild and thus "unused."

THIS IS HOW they have always liked to do things in Alaska. The year was 1958, statehood was still several months away and the plan—called Project Chariot—was to create a major new harbor on Alaska's west coast. No long-term development, this was to be an instant harbor, popped open by exploding three thermonuclear bombs with the cumulative power of 160 Hiroshima blasts. It would be, exulted Edward Teller of the Atomic Energy Commission, the first application of "the great art of geographic engineering." It was now possible, he announced, "to reshape the earth to your pleasure. If your mountain is not in the right place, just drop us a card."

The scale of the undertaking, grotesque by ordinary measures, seemed proportional in Alaska, where small often seems invisible. The time lag between implementation and gratification—30 seconds or so—seemed just about right in a culture shaped by gold rushes and oil strikes. Some wild animals might be vaporized and much of the Northern Hemisphere coated with radioactive fallout, but, heck, as the AEC explained, ground zero was "far from any human habitation." In fact, it was only 30 miles from Point Hope, the oldest continuously occupied settlement in North America.

Most of the state's politicians, newspapers, civic clubs and chambers of commerce all got in line, shook their pom-poms and cheered their lungs out on behalf of geographic engineering. And they did succeed in beginning a new era; it just wasn't the era they had in mind. Instead of the advent of the Big Bang school of landscaping, Alaskans saw the country's first environmental impact study detail the appalling effects of all that radiation on the Arctic. They also saw the creation of the first territory-wide coalition of native people, organized to mount an increasingly effective opposition. And they witnessed the birth of Alaska's environmental movement, giving powerful voice to the idea that some things are worth more than economic development.

Thirty years later, the deeper questions raised by Project Chariot about the future of Alaska have not been resolved. The idea of using nuclear weapons to replace bulldozers is no longer taken seriously, but the notion is very much alive that large-scale, long-lasting harm to the environment is a reasonable trade-off for jobs and prosperity. Meanwhile, the antithesis—that such intangibles as biological diversity, wilderness values and wildlife preservation have become more precious than any potential economic benefit—is also firmly established and gaining strength.

Nowhere are these two conflicting ideas pitted against each other with more apocalyptic intensity than in the current struggle for the soul of Alaska. The conundrum is that while Alaska offers various forms of wealth in staggering abundance—vast deposits of coal and oil, hoards of platinum and gold, huge expanses of timber—the exploitation of such riches often involves annihilating Alaska's other resources. "We're talking about natural resources that are impossible to replace," observes National Wildlife Federation President Jay D. Hair. "You can't put a price tag on them."

To be anywhere in the Alaskan countryside is to be immersed in an experience of sensory and emotional intensity that in most other places on Earth is available only fleetingly. During a starry night on the Midwestern prairie, for example, or a Rocky Mountain sunrise, you might feel for a few minutes the chest-tightening awe, the cosmic vertigo, that in Alaska can continue for hours on end.

You can spend days just looking at Denali (most Alaskans prefer not to call North America's grandest mountain by the official name of that Ohio politician McKinley; they use the Tanaina Indian word meaning "Great One"). This is a mountain so enormous that when authorities built a place where the tourists could gaze at it, they backed off 33 miles. Yet when you have been drunk with the grandeur of it for a whole day, when you take a deep breath and drive away, within minutes you may find yourself in some enormous, empty valley, daubed with the rich palette of varicolored lichens, their soothing splendor pierced by the fierce crystal white of some distant, lesser peak. Then your adrenaline red-lines again. After a day of thus having your spirit stretched so far, your head begins to ache.

One tries, but in the end fails, to ignore the fact that almost everything people have put into this magnificent setting is surpassingly ugly. There seems to be little evidence of architectural planning in Anchorage or Fairbanks, where almost all the buildings are square, sullen and often slightly seedy. The roadsides are littered not only with bottles and hamburger wrappers, but with discarded camping trailers, shacks and bulldozers. In a lonesome gorge overlooking a dancing river, the picture-postcard scene is marred by a plywood sign in helter-skelter red: "Canyon Parking $5." The town of Valdez—a major center of the oil and fishing industries and now of the fight for Alaska's soul—includes as sordid a collection of huts in as spectacular a natural setting as you'll find anywhere on Earth.

Living with such extremes—of expanse and scope, of color and light—may have something to do with the Alaskan mindset and with what has transpired here over the years. The Alaska psyche and history seem to consist of lofty peaks and bottomless pits, with little gentle country in between. A tour of the past is a roller-coaster ride from Gold Rush to bust, from postwar boom to bust, from oil field and pipeline boom to—what next? If the past is prologue, it will be a bust, followed by something big and brash. Observes Celia Hunter, who helped found the Alaska Conservation Society to oppose Project Chariot, "There is no limit to the ways Alaskans will deceive themselves about the panacea effect of huge projects."

Big Oil has been Alaska's panacea since the discovery of the massive oil field at Prudhoe Bay, on the barren shores of the Beaufort Sea, more than 20 years ago. The construction of the 800-mile Alaska Pipeline from Prudhoe to Valdez, at a cost of about $10 million per mile, brought a two-year boom for much of the state. Then the daily flow of two million barrels of oil per day brought prosperity for every Alaskan: not only did oil revenues soon account for 85 percent of the state's income, making a state income tax unnecessary, but dividends from an oil-based trust fund provided annual checks for every living Alaskan. Throughout the 1980s, such dividends averaged about $800 per person per year.

Now there are signs that the long ride may be over. It is bad enough that the price declines of the past few years have devastated Alaska's oil-based economy. The insult added to that injury was the wreck of the *Exxon Valdez* last spring. It was the oil spill that everyone said could not happen, that desecrated the south coast and devastated much of Alaska's billion-dollar fishing industry. Its effects on the fisheries, on the region's wildlife, on the increasingly important industry of tourism, on the subsistence of native Alaskans living as far away as Kodiak Island 350 miles to the southwest, and on the oil business itself have not yet been calculated.

"Everyone in Alaska feels desolated," says Celia Hunter, acknowledged mother of the state's environmental movement. "There is a sense of mourning, and of loss. The fabric of life of the whole area has been changed forever."

Perhaps. But there is still hope in many circles that, unlike gold and World War II, this will be the boom that reinvigorates itself. That hope is based on geologists' belief that the country's largest oil field awaits development a few-score miles east of Prudhoe Bay, on the Coastal Plain of the Arctic National Wildlife Refuge—a Maine-sized piece of real estate located hundreds of miles above the Arctic Circle. The oil companies want to go get the oil.

The industry argues it has demonstrated at Prudhoe Bay and with the pipeline that it can get oil out of the ground without undue harm to the environment or wildlife in the region. More oil, it insists, means more prosperity for Alaska and more petroleum for the United States, with reduced worries about a trade balance knocked askew by oil imports and an energy supply that can be subject to the whims of foreigners.

Environmentalists are horrified at the idea of encroaching on a pristine and sensitive wildlife reserve in order to get, at best, a few months' worth of a polluting fuel that is increasingly wreaking havoc on the atmosphere. They insist that the Prudhoe Bay field, with its air pollution, toxic effluents and by-products, its roads and construction, has done substantial and long-term harm to an ecosystem that does not have the resiliency of warmer climes.

"In fact," says Jay Hair, "very little is known about the cumulative effects of pollutants on the Arctic environment or the wildlife that lives there. We should not even talk about leasing the Arctic refuge until we know more about those effects or, for that matter, until this country has formulated a comprehensive energy policy based on conservation practices and renewable energy resources."

Nevertheless, Congress, unable to resolve the dispute for ten long years, was about to unleash the oil industry on the Arctic National Wildlife Refuge last year. Oil exploration and drilling was also about to commence in southwestern Alaska's Bristol Bay area, site of a $1 billion salmon fishery. But in the aftermath of the *Exxon Valdez* disaster, with its revelations of an endless string of broken laws, broken promises and false assurances, everyone, even in Alaska, was too angry to give the oil companies anything.

Needless to say, however, anger fades, and sooner or later the American public and Congress will have to return to the still-unresolved issue: how much of Alaska's wealth can we go after without destroying Alaska? The answer, if we are successful in finding it, will have significance for the rest of the planet.

But even before we can deal with such fundamental questions as how far can we go, and how much time have we left, we must first dispense with a point of order: who will decide?

Alaskans, naturally enough, think they should have the responsibility. They are the people who need jobs and opportunities, and who live amid the scenery and the wildlife. It is, after all, their state.

Or is it? We the people of the United States bought the place from the Russians, and through our federal government we still own 60 percent of its 315 million acres. Eighty-eight percent of the nation's wildlife refuge system, 67 percent of its national park acreage, and 64 percent of its designated wilderness areas are located in Alaska. Maybe it is we who must decide whether Alaska is to be our national park or national oil tank.

Nor is oil the only source of temptation. The world needs more wood, for example, and Alaska has 129 million acres of forests. Much of it is inaccessible, but within easy reach along the coast of the southeastern Panhandle sprawls the 17-million-acre Tongass National Forest, most of which consists of temperate rain forest—400-year-old Sitka spruce and hemlock trees reaching some 200 feet into the air. Driven by a voracious Japanese market and subsidized by federal mandates that often sell logging rights for prices that do not even approach the cost of building access roads, the timber industry has been clearing this last American rain forest with all the zeal of the much-maligned destroyers of the Amazon rain forest.

With about 300,000 acres of the Tongass cut over, and with 10,000 acres of old-growth evergreens being cut each year, Congress began to take action last year to reduce the subsidies and protect more of the forest as wilderness areas. But the struggle continues between those who claim the right to continue making a living by harvesting trees and those who speak for the brown bears, bald eagles, Sitka black-tailed deer and other creatures of the Tongass—creatures that require vast areas of undisturbed forest for their survival.

These controversies, well established, may soon give way to others just as fervent. There is talk of building a new natural gas pipeline from Prudhoe Bay to bring to market trillions of cubic feet of gas while oil production from the region trails off. There is discussion about the fact that just a bulldozer-blade beneath the North Slope tundra (where you can see tire tracks laid down a generation ago) lie perhaps 130 billion tons of coal, two-thirds of the country's known coal reserves. Real estate development, in temporary remission in an economy devastated by low oil prices, promises to resume its spread if any of the other schemes meet with any success at all.

In each case, the vastness of the country clouds the issue. In a sense, Alaska's enormous, pristine expanse invites its own destruction, because it does not seem possible that a little oilfield here, a tiny housing development there, can possibly do significant harm. Even if drilling is allowed in the Arctic refuge, more than 17 million acres of tundra will still remain untouched. For all the irreplaceable old-growth timber that has been cut in the Tongass, 17 million acres of the rain forest still stand.

However, we have no way of knowing which acre, if removed from a grizzly's range, will make it impossible for the grizzly to survive. Which patch of wetland, if destroyed or crowded, will mean the end of some migratory waterfowl's existence. Or which mile of road will alter the migration of a caribou herd in a way that will turn out to be fatal. We will know for sure only afterward.

For much of the rest of the United States, afterward has already come. And now, everyone agrees, Alaska is indeed the Last Frontier. But does that mean our forty-ninth state represents the last chance to make a fortune in oil, timber, coal or development? Or that it is the last chance to save something precious and intangible that has been lost in many other places?

And so we dream our dreams of Alaska, of unimaginable wealth or untouched wilderness, of flourishing wildlife or thriving developments—dreams as big as Cinemascope and, often, incompatible with one another. We laugh about the Project Chariots of the past, yet cannot quite make out what to do about the Project Arctic National Wildlife Refuge or Project Tongass National Forest that are currently confronting us. Meanwhile, our fragile last frontier could be slipping slowly and irrevocably into the age of unintentional geographic engineering.

Content Considerations

1. What two ideas are in conflict about how Alaska will be used or developed?
2. What is the reasoning for allowing Alaskans to determine how their state will be developed? What is the reasoning for allowing the people of the United States to decide?
3. According to the author, what kinds of development threaten to "spoil" Alaska?

The Writer's Strategies

1. How does the title of this piece reflect its content and tone?
2. Evaluate the author's objectivity in this essay by finding examples of an even and balanced approach. Is his personal view revealed? How?
3. To what extent does this article have a persuasive purpose? What techniques of persuasion does it use?

Writing Possibilities

1. With classmates, explore the idea that Alaska is the last frontier. Consider what a frontier is and what images it evokes. Present your conclusions in an essay.
2. Write an argument to persuade others to agree with your position regarding who should control development in Alaska—the nation or the state itself.
3. Write a letter to a member of the U.S. House or Senate in which you express your position concerning development and land use in Alaska. You may choose to address all types of development or focus on only a couple.

Gretel Ehrlich

The Solace of Open Spaces

Gretel Ehrlich ended up settling in Wyoming after her first trip there as a documentary filmmaker. Now living on a ranch and writing, she spent her first years in Wyoming calving, riding fence, lambing, branding, and herding. In the next essay, from which her 1985 collection of essays takes its title, she focuses on the space rather than the surface of the land.

Space and surface are related, and it may be that the value placed on land depends upon how well its characteristics meet the uses people plan for it. Ehrlich describes Wyoming's rocky and rough surfaces, how those surfaces are used, and how its spaces envelop and define. They stretch to the horizon, and such endlessness seems to affect the people who live there.

Before you start the essay, imagine how space affects people. Then, as you read, think about how much space you need or desire. You might compare the spaces (and surfaces) of where you live to those Ehrlich describes.

It's May and I've just awakened from a nap, curled against sagebrush the way my dog taught me to sleep—sheltered from wind. A front is pulling the huge sky over me, and from the dark a hailstone has hit me on the head. I'm trailing a band of two thousand sheep across a stretch of Wyoming badlands, a fifty-mile trip that takes five days because sheep shade up in hot sun and won't budge until it's cool. Bunched together now, and excited into a run by the storm, they drift across dry land, tumbling into draws like water, and surge out again onto the rugged, choppy plateaus that are the building blocks of this state.

The name Wyoming comes from an Indian word meaning "at the great plains," but the plains are really valleys, great arid valleys, sixteen hundred square miles, with the horizon bending up on all sides into mountain ranges. This gives the vastness a sheltering look.

Winter lasts six months here. Prevailing winds spill snowdrifts to the east, and new storms from the northwest replenish them. This white bulk is sometimes dizzying, even nauseating, to look at. At twenty, thirty, and forty degrees below zero, not only does your car not work, but neither do your mind and body. The landscape hardens into a dungeon of space. During the winter, while I was riding to find a new calf, my jeans froze to the saddle, and in the silence that such cold creates I felt like the first person on earth, or the last.

Today the sun is out—only a few clouds billowing. In the east, where the sheep have started off without me, the benchland tilts up in a series of eroded red-earthed mesas, planed flat on top by a million years of water; behind them, a bold line of muscular scarps rears up ten thousand feet to become the Big Horn Mountains. A tidal pattern is engraved into the ground, as if left by the sea that once covered this state. Canyons curve down like galaxies to meet the oncoming rush of flat land.

To live and work in this kind of open country, with its hundred-mile views, is to lose the distinction between background and foreground. When I asked an older ranch hand to describe Wyoming's openness, he said, "It's all a bunch of nothing—wind and rattlesnakes—and so much of it you can't tell where you're going or where you've been and it don't make much difference." John, a sheepman I know, is tall and handsome and has an explosive temperament. He has a perfect intuition about people and sheep. They call him "Highpockets," because he's so long-legged; his graceful stride matches the distances he has to cover. He says, "Open space hasn't affected me at all. It's all the people moving in on it." The huge ranch he was born on takes up much of one county and spreads into another state; to put 100,000 miles on his pickup in three years and never leave home is not unusual. A friend of mine has an aunt who ranched on Powder River and didn't go off her place for eleven years. When her husband died, she quickly moved to town, bought a car, and drove around the States to see what she'd been missing.

Most people tell me they've simply driven through Wyoming, as if there were nothing to stop for. Or else they've skied in Jackson Hole, a place Wyomingites acknowledge uncomfortably because its green beauty and chic affluence are mismatched with the rest of the state. Most of Wyoming has a "lean-to" look. Instead of big, roomy barns and Victorian houses, there are dugouts, low sheds, log cabins, sheep camps, and fence lines that look like driftwood blown haphazardly into place. People here still feel pride because they live in such a harsh place, part of the glamorous cowboy past, and they are determined not to be the victims of a mining-dominated future.

Most characteristic of the state's landscape is what a developer euphemistically describes as "indigenous growth right up to your front door"—a reference to waterless stands of salt sage, snakes, jack rabbits, deerflies, red dust, a brief respite of wildflowers, dry washes, and no trees. In the Great Plains the vistas look like music, like Kyries of grass, but Wyoming seems to be the doing of a mad architect—tumbled and twisted, ribboned with faded, deathbed colors, thrust up and pulled down as if the place had been startled out of a deep sleep and thrown into a pure light.

I came here four years ago. I had not planned to stay, but I couldn't make myself leave. John, the sheepman, put me to work immediately. It was spring, and shearing time. For fourteen days of fourteen hours each, we moved thousands of sheep through sorting corrals to be sheared, branded, and deloused. I suspect that my original motive for coming here was to "lose myself" in a new and unpopulated territory. Instead of producing the numbness I thought I wanted, life on the sheep ranch woke me up. The vitality of the people I was working with flushed out what had become a hallucinatory rawness inside me. I threw away my clothes and bought new ones; I cut my hair. The arid country was a clean slate. Its absolute indifference steadied me.

Sagebrush covers 58,000 square miles of Wyoming. The biggest city has a population of fifty thousand, and there are only five settlements that could be called cities in the whole state. The rest are towns, scattered across the expanse with as much as sixty miles between them, their populations two thousand, fifty, or ten. They are fugitive-looking, perched on a barren, windblown bench, or tagged onto a river or a railroad, or laid out straight in a farming valley with implement stores and a block-long Mormon church. In the eastern part of the state, which slides down into the Great Plains, the new mining settlements are boomtowns, trailer cities, metal knots on flat land.

Despite the desolate look, there's a coziness to living in this state. There are so few people (only 470,000) that ranchers who buy and sell cattle know one another statewide; the kids who choose to go to college usually go to the state's one university, in Laramie; hired hands work their way around Wyoming in a lifetime of hirings and firings. And despite the physical separation, people stay in touch, often driving two or three hours to another ranch for dinner.

Seventy-five years ago, when travel was by buckboard or horseback, cowboys who were temporarily out of work rode the grub line—drifting from ranch to ranch, mending fences or milking cows, and receiving in exchange a bed and meals. Gossip and messages traveled this slow circuit with them, creating an intimacy between ranchers who were three and four weeks' ride apart. One old-time couple I know, whose turn-of-the-century homestead was used by an outlaw gang as a relay station for stolen horses, recall that if you were traveling, desperado or not, any lighted ranch house was a welcome sign. Even now, for someone who lives in a remote spot, arriving at a ranch or coming to town for supplies is cause for celebration. To emerge from isolation can be disorienting. Everything looks bright, new, vivid. After I had been herding sheep for only three days, the sound of the camp tender's pickup flustered me. Longing for human company, I felt a foolish grin take over my face; yet I had to resist an urgent temptation to run and hide.

Things happen suddenly in Wyoming, the change of seasons and weather; for people, the violent swings in and out of isolation. But good-naturedness is concomitant with severity. Friendliness is a tradition. Strangers passing on the road wave hello. A common sight is two pickups stopped side by side far out on a range, on a dirt track winding through the sage. The drivers will share a cigarette, uncap their thermos bottles, and pass a battered cup, steaming with coffee, between windows. These meetings summon up the details of several generations, because, in Wyoming, private histories are largely public knowledge.

Because ranch work is a physical and, these days, economic strain, being "at home on the range" is a matter of vigor, self-reliance, and common sense. A person's life is not a series of dramatic events for which he or she is applauded or exiled but a slow accumulation of days, seasons, years, fleshed out by the generational weight of one's family and anchored by a land-bound sense of place.

In most parts of Wyoming, the human population is visibly outnumbered by the animal. Not far from my town of fifty, I rode into a narrow valley and startled a herd of two hundred elk. Eagles look like small people as they eat car-killed deer by the road. Antelope, moving in small, graceful bands, travel at sixty miles an hour, their mouths open as if drinking in the space.

The solitude in which westerners live makes them quiet. They telegraph thoughts and feelings by the way they tilt their heads and listen; pulling their Stetsons into a steep dive over their eyes, or pigeon-toeing one boot over the other, they lean against a fence with a fat wedge of Copenhagen beneath their lower lips and take in the whole scene. These detached looks of quiet amusement are sometimes cynical, but they can also come from a dry-eyed humility as lucid as the air is clear.

Conversation goes on in what sounds like a private code; a few phrases imply a complex of meanings. Asking directions, you get a curious list of details. While trailing sheep I was told to "ride up to that kinda upturned rock, follow the pink wash, turn left at the dump, and then you'll see the water hole." One friend told his wife on roundup to "turn at the salt lick and the dead cow," which turned out to be a scattering of bones and no salt lick at all.

Sentence structure is shortened to the skin and bones of a thought. Descriptive words are dropped, even verbs; a cowboy looking over a corral full of horses will say to a wrangler, "Which one needs rode?"

People hold back their thoughts in what seems to be a dumbfounded silence, then erupt with an excoriating perceptive remark. Language, so compressed, becomes metaphorical. A rancher ended a relationship with one remark: "You're a bad check," meaning bouncing in and out was intolerable, and even coming back would be no good.

What's behind this laconic style is shyness. There is no vocabulary for the subject of feelings. It's not a hangdog shyness, or anything coy—always there's a robust spirit in evidence behind the restraint, as if the earth-dredging wind that pulls across Wyoming had carried its people's voices away but everything else in them had shouldered confidently into the breeze.

I've spent hours riding to sheep camp at dawn in a pickup when nothing was said; eaten meals in the cookhouse when the only words spoken were a mumbled "Thank you, ma'am" at the end of dinner. The silence is profound. Instead of talking, we seem to share one eye. Keenly observed, the world is transformed. The landscape is engorged with detail, every movement on it chillingly sharp. The air between people is charged. Days unfold, bathed in their own music. Nights become hallucinatory; dreams, prescient.

Spring weather is capricious and mean. It snows, then blisters with heat. There have been tornadoes. They lay their elephant trunks out in the sage until they find houses, then slurp everything up and leave. I've noticed that melting snowbanks hiss and rot, viperous, then drip into calm pools where ducklings hatch and livestock, being trailed to summer range, drink. With the ice cover gone, rivers churn a milkshake brown, taking culverts and small bridges with them. Water in such an arid place (the average annual rainfall where I live is less than eight inches) is like blood. It festoons drab land with green veins; a line of cottonwoods following a stream; a strip of alfalfa; and, on ditch banks, wild asparagus growing.

I've moved to a small cattle ranch owned by friends. It's at the foot of the Big Horn Mountains. A few weeks ago, I helped them deliver a calf who was stuck halfway out of his mother's body. By the time he was freed, we could see a heartbeat, but he was straining against a swollen tongue for air. Mary and I held him upside down by his back feet, while Stan, on his hands and knees in the blood, gave the calf mouth-to-mouth resuscitation. I have a vague memory of being pneumonia-choked as a child, my mother giving me her air, which may account for my romance with this windswept state.

If anything is endemic to Wyoming, it is wind. This big room of space is swept out daily, leaving a boneyard of fossils, agates, and carcasses in every stage of decay. Though it was water that initially shaped the state, wind is the meticulous gardener, raising dust and pruning the sage.

I try to imagine a world in which I could ride my horse across uncharted land. There is no wilderness left; wildness, yes, but true wilderness has been gone on this continent since the time of Lewis and Clark's overland journey.

Two hundred years ago, the Crow, Shoshone, Arapaho, Cheyenne, and Sioux roamed the intermountain West, orchestrating their movements according to hunger, season, and warfare. Once they acquired horses, they traversed the spines of all the big Wyoming ranges—the Absarokas, the Wind Rivers, the Tetons, the Big Horns—and wintered on the unprotected plains that fan out from them. Space was life. The world was their home.

What was life-giving to Native Americans was often nightmarish to sodbusters who had arrived encumbered with families and ethnic pasts to be transplanted in nearly uninhabitable land. The great distances, the shortage of water and trees, and the loneliness created unexpected hardships for them. In her book *O Pioneers!*, Willa Cather gives a settler's version of the bleak landscape:

> The little town behind them had vanished as if it had never been, had fallen behind the swell of the prairie, and the stern frozen country received them into its bosom. The homesteads were few and far apart; here and there a windmill gaunt against the sky, a sod house crouching in a hollow.

The emptiness of the West was for others a geography of possibility. Men and women who amassed great chunks of land and struggled to preserve unfenced empires were, despite their self-serving motives, unwitting geographers. They understood the lay of the land. But by the 1850s the Oregon and Mormon trails sported bumper-to-bumper traffic. Wealthy landowners, many of them aristocratic absentee landlords, known as remittance men because they were paid to come West and get out of their families' hair, overstocked the range with more than a million head of cattle. By 1885 the feed and water were desperately short, and the winter of 1886 laid out the gaunt bodies of dead animals so closely together that when the thaw came, one rancher from Kaycee claimed to have walked on cowhide all the way to Crazy Woman Creek, twenty miles away.

Territorial Wyoming was a boy's world. The land was generous with everything but water. At first there was room enough, food enough, for everyone. And, as with all beginnings, an expansive mood set in. The young cowboys, drifters, shopkeepers, schoolteachers, were heroic, lawless, generous, rowdy, and tenacious. The individualism and optimism generated during those times have endured.

John Tisdale rode north with the trail herds from Texas. He was a college-educated man with enough money to buy a small outfit near the Powder River. While driving home from the town of Buffalo with a buckboard full of Christmas toys for his family and a winter's supply of food, he was shot in the back by an agent of the cattle barons who resented the encroachment of small-time stockmen like him. The wealthy cattlemen tried to control all the public grazing land by restricting membership in the Wyoming Stock Growers Association, as if it were a country club. They ostracized from roundups and brandings cowboys and ranchers who were not members, then denounced them as rustlers. Tisdale's death, the second such cold-blooded murder, kicked off the Johnson County cattle war, which was no simple good-guy-bad-guy shoot-out but a complicated class struggle between landed gentry and less affluent settlers—a shocking reminder that the West was not an egalitarian sanctuary after all.

Fencing ultimately enforced boundaries, but barbed wire abrogated space. It was stretched across the beautiful valleys, into the mountains, over desert badlands, through buffalo grass. The "anything is possible" fever—the lure of any new place—was constricted. The integrity of the land as a geographical body, and the freedom to ride anywhere on it, were lost.

I punched cows with a young man named Martin, who is the great-grandson of John Tisdale. His inheritance is not the open land that Tisdale knew and prematurely lost but a rage against restraint.

Wyoming tips down as you head northeast; the highest ground—the Laramie Plains—is on the Colorado border. Up where I live, the Big Horn River leaks into difficult, arid terrain. In the basin where it's dammed, sandhill cranes gather and, with delicate legwork, slice through the stilled water. I was driving by with a rancher one morning when he commented that cranes are "old-fashioned." When I asked why, he said, "Because they mate for life." Then he looked at me with a twinkle in his eyes, as if to say he really did believe in such things but also understood why we break our own rules.

In all this open space, values crystalize quickly. People are strong on scruples but tenderhearted about quirky behavior. A friend and I found one ranch hand, who's "not quite right in the head," sitting in front of the badly decayed carcass of a cow, shaking his finger and saying, "Now, I don't want you to do this ever again!" When I asked what was wrong with him, I was told, "He's goofier than hell, just like the rest of us." Perhaps because the West is historically new, conventional morality is still felt to be less important than rock-bottom truths. Though there's always a lot of teasing and sparring, people are blunt with one another, sometimes even cruel, believing honesty is stronger medicine than sympathy, which may console but often conceals.

The formality that goes hand in hand with the rowdiness is known as the Western Code. It's a list of practical do's and don'ts, faithfully observed. A friend, Cliff, who runs a trapline in the winter, cut off half his foot while chopping a hole in the ice. Alone, he dragged himself to his pickup and headed for town, stopping to open the ranch gate as he left, and getting out to close it again, thus losing, in his observance of rules, precious time and blood. Later, he commented, "How would it look, them having to come to the hospital to tell me their cows had gotten out?"

Accustomed to emergencies, my friends doctor each other from the vet's bag with relish. When one old-timer suffered a heart attack in hunting camp, his partner quickly stirred up a brew of red horse liniment and hot water and made the half-conscious victim drink it, then tied him onto a horse and led him twenty miles to town. He regained consciousness and lived.

The roominess of the state has affected political attitudes as well. Ranchers keep up with world politics and the convulsions of the economy but are basically isolationists. Being used to running their own small empires of land and livestock, they're suspicious of big government. It's a "don't fence me in" holdover from a century ago. They still want the elbow room their grandfathers had, so they're strongly conservative, but with a populist twist.

Summer is the season when we get our "cowboy tans"—on the lower parts of our faces and on three fourths of our arms. Excessive heat, in the nineties and higher, sends us outside with the mosquitoes. In winter we're tucked inside our houses, and the white wasteland outside appears to be expanding, but in summer all the greenery abridges space. Summer is a go-ahead season. Every living thing is off the block and in the race: battalions of bugs in flight and biting; bats swinging around my log cabin as if the bases were loaded and someone had hit a home run. Some of summer's high-speed growth is ominous: larkspur, death camas, and green greasewood can kill sheep—an ironic idea, dying in this desert from eating what is too verdant. With sixteen hours of daylight, farmers and ranchers irrigate feverishly. There are first, second, and third cuttings of hay, some crews averaging only four hours of sleep a night for weeks. And, like the cowboys who in summer ride the night rodeo circuit, nighthawks make daredevil dives at dusk with an eerie whirring sound like a plane going down on the shimmering horizon.

In the town where I live, they've had to board up the dance-hall windows because there have been so many fights. There's so little to do except work that people wind up in a state of idle agitation that becomes fatalistic, as if there were nothing to be done about all this untapped energy. So the dark side to the grandeur of these spaces is the small-mindedness that seals people in. Men become hermits; women go mad. Cabin fever explodes into suicides, or into grudges and lifelong family feuds. Two sisters in my area inherited a ranch but found they couldn't get along. They fenced the place in half. When one's cows got out and mixed with the other's, the women went at each other with shovels. They ended up in the same hospital room but never spoke a word to each other for the rest of their lives.

After the brief lushness of summer, the sun moves south. The range grass is brown. Livestock is trailed back down from the mountains. Water holes begin to frost over at night. Last fall Martin asked me to accompany him on a pack trip. With five horses, we followed a river into the mountains behind the tiny Wyoming town of Meeteetse. Groves of aspen, red and orange, gave off a light that made us look toasted. Our hunting camp was so high that clouds skidded across our foreheads, then slowed to sail out across the warm valleys. Except for a bull moose who wandered into our camp and mistook our black gelding for a rival, we shot at nothing.

One of our evening entertainments was to watch the night sky. My dog, a dingo bred to herd sheep, also came on the trip. He is so used to the silence and empty skies that when an airplane flies over he always looks up and eyes the distant intruder quizzically. The sky, lately, seems to be much more crowded than it used to be. Satellites make their silent passes in the dark with great regularity. We counted eighteen in one hour's viewing. How odd to think that while they circumnavigated the planet, Martin and I had moved only six miles into our local wilderness and had seen no other human for the two weeks we stayed there.

At night, by moonlight, the land is whittled to slivers—a ridge, a river, a strip of grassland stretching to the mountains, then the huge sky. One morning a full moon was setting in the west just as the sun was rising. I felt precariously balanced between the two as I loped across a meadow. For a moment, I could believe that the stars, which were still visible, work like cooper's bands, holding together everything above Wyoming.

Space has a spiritual equivalent and can heal what is divided and burdensome in us. My grandchildren will probably use space shuttles for a honeymoon trip or to recover from heart attacks, but closer to home we might also learn how to carry space inside ourselves in the effortless way we carry our skins. Space represents sanity, not a life purified, dull, or "spaced out" but one that might accommodate intelligently any idea or situation.

From the clayey soil of northern Wyoming is mined bentonite, which is used as a filler in candy, gum, and lipstick. We Americans are great on fillers, as if what we have, what we are, is not enough. We have a cultural tendency toward denial, but, being affluent, we strangle ourselves with what we can buy. We have only to look at the houses we build to see how we build *against* space, the way we drink against pain and loneliness. We fill up space as if it were a pie shell, with things whose opacity further obstructs our ability to see what is already there.

Content Considerations

1. In what ways did Native Americans and European pioneers differ in their attitudes toward Wyoming's space?
2. What are the positive effects of space? The negative?
3. Paraphrase the main idea of this essay.

The Writer's Strategies

1. Find examples of figurative language in this essay—especially metaphor, simile, and personification. Does this language distract or engage you? How?
2. What is Ehrlich's attitude toward her subject and audience (tone)? How is her tone expressed?

Writing Possibilities

1. Write about how space affects you; include some anecdotes about times you enjoyed just the right amount or suffered from too little or too much space.
2. Put yourself in the Wyoming Ehrlich presents; choose a persona for yourself and describe how you spend your time.
3. Think of the nature of the land about which Ehrlich writes. Is the land suited for the way it is used? How? How not?

Additional Writing and Research

Connections

1. The articles and essays in this chapter focus on many ecosystems—coasts, high plains, prairies, tundra, forests, rain forests, mountains, polar regions. What do the readings suggest happens when an ecosystem is not considered and treated as a part of the whole biosphere?
2. Compare and contrast the approaches taken in the articles by Sunquist, Sears, and Postel. What beliefs do the authors hold in common? Where do they differ?
3. Using the pieces by Madson, Ehrlich, and Lewis, explore the ways people affect land and land affects people.
4. Each reading mentions human ecosystems that are somehow different from natural ecosystems. Use three of the readings to illustrate the relationship(s) of human ecosystems to natural ones.
5. The essays by Madson and Ehrlich comment upon land as space. Compare the ways they present their views, including the evidence each presents. How are their conclusions different or similar?
6. Using five of the readings in this chapter, determine what considerations must be taken into account in developing a land use policy.

Researches

1. Read about the native peoples who populated America's prairies before European settlement—who were they, how did they live, how did they use the prairie?
2. Sears writes that "to keep land productive, materials must either be recycled or restored" (paragraph 31). Much effort in the United States has been directed toward "restoring" material; that restoration has later proved harmful. Research and write about various soil additives and their effects on living things, including their consequences for humans.
3. Focus on one part of the world that is currently experiencing land degradation. Explore the natural, political, economic, agricultural, and social causes of its desertification.
4. Investigate the kinds and extent of development that have occurred in one coastal area such as Miami or Seattle. Find out what the area was like before development; try to determine how development has affected the coastal ecosystem.

5. Select a place and period of history that interests you. Discover the role of land in the society of that time and place; discover how it was used and the results of that use.
6. Analyze the soil in your area, including an overall view of its geohistory. Then try to determine its quality and whether it is suited to the ways it is used. In gathering information, you may find it helpful to interview farmers and biologists as well as people from chemical companies, environmental organizations, conservation societies, government agricultural agencies, and research laboratories.

2

Species and Endangerment

WITHOUT DOUBT, everyone is familiar with the famous dodo, the hapless and flightless bird of Mauritius clubbed into extinction by passing sailors in the nineteenth century. And most know the story of the thousands of baby seals bludgeoned to death each year for their fur. While the awkward dodo was killed off with scarcely a murmur of protest, a highly publicized campaign to "stop the slaughter" of the cuddly, furry baby seals resulted in the banning by Canada of seal hunting in 1987.

Around the globe in the last ten years, other very visible animals have come under the protection of one organization or another: Elephant ivory has been banned by many governments, and it appears elephants may be saved from the fate of the dodo; rhinos in East and South Africa are being targeted for protection from poachers who seek their horns for sale as aphrodisiacs; whales and dolphins frolic at the top of several conservationists' wish lists.

Indeed, some would say we have gone too far with certain animals: A campaign to save the alligators in Florida has worked so well that news programs often feature irate homeowners who must contend with the reptiles in their backyards. It seems that an organization or governmental body will spring up to save whatever needs saving—condors, wolves, snail darters, or spotted owls. Surely, some group will galvanize action to save a worthwhile animal. And if it is not worthwhile, why bother? So what if another bird species goes down, another fish species goes belly up, or a beetle or two crawl into extinction?

After all, no one can really say how many species of life even exist in the world today—anywhere from 10 million to 100 million, according to which source one consults.

But the numbers are important, it seems. Biodiversity—the sum total of all the plants and animals on the Earth—is now seen to be important in ways few envisioned even a few decades ago.

Extinction—the permanent end of a species, never to exist again—is a natural phenomenon. As climatic conditions change, as habitat varies, as new predators move into the region, some animals lose the race of survival. Eventually it may happen to all existing species; after all, dinosaurs ruled supreme for several million years. As Jared Diamond points out in an essay in this chapter, however, humans have drastically accelerated the natural rate of extinction and are responsible—by destruction of habitat—for exterminating species at a rate that bespeaks potentially cataclysmic results. Within the next century, the UCLA physiology professor estimates, we will have caused the extinction of half the species alive today.

The solution might appear simple.

Why not, for instance, simply capture a few breedable individuals of each species and plunk them into a zoo, to be kept safe from the rapacity of humans—a kind of Noah's ark approach? Or why capture them at all? Wouldn't the world be a better place without some of these creatures, such as sharks? Anyone who saw the movie *Jaws* knows the answer to that. Maybe.

Americans have raised a hue and cry for dolphin-free tuna. The lovable, nearly human creatures that seem to communicate with us shouldn't die in the nets with tuna, to be tossed over the side or possibly end up packed into cans, we have said. But during the rising debate over saving the dolphins, some have started to ask: "What about tuna-free tuna?" Why all this effort to save dolphins and not tuna? And what about sharks? According to Rudy Abramson in his essay in this chapter, they are in danger of being killed off by humans after surviving 400 million years.

On what basis do we decide that elephants and snow leopards and dolphins deserve our protection and others are fair game for progress? Sometimes it is simple expediency, as Peter Matthiessen pointed out in his book, *Wildlife in America*. Vast herds of buffalo—millions of the great, shaggy beasts—were eliminated in the nineteenth century. Matthiessen says that "their disappearance was inevitable. Once the settlers discovered the agricultural potential of the long-grass prairies, and the ranchers bred fat livestock on the short-grass plains further west, the history of these humped, sullen beasts was over." For much the same reason—economics—a battle now rages between loggers and conservationists about the preservation of the spotted owl.

Wolves, coyotes, and grizzly bears, all of which feed on the youngest and weakest of man's domestic animals, have faced a hard fight from farmers and ranchers who have shot, poisoned, and trapped these threats to their livelihood. These animals have been the victims of direct violence by man, and although none of them is now threatened with imminent extinction, their numbers and range have been greatly reduced. Less direct human acts have killed off entire species, altering the natural balance of animals. The introduction of Nile perch into Lake Victoria in Africa as a fishery experiment, for instance, resulted in the extinction of several species in the lake as the perch indulged in a feeding frenzy.

Much has been said and written about the place of humans in the grand scheme of things; many have pondered the accelerating loss of species because of human action. What are the roots of this biocentric orgy of destruction? What are the consequences to future generations of humans as well as other species? Of what benefit to humans are species that offer no known economic or recreational use? Must humans accept the responsibility of deciding the fate of other species?

The writers in this chapter pose and answer such questions about preservation and extinction, make sense out of the conflicting claims about extinctions, and provide a perspective on the importance of other forms of life to humans.

JARED DIAMOND

Playing Dice with Megadeath

Although many of us have been educated to believe that scientists discover and then report truth, the status of what is known changes almost daily. Each new discovery or tested hypothesis confirms, disproves, enlarges, or complicates a host of others. Making knowledge—establishing truth, fact, and relations—is made more difficult when scientists cannot agree on how to define a problem or whether a problem even exists.

Scientists deal more with probability than with certainty, and any two scientists studying the same data may reach opposite conclusions. This is the case with those whose concern is the endangerment and extinction of species— the problem is hard to identify, predictions are difficult to make, and solutions are not easily constructed. Causes and connections are traced and examined; science and history study the past in hopes of repairing the present and preserving the future.

Jared Diamond, a professor of physiology at the UCLA School of Medicine, frequently writes articles of environmental concern. In the following one, published in Discover *in 1990, he examines the possibility that during the next century at least half of the Earth's species will become extinct. The problem he discusses is one with which some scientists disagree; Diamond structures his argument to counter such disagreement. As you read, keep track of the ways in which the author addresses the opposition and the ways in which he reveals how scientists come up with the numbers they use for their predictions.*

ON MAY 21, 1534, the French explorer Jacques Cartier stopped at Funk Island, off the coast of Newfoundland, so that his crew could feed on the meat of the great auk, a large, flightless seabird they could easily capture. After taking their fill, Cartier and his crew went on to explore the St. Lawrence River, and the great auks on Funk Island were safe for the next two and a half centuries—until feather collectors, searching out stuffing for pillows and mattresses, paid the birds a visit in 1785. Over the next half century the demand for the birds' feathers brought a steady stream of collection parties; by 1841 the great auks of Funk Island were gone completely.

So were most of their kin throughout the world. The last great auk on the Orkney Islands, north of Scotland, was killed in 1812; the last on St. Kilda, west of Scotland, in 1821; the last on the Faroe Islands, between Scotland and Iceland, in 1828. Six years later someone bagged the last great auk in Ireland. Finally, on June 3, 1844, two fishermen, Jón Brandsson and Siguror Islefsson, killed the world's last breeding pair at Eldey Rock, an island off the southern coast of Iceland; they also smashed the birds' single egg.

Although it is one of the most famous examples of extinction, the great auk is clearly not the only animal to disappear in modern times. The great auk is not even the only *bird* to become extinct. In North America alone we no longer have any living examples of the spectacled cormorant, the Labrador duck, the Carolina parakeet, and the passenger pigeon. Bachman's warbler, a resident of the swampy woods of the South, may be next on the list; it may already be gone—the most recent definite sighting was in 1977.

Everywhere humans have gone, they have wiped out whole species of birds, mammals, reptiles, fish, and other forms of life. Indeed, we continue to do so on every continent and island we inhabit, except now we act with a technology and a capacity for destruction far greater than that of our forebears. Many of us—scientist and nonscientist alike—find our increased threat to other species alarming. We fear that we have set in motion a wave of extinction that will ultimately undermine the quality, and perhaps even the possibility, of human life.

Not everyone, however, agrees that the risk of mass extinction is real, nor for that matter that it would do us much harm if it in fact occurred. One of the most frequently cited estimates in that debate, for example, is that humans have caused one percent of all bird species to become extinct within the past few centuries. Yet it's easy to imagine that most birds are superfluous as far as human needs are concerned and that we could safely lose, say, ten times more. We might even believe the same of other endangered species, from gray wolves to blue whales—it would be a pity to lose them, but life would go on.

What is the truth of our situation? Is the mass extinction crisis a hysterical fantasy, a real risk for the future, or a proven event that's already well under way? To answer those questions, we need to step back a bit and examine how the numbers bandied about by both sides in the debate are really arrived at. Then we need to know whether the pace of extinctions is rising or falling; we have to compare modern records with evidence dug up from the past. Finally, we need to ask what difference the loss of species makes. If more do become extinct, so what? How much do chickadees contribute to our gross national product anyway?

Birds keep popping up in the estimates for good reason: they are easy to see and identify, and hordes of bird-watchers keep track of them.

As a result, we know more about birds than about any other group of animals, and so they provide a measure of the current destruction.

We know that approximately 9,000 species of birds exist today. Since only one or two new species are being discovered each year, we can safely say that virtually all living birds have been identified. According to the International Council for Bird Preservation (the world's leading agency concerned with the status of birds), 108 species, a little more than one percent of the total, have become extinct in modern times—that is, since 1600, a date that conveniently marks the beginning of scientific classification.

Is this a fair measure of the rate of extinction? The council's list is intentionally conservative: a bird is registered as extinct only after it has been unsuccessfully searched for in all areas where it might conceivably turn up, and after it has not been found for many years. So there is little doubt about the true status of a bird listed as extinct. Can we be equally certain that all those bird species that have not fulfilled the council's rigorous criteria for extinction still exist? For North American and European birds the answer is yes. The presence of hundreds of thousands of fanatic bird-watchers virtually guarantees that no bird on these continents could drift into extinction unnoticed. In many cases researchers have watched a population dwindle to a few individuals and then followed the fates of those last survivors.

For example, the most recent casualty in the United States is the subspecies known as the dusky seaside sparrow, which lived in marshes near Titusville, Florida. As the marshes were destroyed, the sparrows' population shrank. Wildlife agents put identification bands on the remaining sparrows so that they could recognize each individual. When only six remained, they were brought into captivity; unfortunately, they were all males. On June 16, 1987, the dusky seaside sparrow became extinct—with witnesses.

But most tropical countries, where the overwhelming majority of species live, have few, if any, bird-watchers. The fate of many tropical birds is unknown, because no one has seen them or specifically looked for them since they were discovered years ago. Many other species were described from single specimens collected by nineteenth-century explorers who gave only vague indication of the site—such as "South America." We know nothing about the songs, behavior, and habitats of such birds, and even the most dedicated bird-watcher would be hard-pressed to identify one of those birds if he glimpsed or heard it.

Many tropical species therefore cannot be classified as either "definitely extinct" or "definitely in existence," but just as "of unknown status." It becomes a matter of chance which of these species happens to attract the attention of some ornithologist, be searched for, and then possibly be listed as extinct.

Here's an example: The Solomon Islands are one of my favorite birding areas in the tropical Pacific. The International Council for Bird Preservation lists one Solomon bird species, Meek's crowned pigeon, as extinct. But when I tabulated recent observations of all 164 known Solomon birds, I noted that 12 had not been encountered since 1953. At least some of those 12 species are almost surely extinct. Some should have been sighted if they still exist because they were formerly abundant and conspicuous. Some, Solomon Islanders told me, have been exterminated by cats—a predator that missionaries and traders introduced.

Twelve species possibly extinct out of 164 still may not sound like much to worry about—a bit over 7 percent. However, the Solomons are in better shape environmentally than most of the remaining tropical world. These islands have few people, little economic development, and much natural forest. More typical of the tropics is Malaysia, which has had most of its lowland forest cut down.

And in Malaysia we have good records for a group of animals other than birds. Before the advent of massive logging in Malaysia, biological explorers identified 266 fish species in the region's forest rivers. A recent four-year search turned up only 122 of those 266, fewer than half. The other 144 must be extinct, rare, or very local. And they reached that status before anyone noticed.

Malaysia is typical of the tropics in the pressure it faces from humans. Fish are typical of tropical species other than birds in that they attract only patchy scientific attention. The estimate that Malaysia has already lost, or nearly lost, half its freshwater fish is therefore a reasonable ballpark figure for the status of animals and plants in much of the rest of the tropics. Even this estimate, however, applies only to species we have already discovered and described. What about the ones we have never seen or named? Are they too becoming extinct?

Fewer than 2 million plant and animal species have been described, but sampling procedures suggest that the actual number of the world's species may be at least 30 million. Most of them live in rain forests that biologists are just beginning to explore; over 1,500 beetle species, for example, have been found in a single species of tree in Panama.

Humans are now bearing down on this vast, unknown majority of species, and here and there we can glimpse the results. When botanist Alwyn Gentry, for example, surveyed an isolated ridge in Ecuador called Centinela, he found 38 new plant species, many of them strikingly beautiful, confined to that one spot. Shortly afterward the ridge was logged, and those plants were exterminated. On Grand Cayman Island in the

Caribbean, zoologist Fred Thompson discovered two new species of land snails confined to a forest on a limestone ridge. When that ridge was cleared a few years later for a housing development, the snails disappeared.

The accident of Gentry and Thompson visiting those ridges before rather than after they were cleared means that we have names for the now-extinct species that lived there. But biologists don't first survey most tropical areas that are being developed. Humans must have exterminated plants and animals on innumerable other tropical ridges before anyone had a chance to discover them.

In short, then, published lists of species known to be extinct must be gross underestimates of the actual numbers. There is a systematic tendency to underestimate extinctions even among birds, which we know best. For many other forms of life the unknown species—and unknown extinctions—must far outnumber the known.

So far we have counted only species exterminated in modern times, that is, since 1600. But were there no exterminations before 1600, throughout the preceding several million years of human history? To find out if the slaughter is gaining momentum, we need to follow the swath we have cut through the past.

Until 50,000 years ago, when humans first reached Australia and New Guinea, we were confined to Africa plus the warmer areas of Europe and Asia. We reached Siberia 20,000 years ago, North and South America 11,000 years ago, and most of the world's remote oceanic islands only within the past 4,000 years. Over this same period we were dramatically improving our hunting skills and our tools, and developing the technology of agriculture. Also, while expanding over the globe, we were increasing in numbers, from perhaps a few million 50,000 years ago to half a billion in 1600.

Paleontologists have studied many of the areas humans have reached within the past 50,000 years. In every one, human arrival coincided with massive extinctions. After people arrived in Australia, that continent lost its giant kangaroos and other giant marsupials. Around the time humans reached North America, that continent lost its lions, cheetahs, horses, mammoths, mastodons, giant ground sloths, and several dozen other large mammals. In all, counting by genus, 73 percent of North American large mammals became extinct near the time of human arrival. In South America it was 80 percent, and in Australia 86 percent.

We know that people hunted many of these animals. And in areas where meaty mammals were not abundant, people killed other prey. For example, paleontologists have found remains of recently extinct bird species on almost every oceanic island they have explored. New Zealand lost its giant flightless moas when the Polynesians arrived, and Hawaii lost its flightless geese and dozens of smaller birds. Extrapolation to islands not yet explored by paleontologists suggests that 2,000 bird species—one-fifth of all the birds that existed a few thousand years ago—fell victim to prehistoric exterminations. That doesn't count birds that may have disappeared on the continents.

Ever since researchers became aware of these prehistoric extinctions, they've debated whether people were the cause or just happened to arrive while animals were succumbing to climatic changes. But in the case of Polynesia, at least, there is now no reasonable doubt. Extinctions there not only coincided with human arrival but also with a time when climate was stable. Humans did turn up the heat, though—abandoned Polynesian ovens contain the bones of thousands of roasted moas.

To me the evidence seems overwhelming that humans played a role in the earlier extinctions also, especially in Australia and the Americas. In each part of the world in question, a wave of extinction swept over the land after the arrival of humans but didn't appear simultaneously in other areas undergoing similar climatic swings. If climatic changes at the end of the most recent ice age finished off America's big beasts, it's curious that those animals, having survived the previous 22 ice ages, chose to drop dead at the end of the twenty-third, just when human hunters sauntered by.

Moreover, it wasn't just in these newly occupied territories that large animals disappeared. In areas long occupied by humans, a new round of extinctions started 20,000 years ago: Eurasia lost its woolly rhinos, mammoths, and giant deer (the "Irish elk"). At the same time, Africa lost its giant buffalo, giant hartebeests, and giant horses. Like their cousins on the other continents, these big beasts may have been done in by prehistoric humans who hunted them with newer, better weapons.

Evidently, our species has a knack for exterminating others, and we're becoming better killers all the time. The key question for our children, though, is whether the crest of the extinction wave has already passed or whether the worst is still to come.

There are a couple of ways to approach this question. One is simply to assume that the number of tomorrow's extinct species will bear close resemblance to the number of today's endangered ones. Among birds an estimated 1,666 species—18 percent of the total—are now either endangered or at imminent risk of extinction. But this figure is an underestimate for the same reason that one percent is too low an estimate for birds we have already driven to extinction. Both numbers are based only on species whose status caught someone's attention, not on a reappraisal of all birds. The real number of species at risk must be higher.

The alternative approach to prediction is not to look at other species but rather to look at ourselves. Our destruction of others has kept pace with our population and technology, and it is fair to reason that this will continue until our population and technology reach a plateau. But neither shows signs of leveling off. Our population grew from half a billion in 1600 to more than 5 billion now, and it is still growing at close to 2 percent per year. And every day brings new technological advances by which we are changing Earth and its denizens.

If we are to predict the extent of our future devastation, then we must take notice of the mechanisms by which we cause extinction. Of the four leading ones, the first and most direct is overhunting, or killing animals faster than they can breed. This is chiefly how we have eliminated big animals, from mammoths 10,000 years ago to California grizzly bears this past century. And we are not yet finished. We no longer depend on such animals for meat, but we kill elephants for ivory and rhinos for their horns. At current rates of slaughter, not just elephants and rhinos but most other populations of large African and Southeast Asian mammals will be extinct (outside game parks and zoos) in a decade or two.

The second method of extermination is more indirect but no less effective: introducing species to parts of the world where they didn't previously exist. When such newcomers spread, they frequently kill off native species because the victims have no natural defenses. In the United States, for example, an introduced fungus has almost exterminated the American chestnut tree, just as on oceanic islands around the world, goats and rats have exterminated many native plants and birds. Although we might try to keep from introducing pests to new parts of the world, they will inevitably keep pace with human travel and commerce.

Meanwhile, new pests show up disguised as friends. In Africa's Lake Victoria a large fish called the Nile perch is now eating its way through the lake's hundreds of remarkable fish species. The Nile perch was intentionally introduced in a misguided effort to establish a new fishery. It will probably ring up the biggest number of extinctions caused by an introduced predator in modern times.

Destruction of habitat is the third path to extinction. Most species live in only one type of habitat: marsh wrens in marshes, pine warblers in pine forests. Many are even more particular, especially in the tropics. Centinela Ridge is an isolated habitat because it just reaches the clouds; before it was cleared, it was a unique "cloud forest" separated by valleys from similar habitats. When we cut such forests, we eliminate the local species almost as certainly as by shooting each individual. When all the forest on Cebu Island in the Philippines was logged, nine of the ten bird species unique to Cebu became extinct, and that last one may not survive.

The worst habitat destruction is still to come. We are just starting in earnest to destroy tropical rain forests. These forests cover only 6 percent of Earth's surface but harbor at least half of its species. Brazil's Atlantic forest and Malaysia's lowland forest are nearly gone, and most forests in Borneo and the Philippines will be logged within the next two decades. By the middle of the next century the only large tracts of tropical rain forest likely to be standing intact will be in parts of Zaire and the Amazon Basin.

Finally, our fourth method of extermination is by means of an inadvertent domino effect. Every species depends on many others, and often in such complex ways that it's impossible to foresee where any one extinction may ultimately lead. For example, when the Panama Canal created Gatun Lake in 1914 and turned Barro Colorado into an island, no one anticipated that the disappearance of jaguars and other big predators on that island would lead to the local extinction of little antbirds. But it did, because the big predators used to eat medium-size predators like peccaries, monkeys, and coatimundis. When the big predators disappeared, there was a population explosion of the medium-size predators, which proceeded to eat up the antbirds and their eggs.

A realistic projection for the future must take all these effects into account, as well as the knowledge that published lists of extinct and endangered species are gross underestimates. It is likely that more than half of all existing species will be extinct or endangered by the middle of the next century. If the estimate of 30 million for the world's total species is correct, then species are now becoming extinct at a rate of 150,000 per year, or 17 per hour.

Of course, there are people who dismiss the significance of these extinctions. So what if a few million beetle species disappear, they ask. We care about our children, not about bugs.

The answer is simply that, like all species, we depend on others for our existence. We need them to produce the oxygen we breathe, absorb the carbon dioxide we exhale, decompose our sewage, provide our food, and maintain the fertility of our soil.

Then couldn't we just preserve those species we need and let others become extinct? Of course not, because the species we need also depend on other species. The ecological chain of dominoes is much too complex for us to have figured out which dominoes we can dispense with. For instance, if you were the president of a timber company trying to figure out which species you could afford to let become extinct, you would have to answer these questions: Which ten tree species produce most of the world's paper pulp? For each of those ten tree species, which are the ten bird species that eat most of its insect pests, the ten insect species that pollinate most of its flowers, and the ten animal species that spread most of its seeds? Which other species do these birds, insects, and animals depend on?

Even without knowing the answers, you might be willing to take a chance. If you were trying to evaluate a project that would bring in millions of dollars but might exterminate a few obscure species, it would certainly be tempting to run the risk. But consider the following analogy. Suppose someone offers you a million bucks in return for the privilege of cutting out two ounces of your valuable flesh. You might figure that two ounces is less than one thousandth of your body weight, so you'll still have plenty left. That's fine if the two ounces come from your spare body fat and if they'll be removed by a skilled surgeon. But what if the deal requires you to let anyone hack two ounces from any conveniently accessible part of your body, the way much of the tropics are now being razed?

Am I saying, then, that our future is hopeless? Not at all. We are the ones who are creating the problem, so it's completely in our power to solve it. There are many realistic ways we can avoid extinctions, such as by preserving natural habitats and limiting human population growth. But we will have to do more than we are doing now.

If, on the other hand, we continue behaving as we have in the past, the devastation will also continue. The only uncertainty is whether we will halt the juggernaut or whether it will halt us.

Content Considerations

1. Diamond suggests that the numbers of species on extinction lists have been underestimated. Why does he think so? Do you agree with his reasoning?
2. What argument suggests that humans' arrival has caused mass extinction of species in various places at various times?
3. What causes species extinction? What part does the "domino effect" play?

The Writer's Strategies

1. Diamond begins by describing the eradication of the great auk, then comments briefly on the extinction of other birds. Then he states his own position regarding the dangers of extinction. What is his position? Why does he introduce it this way?
2. The fifth paragraph of this article acknowledges opposing arguments.

Why do arguments frequently acknowledge opposing points? How does this acknowledgment affect the writer's credibility?

Writing Possibilities

1. Make a list of technological advances achieved since your grandparents' youth, then speculate on how these advances may contribute to species endangerment and extinction.
2. Human population control is controversial because it involves religious, cultural, and individual beliefs that often conflict with the beliefs of others. Write a paper in which you defend or attack population control as an environmental issue; that is, defend or attack population control as a necessary step in maintaining or regaining Earth's ecological balance.
3. Explain "what difference the loss of species makes" (paragraph 6) to you in your life.

Jake Page

The Owls of Night

Even those who agree that humans have endangered too many kinds of animals and lament the extinction of too many species may not agree on how to save those that remain. They may not even agree about which species should receive the attention of funding available for preservation efforts.

That dilemma is one that concerns Jake Page, a writer who frequently focuses on scientific and environmental issues. In the essay that follows, originally a chapter in his book Pastorale: A Natural History of Sorts, *Page roams among domestic animals and wildly exotic ones, pondering the similarities among them and the possibility of their extinction. Although he does not use it, the term "triage" seems to fit his musings. In a medical sense "triage" describes how treatment is allocated during battles or disasters (and often urban emergency rooms)—those with minor wounds are treated last, and those whose wounds mean certain death are allowed to die. Those who are to receive treatment are ranked according to a system of priorities; attempts to provide treatment are focused on the lives that can be saved, not those for which there is no hope or those in no immediate danger.*

How might the idea of triage apply to preserving species? Write a few quick notes of your ideas, then keep those ideas in mind as you read Page's essay.

SO MUCH of what passes for intellectual discussion these days is about differences and change, but what strikes me as stunning are similarities and continuities.

It is late at night in July—an unusually cool night—and at the far end of the sofa is a 170-pound dog, the female mastiff, who is almost exactly the size and coloration of an Atlas lioness, which happens to be a species that exists now only in zoos. Between me and the mastiff, my wife sleeps, her feet against this furry, warm-blooded body. The television news has just winked out, its message being the strident concern so often heard nowadays—and quite properly—that nuclear fission was the bite of the apple we should not have taken but in any case should immediately and permanently regurgitate. I understand the wish. The thought of a nuclear war is not the sort of thing to make one's heart beat with yeomanlike calm.

It being late, my mind wanders.

I used to love to watch the mastiff as she padded through town, seeing the muscles extend and contract under her tawny coat. A leash law now disallows such behavior in our town, but she can roam, if accompanied, in the meadows and the pastures around town. I love to see her then, the careless power of her haunches, the rise and fall of her huge shoulders as she patrols what she, by sense of smell, considers to be her universe, though shared with other and lesser creatures of her kind. When she runs, she challenges gravity and friction at thirty miles an hour in spurts, and she can make a sudden stop with the certainty and grace given only to the supreme athlete. When I watch her do this I think of lions in their natural state and I also recall schoolboy physics—in particular $F = Ma$—and I am glad I do not need to oppose this creature with my far less compact body.

There are people in town who admire the mastiff less than I, fearing the great size of this Baskervillean hound and forgetting that giants can be gentle. Right now this fearsome giant, whose ancestors, according to Herodotus, could singly bring down and dispatch an elephant, is lying on her back asleep, feet up, her massive neck athwart my wife's ankles, warming them and probably cutting off their circulation. My wife's toes are aligned with those of the mastiff's front paws, both sets of toes so different and yet so miraculously alike.

Both creatures are asleep, a state no one yet fully understands. The wrinkled, dark brow of the mastiff twitches, her feet jerk in the air. A dream. Less important perhaps than when my wife's eyes swerve back and forth under her eyelids. Who dreams what? Is to dream to think in some way? The mastiff seems to think at times. However well intentioned he may have been, "Descartes was a jerk: Anything that so obviously exhibits empathy cannot be a machine. I think it is only fair, in matters of such importance as intelligence, to give dogs the benefit of the doubt until they are proved by some yet to be devised scientific test to be lacking."

And it is fair that my wife and dog and I share a sofa, for we are more alike than different. To be sure, my wife and I are more akin. (Our immune systems both find the dog's hair foreign and to create allergenic reactions, a mindless kind of chemical warfare we choose to ignore, liking the dog.) Indeed, geneologies show that several generations ago, Susanne's and my forebears sprang from the same union, somewhere around here in Virginia, but traced far back enough we are all cousins, a single gigantic clan, relatives even to mastiffs, even to the protozoa of the sea. In the late night, I sit and wonder how we grew so far apart in our notion of mutual responsibility when our blood, in its chemical composition, is so much like the liquid that the earliest cells enclosed within their tenuous walls.

There are two mammals with me on the sofa, endothermic creatures using basically the same system to warm themselves, each other, and me while I sit awake thinking about the news of the world and the chilling suggestions that we are risking—even deliberately summoning—our own doom, an old Anglo-Saxon word that means extinction. Outside, the constellation of Orion, a bunch of thermonuclear reactions with no known physical effect on life or death on this planet, is in its predictable place in the sky, a reassuring sign of permanence. Also outside, a barred owl hoots: Hoo-hoo, hoo-hoo, hoo-hooaw. Night is the owl's affair. Many Indian people think the owl is the sayer of the end. If you hear an owl or, worse, see one, it means a death. I'm against that. I don't particularly care for the dark.

A magazine lies open on the arm of the sofa, open to an article that speaks bravely of evolution and that force's corollary, extinction. It is couched in the cool lingo of statistics, starting back in the Cambrian age, when life existed only in oceans and tide pools, and it proceeds to the present, dealing with the marine organisms. It appears from the article that there has been a terrible problem finding out how many mass extinctions there have been, because the data have been lodged in bins for a genus here, a family there, taxonomically an apples-and-oranges situation. The authors, two paleontologists from Chicago, have now put it all at the family level and have come up with a few altogether discouraging conclusions and a few happy ones.

"First of all, there have been five momentous times of mass extinction when vast numbers of families of (to me) nasty little marine beasts went down the tube." I don't mind thinking about such creatures, but I don't like them underfoot when I ineptly try a little body surfing in the weak waves of the Atlantic. In any event, on five statistically important occasions, lasting merely a few million years, many squishy little marine families bought the farm.

"The data do not tell us, of course," the authors write, "what stresses caused the mass extinctions." But they can be seen in cool little graphs as distinct and abrupt dips against the general curve of "background" extinction, the constant buzz of doom that is apparently always with us. A family here, a family there.

The happy part is that the rate of background extinction is diminishing. If we were still operating at Cambrian rates of extinction, we would have lost 710 families of marine organisms more than we have in fact lost, out of an approximate 3,300 families. The rate has dropped from about 4.6 to 2 families per million years. But "a decrease in extinction rate is predictable from first principles if one argues that general optimization of fitness through evolutionary time should lead to prolonged survival. This is speculative but worthy of further consideration."

Indeed.

It is nice to think of evolution as generally progressive and not just random change in a random world. Dr. Pangloss would be delighted. The sanguinity of the Chicago scientists is really quite remarkable when looked at in the light of evidence that by the year 2000 we may well have lost twenty percent of extant species—1 or 2 million plant, animal, and microbial species—through destruction of their habitat. The rain forests are going. By the turn of the century they—the most diverse and richest habitat in the world, with the possible exception of coral reefs—may be represented in only one or two places around the equator.

Closer to my home, Chesapeake Bay, the largest estuary on the continent, spawning ground for a prodigious quantity (maybe a majority for the Atlantic) of shellfish, is threatened with continuing pollution and death. Maybe we are producing the stresses needed for the next great interval of extinctions of marine life. Certainly the most related (to us) forms of sea life are in urgent danger: the whales.

I remember once, on a strand of beach along North Carolina's Outer Banks, I stole my daughter's air mattress right out from under her in the surf. Over her justified protest, I used it to go straight out to sea. For several days we had watched a school of dolphins (or is it a pod?) about a half mile out, head north one day, south the next. On this day they were heading north, somewhat closer to shore.

I set a course that would intersect the dolphins' and paddled desperately toward the rendez-vous. On they came through the calm water, arching, dark, bigger than I had thought, and a freak wave caused me to slide off the mattress. Terrified, I scrabbled back aboard like a crab, just as a detail of three dolphins broke off and speeded directly at me. One after the other, each leaped out of the water and plunged back in about three feet from my head, racing toward the shore, where they danced briefly in the surf. Suddenly there were dolphins all around me— eight-footers, I guessed, maybe a hundred of them—in an area about half the size of a football field. They leapt and plunged, eying me, churning up the water in some wild sort of ballet to the rasping accompaniment of their own watery snorts. Just as suddenly, I was not afraid.

After some five minutes, as if by signal, they resumed their northward journey and, as I recall it, I shouted, imploring them not to leave me, and paddled futilely after them a few yards. But they went off to some unknown place, flashing arcs, vanishing instants of quicksilver spray.

Such impulsive forays, I learned later, are not considered wise beach behavior. Nevertheless, the experience recurs from time to time in my mind just as the melody of a song can show up for no reason on a given morning, a melody that was popular, say, when one was thirteen and fell hopelessly in love with an incomprehensibly wondrous sixteen-year-old. We have all fallen in love at one time or another with what can best be thought as the unattainable "other."

Romance, of course, is not universally accepted as one of the appropriate modes of scientific discourse (though I have never met a cetologist who wasn't crazy about his subjects or, for that matter, an astronomer who found stars unexciting). But ever since my miniadventure on the Outer Banks I've had a keen sense of kinship with dolphins and *their* kin, and I am inclined to hear of people who consider these animals as "resources" with an emotional wince.

This happens to have been the philosophy of the International Whaling Commission when it was set up after World War II to conserve and manage the whale "fishery," the better that the world could use this resource. (Wince.) The commission has done its best, what with conflicting scientific data about the number of whales in the ocean and their ability to bounce back. Japanese scientists, for example, have an entirely different way than Americans of estimating the number of sperm whales, using the whale's age at sexual maturity, rather than the method that involves the length of the whale.

In any event, the commission has suggested a moratorium on all whaling that has been met with outrage on the part of the whaling nations (notably Russia, Japan, and Norway, a strange bedfellowship), and their outrage has been met with comparable outrage by environmentalists who claim, with what I consider proper thinking, that whales have rights. I think one of the first rights whales should be accorded is that they will not be used by *Homo sapiens* in a shabby game of geopolitics and greed.

There is plenty of room for scientific disagreement about whales, precisely because not very much is known about them. Whaling has long been thought of in terms of fisheries, and not only is the science of fisheries pretty inexact, but its principles as such cannot be expected to apply to highly socialized and communicative mammals that simply happen to live in the sea. What is the effect of clearing the dominant males out of a society of whales? No one knows, but one thing is certain: Whale populations do not tend to bounce back even after decades of total protection. Only the California gray whale among such protected species has increased its numbers in any significant way.

Then what is to be done if scientists disagree on fine points, if international agencies are unable to enforce their recommendations (which is unhappily the case)? Perhaps it is time for nonwhaling nations to think of carrots rather than sticks, to gain public support within the whaling nations for these creatures who seem brainy, affective, and even capable of mourning their own. Or perhaps we must simply sit by and watch, hoping that whaling as an industry will soon collapse from overdepletion and bad economics *before* the whales have *all* become vanishing instants of quicksilver spray, memories of almost attainable friends, never again to be seen.

Whales seem perfectly adapted to their world, but for the parasitism of man—something we could learn to restrain, especially in the oceans. On the other hand, humanity will always have an unavoidable effect on the land. At its best, this impact is beneficial. From my backyard I can see parkland (the yard), hedgerows, pastures, croplands, and wooded mountains, a splendid diversity of habitats that support more species of plants and animals than, in its pristine forested state, this land once provided for. The people who have farmed this land, the people who came before me on my acre, did things right.

There's an old dam on the creek, made of boulders and now falling down enough to create a waterfall and a swimming hole. Farther up the creek is a beaver lodge and the nest of a red-tailed hawk. Engineering of various sorts and all for the best. The land prospers; food chains thrive; and circling above it all, by day, are the turkey vultures, which remind me of another problem with wildlife conservation.

Up close turkey vultures are so ugly that you wonder if someone had not designed them to gross people out. In the air, they make you nearly weep with admiration for their way of sailing effortlessly through the sky. On a good day there can be as many as fifty turkey and black vultures soaring, silent and graceful, behind my house, reminding me of the marvels of aerodynamics and also of several of my own limitations. Being carrion eaters they offend the sensibilities of some people, which is silly, because getting rid of carrion is no less important a function than waste management, and it's not just vultures that do it either. Eagles eat carrion quite often and no one thinks they're gross.

On the other hand, some people worry a lot about vultures and that is because one of their kind seems about to vanish from the planet. This is the California condor, the biggest of all flying birds in America. It lives in but two counties now, Ventura and Santa Barbara, requiring great canyons and vast spaces. It feeds on, among other things, dead cows supplied by the government. There are less than thirty left. They breed irregularly. The prognosis is bad.

Recently there have been much publicized attempts to capture some of the birds and breed them in captivity, and a baby condor's face soon adorned the newspapers across the land. This, we are told, is perceived as a noble initiative on the part of an otherwise greedy and rapacious culture, an example to be held up like a marvelous prism of morality to shame a heedless society into paying attention to the environment.

But when the condor was at its peak, a million or so years ago, during the Pleistocene, its diet was made up of creatures like dead mastodons. There are no more mastodons. Extinction, like death, is an unavoidable feature of life. The great mammals, the disposal of which became the business of these great vultures, are no longer with us, even in Santa Barbara County, except as fossils. "Maybe it is time to rethink wildlife conservation and forget about making spectacular last-ditch stands to rescue living fossils for a world they no longer fit."

The condors arose 65 million years ago and wound up in a particular niche. That niche is vanishing. People happen to be the dominant species on earth now and, like it or not, the niches have changed. We don't have to be so heedless about the environment—quite the contrary—but we are, after all, here. We evolved into our present form along with a lot of other creatures long after the appearance of the condor, and we produced a new environment, and that is something we can neither deny nor altogether undo. What good is a zoo for a condor or an Atlas lion?

Several environmentalists of considerable stature have asked if we cannot save vultures—if *even* the vultures are dying—what hope is there for all of us? To the contrary, there are two species of vultures soaring daily over the pasture behind my house, and they seem to be doing fine in *our* kind of landscape—farmland, forest, town, and city, even deserts. Anyone who wants to see turkey vultures can see plenty of them in the nation's capital, plying the carrion trade with exemplary skill in the city's still tree-lined streets and the minicanyons of Rock Creek Park. There are, it must be said, some apartment dwellers who object, because the vultures roost by night on their parapets, resembling impatient gargoyles and providing the residents with nerve-racking omens.

But the vultures are there, alive. So are bluebirds and vireos. You can find trout lilies within the city limits and a vast variety of living forms, most of them post-Pleistocene in origin. Must we be responsible for everything? Can't we distinguish?

Perhaps wildlife conservation should be based on a notion of modernity—the idea of using limited resources to preserve the world we grew into and on which we have made our mark. No one needs to kill a whale. But how many acres of marshland could have been bought up and preserved with all the money poured down the unwilling craw of the California Condor, how many bluebird boxes built and deployed? How many trees planted in Detroit, how much pollution halted in Chesapeake Bay? Clearly, as a species, we need a canary in the mine, but the condor is not that canary. Think instead of Henslow's sparrow, a modest bird that evolved concurrently with us and prefers damp pastures and land that has been ignored, of which there is little to be sure.

Ignoring things creatively is a virtue we have yet to learn as a civilization, but if we are so selfish as to be unable to retain a few ignored places on the planet where Henslow's sparrow can hang out successfully and where turkey vultures can swing in the sky, then we have truly had it.

And of course we are surrounded with omens. The apartment dwellers in Washington think the vultures are bad omens. The Indians think owls are the same.

A few miles from where I sit, there is a bunker built into the rock and equipped with vast quantities of communications apparatus. It is one of the several places the President and other members of the government can go in the event of a nuclear attack and there, in the confines of the Blue Ridge Mountains, our leaders can preside however they think it wise over the ultimate extinction of a nuclear winter or two. Tonight a dog and a woman sleep peacefully on the sofa and insects flutter at the screens, seeking my man-made source of light. I dearly hope the people I listened to on television earlier tonight and the people around the world who speak different languages and dwell upon the differences among us can pause to consider the miracle embodied even in so insignificant a creature as my mastiff—so like me and my wife and children—when they make their decisions.

I remind myself that the sun is shining on the other side of the earth. I don't feel much like sleeping. I will try to sit up through the night. My intention is to outlast the hooting of the owl and make sure that the next big light I see in the sky is the morning sun. There are many human cultures which believe in the need to pray for the return of the sun, so I have precedent.

Content Considerations

1. What is the difference between "mass extinction" and "background extinction"? What can be done to stop both kinds?
2. What are the author's views on saving the California condor?
3. What is the "notion of modernity" on which preservation attempts could be based? Were it implemented, how would it affect efforts to save species from extinction?

The Writer's Strategies

1. The author begins this essay by saying, ". . . . what strikes me as stunning are similarities and continuities." What similarities and continuities hold this essay together?
2. What is Page arguing for? How is his argument constructed?

3. This essay juxtaposes domestic and exotic creatures. Why? How does this juxtaposition fit the argument of the essay?

Writing Possibilities

1. Write about animals that are "so ugly you wonder if someone had not designed them to gross people out" or about animals that "make you nearly weep with admiration" (paragraph 26).
2. As Page points out, humans have created the current environment. In an essay, explain what parts of it you think we should attempt to fix or correct.
3. Construct an argument for the way you think we should go about deciding which species to save from extinction. After you have read your classmates' papers, work into your own an acknowledgment of opposing viewpoints.

RUDY ABRAMSON

Sharks Under Attack

Saving sharks from extinction doesn't carry the same glamour that saving whales, California condors, black rhinos, or snow leopards does. Sharks are not pretty, according to most definitions. Their eyelids roll up, and their jaws drop, sending sharp rows of teeth into their prey. These creatures inspire fear and loathing as often as awe. Yet many species of sharks are now the target of preservation efforts.

What has endangered sharks is what endangers most species: human intervention in ecosystems, mostly for profit of some kind or another. Yet the extent of their endangerment is not clear. Because sharks live and roam in oceans (and an occasional river), their presence—or absence—is not always evident. What has become evident, however, is that thousands of sharks are caught each year. Some are sold; some are killed and discarded.

In an article syndicated to newspapers in the summer of 1990, Rudy Abramson reports on diminishing Atlantic Ocean shark populations. Before you read the article, try to predict why the shark population is shrinking; recall what you know about food chains to predict how sharks fit into the ocean's ecosystem.

AFTER SURVIVING FOR 400 MILLION YEARS and even acquiring a modern-day reputation as indestructible killing machines, sharks have come under serious attack by a superior predator.

Man.

In the ocean, sharks reign at the apex of the predatory system. But as humans have added shark's fin soup and shark filets to more and more menus, the predators increasingly have become the prey.

The business of shark fishing has exploded, and experts say sharks were caught at an unsustainable rate throughout the '80s.

From 500 tons in 1980, when shark fishing was officially encouraged, the commercial catch grew to 7,500 tons in 1989. When recreational fishing is counted—along with sharks that are killed and thrown back into the sea—the National Marine Fisheries Service calculates that the total shark take last year was almost 25,000 tons.

Even more ominous, American shark catches are only the tip of the problem. Five years ago, the United Nations' Food and Agricultural Organization estimated the worldwide catch of cartilaginous fishes totaled

1.3 billion pounds—enough to circle the earth with sharks four times, according to Samuel H. Gruber and Charles Manire of the University of Miami.

Harry Upton, fisheries program director at the Center for Marine Conservation in Washington, says the dramatic growth of shark fishing in the United States means that the Atlantic populations of mako, hammerhead, blacktip and thresher are all facing "imminent collapse."

Gruber, considered one of the nation's top authorities on sharks, agrees. Unless there is a dramatic turnaround, he warns, many species will be pushed to the brink of extinction within 20 to 30 years.

Some already have been.

Commercial volumes of the soupfin shark were wiped out in the 1940s. Overfishing reduced the Scottish and Norwegian stock of spiny dogfish to levels at which they could no longer be caught in profitable numbers—a situation that experts describe as "commercial extinction." And in the 1960s, fishing fleets nearly eradicated the porbeagle shark in the Atlantic.

Commercial shark fishing developed in earlier times because shark livers were a rich source of Vitamin A. Scientists long ago developed synthetic vitamins, but the exploitation of sharks has been propelled faster by new forces.

The warning signs are dramatic:

- Around the Florida Keys, only a few years ago, lemon sharks could be found by the hundreds. Now they are hard to find at all. Most have been slaughtered for crab bait.
- Around Costa Rica's Cocos Island, hammerhead sharks, which feed on coral, have been decimated for their fins.
- In the past several years, the widening market for fins of nearly any species has produced an upsurge in "finning"—a practice in which the fisherman catches the sharks, slices off their valuable fins and throws them back into the ocean to die.

By Gruber's estimates, the take of shark fins in waters around Florida has skyrocketed—from a mere 1.5 tons in 1986 to 40 tons just two years later.

"This works out to over 100,000 individual animals," Gruber reported late last year, "and in many cases only the fins are saved. In some grotesque cases, the shark is returned to the ocean alive without fins, to starve to death."

But federal protection for sharks is on the way.

Last October, the National Oceanic and Atmospheric Administration produced a draft management plan that for the first time would impose limits on commercial and recreational catches of 38 species in the Atlantic. It would require shark fishermen and dealers in shark meat to obtain permits. And it would seek to control finning by limiting the buying and selling of fins to a proportion of the number of carcasses.

The initial proposal would set an overall annual limit of 16,250 tons and would create a federal oversight team to monitor compliance.

Although the final plan was to have been published last February and put into effect in June, it is yet to be completed—because officials have not been able to agree on what the limits ought to be.

As an interim step, conservationists have asked Commerce Secretary Robert A. Mosbacher to take emergency steps while the protection plan is being completed. The Washington-based Center for Marine Conservation called on Mosbacher recently to put the recommendations on finning into effect immediately and to limit recreational shark catches to one per vessel.

Similar pleas came from the University of Miami's Rosenstiel School of Marine and Atmospheric Sciences and the National Wildlife Federation. "This is a typical case of federal and state governments responding after a fire is out of control," said George Burgess, senior biologist at the Florida Museum of Natural History.

"If we wait for hard data to come in, the sharks will be gone," Burgess said. "Decades, not years, would be required for them to recover. My personal opinion is that if we impose stringent limits now, the populations could recover in a reasonable amount of time. But if we wait another year, the crash will have occurred."

What makes the threat of the population collapse so critical is that sharks, unlike other fish, reproduce slowly. They have long gestation periods and give birth to relatively few young. In some cases, they take years to reach sexual maturity.

But there is more at stake here than protecting creatures merely for their intrinsic worth.

Sharks reign at the apex of marine life—350 species of them, from 35-foot whale sharks and the fearsome great white sharks of the movie "Jaws" down to 8-inch cigar sharks. If their population were to collapse catastrophically, it could spark a population explosion among the predators beneath them in the chain, setting off a ripple effect that could reach major consequences.

Scientists also say they can learn from laboratory studies of shark physiology—if the fish are available.

"Without access to research specimens," Gruber said, "we may never understand why sharks are so resistant to cancer, how a serious wound heals in less than 24 hours, or why a lacerated cornea, which in any other creature would permanently cloud over, remains clear, functional and rapidly heals."

Marc Agger, a 32-year-old entrepreneur who buys shark fins from 300 boats ranging over the Atlantic and sells them to upscale Oriental restaurants around the world, agrees it is time for regulation.

Agger, who has a master's degree in business from Yale, was in mergers and acquisitions on Wall Street before entering the shark fin business, operating in a smelly warehouse in the old Brooklyn Navy Yard.

Agger's company buys, dries and sells more than 100,000 pounds a year of fins from more than a dozen species of sharks—makos, blues, lemons, tigers, blacktips and muskies. The cured shark fins fetch $10 to $30 a pound, and restaurants use them in soup.

Even Agger says the government needs to place some limits on the business. "Fishing right now is like the Wild West," Agger said. "It's like, 'Let's go out and shoot a buffalo for dinner.'"

Content Considerations

1. What has caused shark populations to decline in the Atlantic Ocean?
2. For what reasons are conservationists trying to prevent further declines?
3. Summarize the shark management plan proposed by the National Oceanic and Atmospheric Administration.

The Writer's Strategies

1. How does the article's beginning reveal the tone the article takes?
2. Each quotation is preceded by a statement that emphasizes or summarizes the quotation's focus. How do such statements help the reader?

Writing Possibilities

1. How might the shark's reputation as a malevolent creature affect preservation efforts?
2. Discuss the problems that might arise in efforts to draft and enforce a "management plan that . . . would impose limits on commercial and recreational catches" of sharks (paragraph 16).
3. Speculate on how the extinction of some or all shark species would affect you, the economy, or ocean ecosystems.

MICHAEL MILSTEIN

Unlikely Harbingers

A harbinger, in this piece, is a creature that suggests or foreshadows what is to come. The creatures you will read about are some whose numbers are declining. Such diminishing populations suggest that portions of the environment have become so hazardous that they cannot sustain many varieties of life. The worry, of course, is that these creatures are only the beginning, that they as well as other and larger species will eventually disappear from the Earth because their habitats have been polluted or destroyed.

The harbingers of the possibility of wide and massive extinction are, in this case, amphibians. Around the world a mystery has descended: Where are the frogs? Where are the toads? The Salamanders? Scientists are trying to discover how such a disappearance occurred so suddenly and so widely. In the following article from National Parks, *Michael Milstein, once a park ranger and now a reporter, explains what scientists have found and what they suspect; he returns us to the question of why these species matter.*

Before you begin reading, think of why the world needs frogs and toads. Also, try to account for their disappearance.

HIGH IN CALIFORNIA'S SIERRA NEVADA RANGE, just north of Yosemite National Park, lies an isolated alpine lake where Lawrence Cory could always find frogs. A biology professor at St. Mary's College in northern California, Cory spent nearly a decade in the 1960s studying mountain yellow-legged frogs, abundant throughout the glacier-tilled Sierra. Whenever he ventured to Koenig Lake, it was teeming with many hundreds of the speckled creatures hopping about, and thousands of their young tadpoles basking at the edge of the water.

In 1988, Cory returned to Koenig Lake after almost 20 years. This time, it was eerily still and silent. This time, there was not one frog or tadpole to be found. Quickly, he checked a half-dozen other lakes that had been astir with frogs in the past, and discovered them vacant, too.

"It was the first time I had an emotional reaction to a change in the environment. I was shocked," Cory recalls. "It's like being in downtown San Francisco at rush hour and not seeing any cars. Something's got to be wrong."

Around the nation and the world, scientists are now uneasily echoing that sentiment. Prolific frogs, toads, and salamanders are mysteriously disappearing even in untouched national parks and wilderness areas. They are disappearing from the pure, high mountains of the Northwest and the arid lowlands of Arizona, from the Australian interior and the green, tropical hills of Costa Rica.

In what could be a potent omen of human-wrought environmental damage, lakes, ponds, and streams once melodic with frogs' deep, intimate mating calls have now fallen deadly quiet. This strange silence may mean more than the simple loss of a familiar childhood playmate, or an expected boom in swarms of mosquitoes once kept in check by amphibian appetites.

In February, David Wake, a biologist at University of California (Berkeley), led a hastily convened National Research Council conference to discuss worldwide amphibian declines. "There's no question this is the result of human meddling," he now concludes. "Frogs dying off may just be an early sign of the harm we're doing to the environment."

While scientists know that many once-common amphibians are becoming extinct or are already extinct, no one knows exactly why. No single theory explains all of the missing species. There are few corpses in this mystery, and fewer clues.

Many of these ecologically sensitive creatures are clearly gone because their wetland homes have been wrecked by trampling cattle, uncontrolled logging, or other human development. Inescapable acid rain and snow is also a possibility, as are pesticides, drought, and competition from exotic species. Most alarming, though, are inexplicable, simultaneous amphibian declines in America's protected, relatively pristine national parks and wilderness areas.

"These are some of the last places you would expect to see such common creatures becoming extinct before our very eyes," says Andrew Blaustein, an Oregon State University herpetologist who has watched several species of western frogs vanish from much of their homeland. "If they're dying in undisturbed wilderness, it proves nothing, no place is safe anymore."

Uniquely adapted to altitudes above 6,000 feet, the Yosemite toad, for example, has largely disappeared from its namesake park. Completely gone from Yosemite now, yellow-legged frogs have also declined in Sequoia and Kings Canyon national parks, farther south in the Sierra. Once common within Oregon and the Northwest's wild Cascade Range, amphibians such as the western spotted frog and the Cascade frog are now seen there rarely.

Early settlers near Mesa Verde National Park in southern Colorado reported thousands of amphibians along the Mancos River. Although spadefoot toads are still common, leopard frogs have not been seen for years. In and around northern Colorado's Rocky Mountain National

Park, boreal toads (close relatives of Yosemite toads) and leopard frogs are gone from up to 90 percent of their former haunts. Tiger salamanders and chorus frogs are also missing from some sites.

Such swift declines in far-flung locales do not bode well for a class of animals that has persisted in generally the same form for more than 75 million years. Today's amphibians are descendants of the first creatures to evolve from a waterborne existence and successfully adjust to life on land. They lost ground when dinosaurs became dominant, but persevered by plying the damp zone straddling air and water.

In bridging that gap, ancient amphibians became the conduit onto land for all modern birds and mammals, including people.

Since their intermediate anatomy is vital in understanding more advanced animals, frogs are examined in classrooms everywhere. Studying humans without comprehending amphibians, biologists say, would be like studying U.S. history without considering the American Revolution.

"When you look at a toad or frog, you're looking at a representative of the earliest, most primitive forms of life on land," explains David Martin, a biologist at San Jose State University. "If they hadn't made that jump from the water, none of us would be here."

Each year, just as snow melts and small ponds begin to brim with runoff, amphibians emerge from their winter's sleep in soggy burrows or rotten logs. Then, they look for romance. Like sweet-talking Romeos, these familiar harbingers of spring voice loving croaks to help them lure available mates.

As females lay hundreds of eggs in the water of a still pond, males then douse them with a milky cloud of sperm. Salamanders differ only in that females fertilize their eggs internally, after gathering male sperm from the pond bottom. They later attach the eggs, as do most frogs and toads, to submerged plants, rocks, or logs.

Broods of frogs, toads, and salamanders develop in two distinct stages from long strands of eggs or from softball-sized, jellylike egg masses. Young, gilled tadpoles that emerge from the egg mass must stay immersed until they mature into adults, when they can also survive on land.

In adapting to their niche, adult amphibians have evolved the unique ability to breathe through their moist skin. Thus, they can absorb oxygen either above or under water. On land, they can also use humanlike lungs.

Such special devices, which allow amphibians to live in two environments, also make them overly vulnerable to ravages in each. Their permeable skin can easily absorb toxins from both air and water. Absorbed toxins can then be passed on to eagles, herons, raccoons, and others that eat frogs.

Because tadpoles are low on the food chain and adult amphibians are much higher, amphibious species are affected by environmental vagaries at both stages. That exposes their kind to a broad spectrum of problems.

Amphibians are, therefore, notable "indicator" species, biologists say, a vibrant thread in the environmental fabric. Like the proverbial canary used to detect invisible but deadly fumes in a coal mine, frogs are telling bellwethers of wide environmental change.

Yet when the creatures began disappearing sometime in the early 1970s, few really noticed. Slimy, uncomely amphibians rank low in the hierarchy of wildlife research compared to money-making game animals such as elk, bears, or deer. So, frogs and their kind were never much studied to begin with. "People have kind of ignored things like frogs and toads because they're not charismatic species," says David Stevens, a biologist at Rocky Mountain National Park in Colorado, where surveys have shown that frogs and toads have disappeared from alpine ponds.

Federal researchers will continue on a shoestring budget at Rocky Mountain this summer, after finding one surprisingly healthy toad community last year. They hope to discover what conditions have allowed this one population to survive.

Biologists trying to explain the amphibian eclipses are shackled by a dearth of scientific data. To document declines, many have returned to ponds where species were recorded decades ago. But accurate records are rare, since there was never a need for them—until now.

"All our frogs were so common a couple decades ago that nobody even bothered to write them down," Yosemite wildlife biologist Jeff Keay says, lamenting the lack of information. "You wonder how many other critters are having the same problems."

To confuse matters, amphibian species, in some areas, are still doing fairly well. And, scientists could explain earlier declines in terms of natural boom-bust breeding cycles typical of frogs and toads.

Until scientists at the first world herpetology congress in England last fall finally realized that they were all seeing similar declines, some researchers had figured these die-offs to be just routine, local events.

"We never knew whether or not to make anything out of it," remembers Marty Morton. A biologist at Occidental College, California, Morton studied Yosemite toads through the 1970s with colleague Cynthia Kagarise Sherman. As they were finishing their work, they noticed a sharp loss of toads where there had been hundreds breeding each spring.

"Now," Morton muses, "we may have some of the only good information on a dying species."

The pair expects to return to Yosemite this summer to compare past and present toad numbers, but none of the small, olive-green residents have been noted in the park for the past two years.

Outside of the West's remaining wildlands, perhaps the most visible but least appreciated threat to amphibians is the outright loss of their habitat. Much of their matchless wetland home has been flooded or dried out by giant reclamation projects such as those that have damaged the Everglades. Even more has been razed and paved to make room for human homes and progress.

Little-publicized federal research has shown clearcut logging even more devastating to forest amphibians than to the renowned spotted owl, an icon of battles over virgin Northwest forests. By changing the forest climate and dirtying streamside habitat, clearcutting—an economical method of felling vast tracts of timber at once—may isolate and extinguish longtime amphibian residents.

Or, habitat destruction can be more subtle. Biologist Martin has seen simple disorder, such as vehicle traffic or lumbering cattle, throw off the Yosemite toad's careful mating rituals and interrupt their sonorous croaks. At many lakes under heavy fishing and recreational pressure, toads are few and far between.

Even within their habitats, amphibians are at risk. Trout and other fish planted in mountain lakes high in the Sierra and Rockies feed voraciously on amphibian eggs and tadpoles. A century ago, no native fish lived in any of Yosemite's 339 lakes—now about a third contain rainbow, brook, and other exotic trout. National park managers still stock some lakes to please anglers, although they are now reevaluating that policy.

Possibly the most ominous specter hovering over world amphibians is the airborne poisons hidden in acid rain and snow. Prevalent for decades in the eastern United States and growing worse in the West, acid precipitation results when industrial and automobile sulfur and nitrogen exhaust binds with atmospheric elements to create sulfuric and nitric acids. These acids then fall back to earth cloaked in raindrops and snowflakes.

Some amphibians, and especially their waterlogged eggs, have proven extremely sensitive to increased acidity. When exposed in a laboratory to acid amounts that were only slightly higher than is normal, some eggs died, while other eggs grew into deformed tadpoles.

Acidity in the water can also slow the growth rates of young amphibians, keeping these creatures from maturing before the water in their temporary pond refuges evaporates during the heat of the summer.

When exposed to caustic moisture, frogs' sperm is even more vulnerable than eggs. Since frogs fertilize their eggs in open water, any acid present could kill their sperm instantly, rendering males impotent.

"All the way around, acidity is extremely detrimental," says Benjamin Pierce, a Baylor University herpetologist and one of the first to study effects of acid on amphibian breeding. "These frogs can't help but expose themselves to it simply by the way they live."

Most precipitation falls in mountain ranges as snow. When the snow melts, it then fills the small ponds where amphibians breed and lay their eggs.

Even at areas such as Sequoia and Kings Canyon national parks, where unusual acid levels are uncommon, researchers have detected a fast, concentrated "pulse" of acidity flushed out of melting snowbanks in a matter of hours. This acidic pulse washes into ponds and lakes just as frogs and toads begin mating.

"It was coming right at the most crucial and unprotected time in their lives," says U. C. Berkeley biologist John Harte, who has come closest to linking acid pulses to a threefold decline of tiger salamanders in one watershed in the Colorado Rockies. "It's not yet as clear-cut an effect as smoking and lung cancer, but it's there."

In the East and at lower elevations, plentiful limestone and thick soil help absorb acidity like Pepto Bismol soothes stomach pain. Some eastern amphibians, such as the pine barrens tree frog in New Jersey and the Carolinas, are even tolerant of their boggy, acidic habitat. Scientists theorize these frogs may have evolved in the inhospitable habitat to avoid undue competition.

Such unique adaptations as well as buffering soils and stone may have kept amphibians from disappearing in the East—so far.

In the West, younger, higher mountains are carved from impenetrable granite, covered with little buffering soil. It is an ecosystem intolerant of the rapid changes humans have wrought in just a century. Neither the western environment nor its wild residents, including frogs, fish, and other species, can handle the influx of acids, overused pesticides—the diverse assortment of pollutants that have appeared.

Weather patterns that deliver acid rain seem to coincide with many amphibian declines. In the late 1960s, biologist Cory found residues of the pesticide DDT in yellow-legged frogs at some of the highest elevations in the Sierra.

Conversely, amphibians in Yellowstone and Grand Teton national parks, geographically less exposed to air pollution, still seem in good shape.

Acidity does not explain all the lost amphibians. Once the most abundant species in Oregon's lush Willamette Valley, where no measurable acid rain has been found, the red-legged frog is now extremely rare. Western spotted frogs, once common throughout the mountainous West, are now extinct over nearly half their former range and are a new candidate for the federal endangered species list.

In California's Calaveras County, native frogs have been either eaten by people or exotic eastern U.S. bullfrogs, caught by scientists or youthful collectors, or killed by toxins in mining waste. Now the notorious jumping frogs of Calaveras County—famous around the world—exist only in the tales of Mark Twain.

While acid snow is present in the southern Rockies, the Fish and Wildlife Service–NPS survey that included Rocky Mountain National Park found similar losses of boreal toads and leopard frogs in southern Wyoming. Acid levels measured there were not low enough to disturb them.

"There is no smoking gun to explain this right now," says Bruce Bury, one of the two U.S. Fish and Wildlife Service biologists who conducted that survey. "It could turn out to be very simple, or it could be something bigger than we realize."

There are a few reassuring signs. In a wilderness section of Michigan's Isle Royale National Park, David Smith, a Williams College research scientist, watched numbers of chorus frog young plummet from 20,000 to less than 5,000 in the early 1980s. They have since recovered to around 13,000 tadpoles last year, but these fluctuations are not well understood.

In Sequoia National Park, biologists will try to reintroduce now-absent foothill yellow-legged frogs to lower elevations, and see if they survive. Some species, such as the Pacific tree frog in California, seem to be holding their own. They may be less sensitive to acid rain, or it might be something else altogether. Frog eggs, for instance, can also be ruined by ultraviolet light making it through the Earth's failing ozone layer.

What is most urgently needed now, biologists say, is research to prove exactly what is killing frogs and their kin. Backcountry rangers working in several national parks, including Canyonlands, Yosemite, and Sequoia, will be on the lookout for healthy amphibian populations this summer.

Scientists such as Cory, Martin, and Morton can then compare areas of healthy populations to those where amphibians have declined or disappeared altogether.

Maybe something can be done to rectify the damage, even if it is only to halt subsidies of exotic, frog-gobbling trout. Or, it may already be too late. Amphibians may join the 15 percent of the world's animals expected to be extinct in the next 30 years.

"A lot of what we know is only anecdotal bits and pieces," says UCLA biologist David Bradford.

Last summer, Bradford found frogs in only one of 46 Sequoia National Park ponds where he had studied them in the 1970s.

"You'd expect people to disappear before frogs, because of their proven adaptability," he says. "When you look at some of these crashes together, though, it is extinction. It's for good."

Content Considerations

1. Why are amphibians especially vulnerable to their environment? What are the consequences of their vulnerability?
2. What is an "indicator" species? How useful is the information it provides?
3. Why aren't amphibians in eastern ecosystems disappearing as rapidly as those in the West? What does the situation in national parks suggest about amphibians elsewhere?

The Writer's Strategies

1. Milstein begins this article with a description of a place as it existed in the 1960s, then quickly contrasts that description with what a researcher found in 1988. In what ways is this method of introduction suitable? How does it prepare you for the tone and content that follow?
2. One method of persuading others is by an appeal to authority, by giving evidence from the experts. What experts are cited in this article? How credible do you find them? How does the reader decide to trust the authority or expert?

Writing Possibilities

1. According to the author, "frogs are examined in classrooms everywhere" (paragraph 14). Write about such an examination.
2. With classmates, brainstorm a list of questions this article provokes, then write about those questions and how answers can be found.
3. Recall amphibians featured in stories, books, pictures, and films. Then imagine a world in which amphibians have become as extinct as dinosaurs. How might they then appear in literature, art, and film?

Margaret Knox

Africa Daze Montana Knights

Humans live by the Earth and what they take from it. The current attention to ecology results from a fear that humans have taken too much, have not replenished what they have used, or have taken in ways that ruin the Earth beyond repair. Whether they reside in cities or in the countryside, humans survive by depending on the Earth. Many go beyond survival and enjoy wealth.

Preserving a species for the entire Earth and its population usually means that some group of people must change their attitudes and their practices. They must quit farming or farm differently, quit hunting or hunt differently, quit manufacturing or manufacture differently. They must consume less or consume differently, pollute less or learn to clean up what they have dirtied. On the issue of preserving species, some group sees its interest losing out to those of other groups.

The interests are many—businesses, farmers, hunters, consumers, the poor and the rich. At times preservation may put profit at stake, at times it threatens the traditional means of survival for large or small groups of people. The preservation of species has economic repercussions.

Margaret Knox, a writer who now lives in Montana after a long stay in Africa, explores such interests and repercussions in the next essay. Before you read, write briefly about the standards you would use in deciding whose interests to protect in a preservation question.

ONCE UPON A TIME, conservation was a gentleman's game. Over stem-glasses of sherry and fat cigars, in the drawing rooms of Europe and New England, hunters and nature lovers lobbied for the preservation of Yellowstone National Park and claimed the vast Serengeti as a perpetual safari-land. They forged agreements, drafted laws and organized scientific expeditions. Their passions flowed eloquently through delicate quill pens. By and large, conservation work was a polite and decorous undertaking.

No more. As the globe shrinks and human populations explode, conflicts over the last unpolluted, unpaved and untilled patches of soil are becoming ever more desperate and violent.

The conflicts aren't only between those who would leave nature alone and those who would eat it, wear it, carve it or graze it. Even those who call themselves conservationists can't agree on the most fundamental questions: whether ranching is acceptable, whether hunting is appropriate, whether wildlife should be managed at all. People who have different ideas about how to protect nature are coming to blows. The language of preservation has become the language of war.

Fighting wars over scarce resources is not new: today, in a world of five billion people, land is one of our scarcest. Most of us, when we think about it, would like to preserve vast tracts of pristine wilderness. Unfortunately, with so many people to feed, human needs for land often take precedence, especially in underdeveloped and developing countries. Overgrazing, deforestation and erosion result, tragically leaving ever less land for ever more people. People attack the earth with hoe and hoof and fire, and nature fights back with dust and flood and famine.

The warfare isn't just people fighting nature, but people fighting people on behalf of nature. Helicopter-and-tommygun combat teams patrol through an African valley in search of poachers, and animal-rights activists physically assault hunters in Montana. Today's environmental crises push the conservation debate beyond the petition toward the tree spike. The rhetoric is heating up as well, to the point where even those who fundamentally share interests—loggers, say, and those who oppose rapacious, job-threatening clearcuts—are at each other's throats.

I am troubled. Part of me—my instinctively pacifist, consensus-seeking side—recoils from seeing the movement to defend nature become shrill to the point of violence. But another voice tells me nature is in such peril that, to twist Barry Goldwater, extremism in defense of nature is no vice. And that scares me.

It is bad enough that the earth and its defenders are at war with those who would rape it for profit. But even supposed allies, those fighting on the side of the earth, are increasingly vicious with one another. Again and again conservation groups, each thinking it holds the franchise on ultimate wisdom, vilify and sabotage each other, fracturing the overall "environmental movement" and wasting precious energy that could better be spent working on the earth's behalf. The passions that have built up over the environment in the past few years give a sense of the magnitude of the crisis behind them.

The war between human and wildlife interests is being fought in Zimbabwe with machine guns. The humans are by and large citizens of Zambia, which lies just north across the Zambezi River from Zimbabwe. If ever a people were economically desperate, it is the people of Zambia, whose monoproduct economy collapsed with copper prices in the 1970s and who have been driven to poverty so extreme that their *commercial export* farms have reverted to oxen. The Zambians are what we think of when we pity the destitution of the third world, enormously deserving of our sympathy.

Yet they are largely responsible for slaughtering the world's last viable herd of black rhinoceros, which lives in the national park on Zimbabwe's side of the river. It's not that the Zambians are evil or vicious; they're hungry, and the proceeds from a single rhino horn can feed a family for months. What's more, the horns are easy to get. Rhino are curious and nearsighted, and it's a relatively simple matter to get close enough to drop one with a single shot. Once the animal is down, the horn—which isn't bone at all but rather densely packed hair—can be sheared off the nose in seconds with a knife and spirited across the river before daybreak.

To look at the rhino from the Zambians' perspective: here's an animal walking around with thousands of dollars on its nose, an animal that doesn't do anybody any good anyway, except (again from the Zambians' perspective) as an attraction for rich tourists. Should Zambian families starve so tourists can take pictures of rhino?

Zimbabwe, which is relatively prosperous and can afford the "luxury" of wilderness and wildlife, has moved many of the beasts out to the safety of game farms farther from the river. But ultimately President Robert Mugabe decided the rhino has a right to live where it lives and that it does Zimbabwe good to have wild rhino in the Zambezi Valley. In 1985 he ordered "Operation Stronghold," which is just as military as it sounds. Today, anyone found armed within the valley is considered a poacher and is shot on sight.

In the past five years, more than 60 poachers and at least one scout of the Zimbabwe National Parks Department have been shot dead in firefights. "The poachers smear their bodies with juju (magic oils) to make themselves invisible," sighed Peter Matoka, when asked about it in his posh home in Zimbabwe's capital, Harare. Matoka, then Zambian High Commissioner to Zimbabwe, spent a good deal of his time repatriating the corpses of his countrymen killed in the rhino wars.

At first the Zambian parliament was outraged: Why should the life of an animal take precedence over that of their people? Why was a misdemeanor like poaching being punished with death?

To answer that, you have to step back. We are in a crisis of extinctions. During the past 600 million years, the natural rate of extinction averaged one species a year. We now witness one to three extinctions every day. Scientists predict by the early 21st century we will be losing several hundred species a day. The rhino, a marvelously weird-looking beast, has become a symbol of the effort to slow the tide of extinctions. Its low-slung head, nose-mounted weaponry and armor of thick, nearly hairless, skin evolved 55 million years ago and is an exhilaratingly vivid link to prehistory. As many as 65,000 black rhino roamed the African bush in the early 1970s, but the Middle Eastern and East Asian markets for rhino horn have whittled the species to about 600 wild beasts and their last remaining habitat to Zimbabwe's narrow and besieged Zambezi Valley.

After a few weeks of raucous protest on the part of his Parliament, even Zambian President Kenneth Kaunda decided the Zimbabweans were right. Throughout Africa, every less extreme method of protecting the black rhino in its natural habitat has flopped. The Kenyans have resorted to electrified fences and round-the-clock guards to protect their last 200 black rhino. Even the Tanzanian government, which once forcibly moved several villages out of the path of a wildebeest migration, was unable to curb the rhino poaching.

The world pays lip service to its love of wildlife—at least the large, spectacular mammals. But Asian pharmacists pay cash—astronomic prices for rhino horn to grind into medicine. Yemeni men lay out more than 1,000 for a carved rhino-horn dagger.

Steve Edwards looks more like a green beret than a game warden. At his Zambezi Valley base camp after a weekend patrol last year, his khaki chest was crisscrossed with web gear and his safari-vest pockets bulged with hand grenades. A battered automatic rifle was never far away. Edwards, a trained naturalist, wasn't in the mood to talk about the obsessive nesting habits of weaver birds or the water-retaining wonders of baobab trees. His thoughts ran to stalking human prey, ambushing the enemy and counting bodies—of rhino and Zambians. His stories balanced the gruesome details of rhino deaths against graphic accounts of victories over poaching gangs.

One hair-raising story sticks in my mind: An anti-poaching patrol found a poachers' camp and planted a landmine under the fire ring. When the poachers returned and started a fire, the heat set off the mine. "We came back and found a biltong tree," chuckles Edwards. Biltong is beef jerky. Is this the modern environmentalist?

As a newcomer, it was difficult putting Edwards' fighting words together with my Edenic surroundings. An orange sun was simmering into the river upstream from us. Two elephants splashed ashore on the bank nearby. Fireflies blinked. And we sat talking about triangulated gunfire, Claymore mines and the biltong tree. The wilderness has become an armed fortress.

Poaching, of course, isn't the only way human needs encroach on the Zambezi Valley. There is a danger that this lush valley will be snatched out from under not only the rhino, but all its other wondrous inhabitants from the corn cricket to the elephant. The Zimbabwean government, continuously strapped for foreign exchange, last year gave Mobil Corp. exploration rights throughout the valley—including a national park and United Nations-designated World Heritage Site. On the other side of the country, Zimbabwe has thrown up a ten-foot storm fence at the Botswana border to keep wildebeest from spreading hoof-and-mouth disease to cattle. At the end of every drought, news photos show parched wildebeest dying against the fence just yards from a Zimbabwean water hole.

Wildlife and domestic livestock have hurled diseases back and forth like spears since rinderpest brought by Indian cattle in the 1890s nearly wiped out the wildebeest. If hoof-and-mouth works against wildlife, another pest works in favor of the Zambezi Valley. When some Zimbabwe government officials see the Zambezi Valley, they see potential grazing land for thousands of families. Land-poor cattle grazers also are pressing their noses against these public lands. Their pleas for more range can't be ignored by a government whose revolutionary slogan 10 years ago was "Land for the People."

Right now, all that is preventing the human and bovine invasion of the Zambezi Valley is the native tsetse fly, which is no trouble to thick-skinned wildlife, but whose diseased bite wreaks havoc on domestic livestock. The government is paradoxically willing to kill poachers in the valley but also experiments with tsetse fly traps to open it someday to grazing. So far, conservationists have managed to halt a full scale tsetse fly eradication program at the lip of the valley. But whenever Edwards finds traps, he destroys them for the same reason he hunts poachers—to preserve the Zambezi wilderness. "We can kill people all day long," he grins, "but the only thing that's really going to save this valley is the tsetse fly."

The Zimbabwean anti-poaching campaign may seem a far-off, dire measure of a desperate third-world government. But human pressures on American wildlands are growing, too. And with it grows the intensity of rhetoric and action on nature's behalf. The pressures here come not from hungry peasants but from corporations promising people jobs, food and a fast fix for foreign debt.

Already, nearly three-quarters of our nation's forests are in private hands; only 5 percent of native virgin forests remain. About 80 percent of all public lands in the west, including wilderness areas and national monuments, are leased for grazing. And while some conservationists continue ponderous legal efforts to protect these lands, others have lost patience. They see their cause—biological diversity for the long-term benefit of the earth—as too urgent and complex for easy persuasion.

In the name of the spotted owl, they have resorted to tree spiking in the old-growth forests of the Northwest. In the name of streambanks, they have sabotaged ranch equipment in Nevada. Their violence—for there is no other word—is a double-edged sword. A tree spiking or a destructive protest makes the evening news in a way a march or a speech to the legislature doesn't. Issues that might otherwise be ignored by the media are thrust before the public for scrutiny. But as in Zimbabwe, the violence can repel people who might otherwise be sympathetic. The protests of Kenneth Kaunda's government to the shooting of Zambians sound piteously feeble beside the uproar over eco-sabotage in this country. I got a fast lesson in the dangers of eco-polarization the night I first arrived in Montana.

My husband and I had stopped for the night at the Rocky Knob Inn, a rustic logger's bar and motel far down the Bitterroot Valley at the bottom of Lost Trail Pass. It was snowy, and dark, and both of us were exhausted from wrestling our heavily-laden truck through the pass. We wanted only to have a quick dinner and go to bed.

At the bar stood a Central-Casting logger: six-foot-five and broad as an old-growth redwood, he had a drooping mustache, sweat-stained Stetson and a red bandana around his neck. His voice was rough and his hands rougher: a toothpick danced at the corner of his mouth as he talked. He must have given us his name, but I remember him only as "Hoss," a cheerful, generous ambassador of Montanan goodwill.

We told him we were on our way to a new home in Missoula, and he asked what brought us. I'm an environment writer, I told him, a mistake I won't make again.

"Environmentalist?" he snarled, his friendliness shattering at once. "You a flower-sniffer? You a *tree-spiker?*" For him, environmentalist means tree-spiker, end of story.

And he isn't the only one. Throughout western Montana signs are posted on houses and stores and bumpers: We Support the Timber Industry. Or, even more tragically, This Family Supported by Timber Dollars. To those families, anyone who questions the wisdom of clearcutting our national forests is a "flower-sniffer" or a "tree-spiker."

Yet if anyone should be making common cause with the "flower-sniffers" who oppose clearcutting, it is the families who are supported by timber dollars. If the logging continues at its present pace, there won't be any timber dollars in Montana in a generation because there won't be any timber. The timber debate in Montana has grown too hot for its own good. Too much shouting is being done, and too much name calling. Tree spikes *have* saved stands of old-growth forest, and they *have* thrown the media spotlight on the rape of our forests. But they carry a price that the conservation community must consider, just as those dead Zambian poachers do.

In March, an incident of eco-sabotage both amplified and clouded a complicated tug of war between natural and human imperatives. The setting was Hebgen Lake, just outside Yellowstone Park.

Within the park, the National Park Service can claim credit for one of the century's great conservation victories: saving the bison. Only 500 remained after the carnage of the late 1800s; today more than 2,000 roam the park. So successful has the Park Service been that the herd regularly overflows Yellowstone. Through forests, down highways and along old railbeds, the shaggy humpbacked matriarchs annually lead their herds across the park's artificial boundary and toward an easier life on ancient grazing grounds claimed by cattle ranchers, the ground on which humans and nature today are clashing.

More than half of Yellowstone's bison are infected with a disease called brucellosis, which causes abortions in cattle and a raging fever in people who drink unpasteurized milk. Though it hasn't been conclusively proven, it is likely the bison can easily pass the disease to their domesticated cousins. So as Montana ranchers—who have spent more than $30 million ridding their herds of brucellosis—watch the bison pouring out of the park, the human-versus-wilderness debate stokes up as hot as in southern Africa.

The U.S. Department of Agriculture, whose mission is to protect agricultural, not wildlife, interests, would like to inoculate the Yellowstone bison and sacrifice the park's essential wild nature to protect the ranchers. The Park Service so far has refused, saying the bison aren't a problem *inside* the park but only *outside*. Therefore, the Park Service's reasoning goes, they should be dealt with outside the park. Montana agreed to do just that, passing a law five years ago that allows lottery-picked hunters to shoot the bison as they cross the line.

This year, enraged Earth First! and Fund for Animals protestors tried to disrupt a hunt on Montana's Hebgen Lake, and by the time it was over two of them were up on assault and other charges for lashing one hunter with a ski pole and smearing bison blood on the face of another. The newspapers and TV news had a field day.

The outcome: Montana's hunters have been conditioned to think of anyone opposed to the bison shoot as a pole-wielding, blood-smearing extremist. Those opposed to the hunt are confirmed in considering the ranchers greedy spoilsports willing to contract-murder defenseless bison. And almost everybody has one beef or another with the state fish and wildlife department, the USDA or the Park Service. Once again, anti-wilderness politicians such as Republican Representative Ron Marlenee of Montana are trying to capitalize on the public's image of "environmentalists" as a small bunch of shrieking loonies. This time, the "flower sniffers" committed an unpardonable sin: They wrecked a good day's hunting for some red-blooded Montanans.

In some ways, the real issue—what is the best way to balance the interests of the bison and the ranchers—was lost amid the shouting. A rational solution seems further off than ever.

On the other hand, without such a dramatic, telegenic protest, the public's interest might not have been aroused. Nobody will ever shoot a bison in Montana again without raising public debate about the wisdom of the hunt policy.

The new wave of high passions and dramatic acts has been undeniably useful in bringing our environmental crisis to the public eye and holding it there. It's also clear now that militant actions should be carefully chosen. Hot words and bursts of violence not only alienate the conservation community from those it would like to recruit, they also can fracture the community itself. If the rhino campaign in Zimbabwe falls apart, it won't be because the rest of the world objects to it. It will be because the agencies running it can't agree to get along with one another.

Nine rhino fell to poacher's guns over the Christmas holiday in 1988 when the anti-poaching campaign was effectively shut down by a turf war between the Zimbabwe national police and the parks department, both of which are ostensibly on the same side. The leader of the parks department's anti-poaching campaign and two of his men killed a poacher, who turned out to be the nephew of an important district official. The police, long bitter over their second-fiddle role in the anti-poaching campaign, seized the chance to arrest the parks men, and the anti-poaching war slid into a week of inaction. The charges against the parks men were finally dropped, but not before nine precious and unguarded rhino—1.5 percent of the total herd—were killed by poachers.

Perhaps no conservation organization has been as committed to protecting endangered species as the Worldwide Fund for Nature, formerly the World Wildlife Fund. It provided the Zimbabwe rhino campaign's only helicopter, which the game wardens considered absolutely essential in the vast tangle of the Zambezi Valley. Yet in May 1989, the group yanked its funding for the chopper because it wasn't pleased with the Zimbabwe government's paperwork.

"The parks directorate is too bloody disorganized," said the fund's ecologist Raoul du Toit. "The helicopter was a holding action until the agency could come up with a long-term approach."

At the present rate of poaching, the rhino population was barely holding its own and the conservation group wanted the government to consider concentrating its meager forces on the densest herd. The government, bent on protecting the whole valley, never found time to plan alternative strategies. "It's such a shame," commented Dick Pitman of the Zambezi Valley Society, a conservation group. "We're losing the helicopter to what appears to be a personality conflict."

Although the hunting controversy in the United States can't quite be described as a personality conflict, it does hinge on deeply personal moral issues not directly related to the cause of biodiversity and the survival of species. The hunting community can call up volumes of statistics on species they have helped save and land they have protected from development. Hunters were among the most vociferous advocates for setting aside Yellowstone in 1872. And though rapacious hide-hunters had by then nearly wiped out the bison, it was another hunter, Blackfoot Indian Walking Coyote, who saved the little herd that grew into today's Yellowstone bison.

But here, as in Zimbabwe, people who profess the same allegiance to wilderness shriek at each other. To read pamphlets from hunting and anti-hunting lobbies is to read almost the exact same words. "We now know that the web of life is of enormous importance to our own existence and that there is an urgent need to preserve what is left of the natural world," says the anti-hunting group Friends of Animals.

"Wildlife conservation will not take care of itself. The daily loss of habitat is still a constant threat," says the hunting advocacy group Wildlife Legislative Fund of America. Both sides decry the loss of natural habitat. Both sides profess an almost religious veneration and respect for wildlife. Both sides lobby hard to protect wildlands from development.

Yet with so much in common, the hunting and anti-hunting lobbies have deionized each other to the point where the differences in their philosophies, not their similarities, get all the attention, such as in a recent cover story on the hunting controversy in *U.S. News and World Report*. "These bloodthirsty nuts claim they provide a service for the environment. Nonsense!" Cleveland Amory of the New York-based Fund for Animals told *U.S. News*.

"The anti-hunters are invariably long on name calling and short on scientifically based suggestions for solving complex wildlife conservation problems," countered George Reiger, conservation editor of the hunting and fishing magazine, *Field and Stream*. Hunters point to the Pitman-Robertson Act, which established an excise tax on guns in 1937 when frantic burning, draining, plowing and grazing were fast shrinking the habitats of birds and mammals. The tax has poured nearly $2 billion into training state fish and wildlife experts, counting animal populations and setting aside habitat. Some animals have their own fan clubs, private hunters' groups such as Ducks Unlimited and the Elk Foundation which donate millions to protect their chosen prey. Above all, hunters say their sport keeps thousands of people in touch with nature.

Dan Jacobs talks lovingly of the Montana back-country he hunts and fishes, and of the sports he practices. Jacobs, a 39-year-old air-ambulance nurse, hand-loads his own bullets. He target shoots to keep up his skill—says he gave up bow hunting after badly wounding a bear and having to shoot it with a pistol. As much as anyone, he loathes the sight of careless, insensitive and drunken hunters who wound and lose animals. Jacobs earns his living, after all, relieving pain.

When Jacobs set out with two other hunters to kill a bison at Hebgen Lake in March, he was wearing a t-shirt emblazoned "Keep Montana Wild." To hear him talk about his role, he honestly thought he was doing the beasts a favor. Hunting, to Jacobs, fills an important ecological niche. "They starve up there," said Jacobs. "That's the alternative."

Lee Dessaux was so enraged when Jacobs dropped a half-ton cow bison, he whacked him again and again with a ski pole, then jabbed Jacobs in the shoulder with its tip. "Compared to what happened to the buffalo, pokin' that guy in the ribs was nothing," Dessaux, a 25-year-old Earth First!er, said afterwards. "I think passionate expression is quite appropriate in light of what happened."

Never mind Jacobs' t-shirt and eloquent waxing over the beauties of wildlife. Hunters, to Dessaux, are just people who like to kill things. And he isn't alone. According to one Yale University study, a third of all Americans favors a total ban on hunting, and more than half think hunting for sport and trophies is wrong. Anti-hunters and hunters don't yet go at it with nearly the ferocity or firepower of Zimbabwean game wardens and poachers. But officials in a wildlife refuge in Connecticut recently canceled a hunt after an angry sportsman loosed a load of birdshot toward a crowd of demonstrators.

Again, it's hard not to wish that the hunting and anti-hunting camps—both of which have demonstrated their love of wildlife and habitat—would advance their dialogue beyond the clamor of moral outrage and consider the similarities of their positions. A lot of good pro-environment energy is being spent backbiting within the broad population of nature-lovers, while wetlands are draining, rainforests are burning, auto-emissions standards are being relaxed and the atmosphere heats up.

It's unclear to me what was accomplished at Hebgen Lake. But it's clear that the era that swelled Earth First!'s impatient membership to

10,000 has been one of unparalleled greed in the United States and unprecedented environmental degradation worldwide. When the frock-coated conservationists of the past century were maneuvering the first national parks through Congress, the South Pole hadn't even been visited, much less an ozone hole discovered above it. The world's population has tripled in the past century and industrial production has grown fifty-fold. The desperation of Zambian peasants, the equal desperation of the effort to save the rhino and the audacity of American eco-militants only reflect the level of crisis into which we have plunged our earth. If we're surprised at how violent the struggle to share and save the earth has become, we'd best remember: Tempers rise with the stakes.

Content Considerations

1. Knox writes that one of the pressures in environmental decisions is the conflict between human need and diminishing wilderness and wildlife. How does she weigh the conflict?
2. Opposing factions (such as hunters and anti-hunting groups) often share goals regarding wildlife. How can this be? What is the essential issue?
3. What does the author conclude about "extremist" environmental groups?

The Writer's Strategies

1. The writer begins this piece with a preservation issue in remote Africa, then shifts to the American West. How are the situations of the two places comparable?
2. In what ways does the writer show the interrelatedness of conservation issues?

Writing Possibilities

1. Construct opposing viewpoints of two people who both claim to be speaking in defense of wilderness or wildlife; present their views in a paper.
2. Write a persuasive essay in which you present your ideas about when and where hunting is acceptable.
3. With a partner, think of a situation in which you might take violent action to preserve something or to effect change; write about the situation and why your action would be justified. Or, if you believe that violent measures are never justified, explain that position and the alternatives you support.

LEWIS THOMAS

Natural Man

In Lives of a Cell *(1971), the book from which this essay was taken, pathologist Lewis Thomas explores and reflects upon biology, humans, nature, human nature, language, and how life and its parts connect. In the essay that follows he focuses on how mankind has viewed its connections to nature. Over time these views have changed, as have the practices that precede and follow such changes.*

The ways in which we see ourselves connected to nature—master, servant, partner, separate and unrelated entity—will affect what we do to and on the Earth. Some connections or relationships may be more harmful than others to Earth, and some may impoverish both human minds and lives. Others may allow both people and Earth to flourish.

As you read Thomas's essay, think about where you place the human species. Above all other species and master of them? Above all other species but caretaker of them? Among other species in a symbiotic dance of survival? Somehow in partnership, giving and taking what is needed?

THE SOCIAL SCIENTISTS, especially the economists, are moving deeply into ecology and the environment these days, with disquieting results. It goes somehow against the grain to learn that cost-benefit analyses can be done neatly on lakes, meadows, nesting gannets, even whole oceans. It is hard enough to confront the environmental options ahead, and the hard choices, but even harder when the price tags are so visible. Even the new jargon is disturbing: it hurts the spirit, somehow, to read the word *environments,* when the plural means that there are so many alternatives there to be sorted through, as in a market, and voted on. Economists need cool heads and cold hearts for this sort of work, and they must write in icy, often skiddy prose.

The degree to which we are all involved in the control of the earth's life is just beginning to dawn on most of us, and it means another revolution for human thought.

This will not come easily. We've just made our way through inconclusive revolutions on the same topic, trying to make up our minds how we feel about nature. As soon as we arrived at one kind of consensus, like an enormous committee, we found it was time to think it through all over, and now here we are, at it again.

The oldest, easiest-to-swallow idea was that the earth was man's personal property, a combination of garden, zoo, bank vault, and energy source, placed at our disposal to be consumed, ornamented, or pulled apart as we wished. The betterment of mankind was, as we understood it, the whole point of the thing. Mastery over nature, mystery and all, was a moral duty and social obligation.

In the last few years we were wrenched away from this way of looking at it, and arrived at something like general agreement that we had it wrong. We still argue the details, but it is conceded almost everywhere that we are not the masters of nature that we thought ourselves; we are as dependent on the rest of life as are the leaves or midges or fish. We are part of the system. One way to put it is that the earth is a loosely formed, spherical organism, with all its working parts linked in symbiosis. We are, in this view, neither owners nor operators; at best, we might see ourselves as motile tissue specialized for receiving information—perhaps, in the best of all possible worlds, functioning as a nervous system for the whole being.

There is, for some, too much dependency in this view, and they prefer to see us as a separate, qualitatively different, special species, unlike any other form of life, despite the sharing around of genes, enzymes, and organelles. No matter, there is still the underlying idea that we cannot have a life of our own without concern for the ecosystem in which we live, whether in majesty or not. This idea has been strong enough to launch the new movements for the sustenance of wilderness, the protection of wildlife, the turning off of insatiable technologies, the preservation of "whole earth."

But now, just when the new view seems to be taking hold, we may be in for another wrench, this time more dismaying and unsettling than anything we've come through. In a sense, we shall be obliged to swing back again, still believing in the new way but constrained by the facts of life to live in the old. It may be too late, as things have turned out.

We are, in fact, the masters, like it or not.

It is a despairing prospect. Here we are, practically speaking twenty-first-century mankind, filled to exuberance with our new understanding of kinship to all the family of life, and here we are, still nineteenth-century man, walking bootshod over the open face of nature, subjugating and civilizing it. And we cannot stop this controlling, unless we vanish under the hill ourselves. If there were such a thing as a world mind, it should crack over this.

The truth is, we have become more deeply involved than we ever dreamed. The fact that we sit around as we do, worrying seriously about how best to preserve the life of the earth, is itself the sharpest measure of our involvement. It is not human arrogance that has taken us in this direction, but the most natural of natural events. We developed this way, we grew this way, we are this kind of species.

We have become, in a painful, unwished-for way, nature itself. We have grown into everywhere, spreading like a new growth over the entire surface, touching and affecting every other kind of life, *incorporating* ourselves. The earth risks being eutrophied by us. We are now the dominant feature of our own environment. Humans, large terrestrial metazoans, fired by energy from microbial symbionts lodged in their cells, instructed by tapes of nucleic acid stretching back to the earliest live membranes, informed by neurons essentially the same as all the other neurons on earth, sharing structures with mastodons and lichens, living off the sun, are now in charge, running the place, for better or worse.

Or is it really this way? It could be, you know, just the other way around. Perhaps we are the invaded ones, the subjugated, used.

Certain animals in the sea live by becoming part-animal, part-plant. They engulf algae, which then establish themselves as complex plant tissues, essential for the life of the whole company. I suppose the giant clam, if he had more of a mind, would have moments of dismay on seeing what he has done to the plant world, incorporating so much of it, enslaving green cells, living off the photosynthesis. But the plant cells would take a different view of it, having captured the clam on the most satisfactory of terms, including the small lenses in his tissues that focus sunlight for their benefit; perhaps algae have bad moments about what they may collectively be doing on the world of clams.

With luck, our own situation might be similar, on a larger scale. This might turn out to be a special phase in the morphogenesis of the earth when it is necessary to have something like us, for a time anyway, to fetch and carry energy, look after new symbiotic arrangements, store up information for some future season, do a certain amount of ornamenting, maybe even carry seeds around the solar system. That kind of thing. Handyman for the earth.

I would much prefer this useful role, if I had any say, to the essentially unearthly creature we seem otherwise on the way to becoming. It would mean making some quite fundamental changes in our attitudes toward each other, if we were really to think of ourselves as indispensable elements of nature. We would surely become the environment to worry about the most. We would discover, in ourselves, the sources of wonderment and delight that we have discerned in all other manifestations of nature. Who knows, we might even acknowledge the fragility and vulnerability that always accompany high specialization in biology, and movements might start up for the protection of ourselves as a valuable, endangered species. We couldn't lose.

Content Considerations

1. According to Thomas, ideas about how humans relate to nature have changed several times over the years. Summarize the changes.
2. In the last change, Thomas says, we humans began to think of ourselves as masters. Explain what he means. What evidence do you see that people today think of themselves this way?
3. Finally, Thomas says that "Perhaps we are the invaded ones, the subjugated, used" (paragraph 12). What does he mean? Could he be right?

The Writer's Strategies

1. In paragraph 12 the author suggests that maybe what he has been explaining isn't at all true, that perhaps the opposite is true. How does this reversal affect the reader's understanding? What is the author's true position?
2. Thomas compares human placement within the ecosystem with that of the giant clam. Explain his analogy, then determine how it shapes the comments of the essay's last two paragraphs.

Writing Possibilities

1. Imagine yourself as a "Handyman for the earth" (paragraph 14). Explain your job.
2. With a partner, discuss the changes in thought and belief that Thomas describes, then link each thought and belief to the kinds of action you both would expect from a person who espoused it. In an essay, write your conclusions.
3. Examine the idea that the earth is using humans; explain how this could be so.
4. Discuss whether humans are, in any sense, an endangered species.

Additional Writing and Research

Connections

1. Although the readings in this chapter explore several kinds, views, and causes of species endangerment, they also make some mention of economic considerations. Using information from all of the readings, explain how economics is a factor in species endangerment and preservation.
2. Both Milstein and Diamond approach the topic of extinction from a scientific viewpoint. In what ways are their views similar? How does each writer establish the validity of his view?
3. Select some of the comments made by people interviewed and quoted by Knox, Abramson, and Milstein, then connect these comments to Thomas's ideas regarding human relationships to nature. How do the interviewees' comments indicate their beliefs of how humans relate to nature?
4. Page addresses the beauty of wildlife. What are his beliefs? What part might aesthetics play in species preservation for this author?
5. Using any four of the readings in this chapter to support your position, explain the social attitudes and practices that may hasten species extinction.
6. Isolate the reasons all six readings give for caring about species endangerment and extinction. Using the reasons that are valid for you, explain your own level of concern.

Researches

1. Locate several lists of endangered species, then compare them to see how much agreement exists among the organizations and agencies that compile such records. Determine also how such lists are compiled. What can you conclude from the comparison?
2. Find out how the arrival in America of European explorers and settlers affected indigenous animal populations.
3. Limiting your focus to the last two hundred years, discover some specifics about how fashion relates to species decline or extinction.
4. View several wildlife programs or examine several wildlife publications to determine the extent to which they feature animals that are dirty and blemished or animals that are clean and perfect. What do the results of this survey suggest to you?

5. Locate a copy of the Endangered Species Act; read about the arguments that preceded its passage and those that continue afterward. How is the act executed and enforced?
6. Examine the aesthetic experience presented in several poems about wildlife. Describe the aesthetic experience of reading and making meaning of the poem.
7. Learn more about the factors that influence how humans make judgments about animals and their relative worth. Find information about how humans sometimes personify the characteristics of certain animals, attributing to them qualities that are admired or despised.

3

Controlling the Rivers

MARK TWAIN, perhaps the most famous of the nineteenth-century riverboat pilots, knew intimately the bends and riverbars, the snags and shallows, of each mile of the Mississippi between St. Louis and New Orleans. Twain was also keenly aware of the capriciousness of the river, of the vagaries of nature that weekly or daily could change the very course of the mightiest river in North America. At times the river shortened the distance between the cities by cutting off large loops, often stranding ports miles away from it while flooding farms and towns that had once been far distant. In his *Old Times on the Mississippi,* Twain treats with humor the phenomenon of the constantly changing river:

> In the space of one hundred and seventy-six years the Lower Mississippi has shortened itself two hundred and forty-two miles. That is an average of a trifle over one mile and a third per year. Therefore, any calm person, who is not blind or idiotic, can see that in the Old Oolitic Silurian Period, just a million years ago next November, the Lower Mississippi River was upwards of one million three hundred thousand miles long, and stuck out over the Gulf of Mexico like a fishing-rod. And by the same token any person can see that seven hundred and forty-two years from now,

the Lower Mississippi will be only a mile and three quarters long, and Cairo [Illinois] and New Orleans will have joined their streets together, and be plodding along under a single mayor and a mutual board of aldermen.

What Twain found humorous others have found less amusing. Concentrated effort has been directed at channeling rivers across America, at levying their banks to control floods, at damming their courses to provide hydroelectric power and lakes for recreation, at harnessing, controlling, and taming the flow of the waters.

Seen from an airplane or space shuttle, many rivers still writhe across the landscape of earth like serpents, coiling, twisting, sidling from interiors, always seeking oceans. In some years, in some seasons, these most vital of earthscape features rise from their banks and spread across the land, resembling vast lakes more than meandering rivers.

Human history has a long and close association with rivers: In Mesopotamia and China and Egypt, rivers and their annual soil-enriching floods were the wellsprings of nascent moves from hunting and gathering to agriculture, the foundation of all civilization. Rivers often contribute to a nation's sense of self, to a national character. The mysterious, powerful Amazon undulates through the dark rain forests of interior Brazil; the ancient Nile spreads fanlike across the delta of Egypt; the Thames flows placidly by Big Ben. The Niger, the Mississippi, the Volga, the Rhine, the Seine, the Yalu—each is intimately linked with its region's history.

In the United States rivers provided the paths for exploration into the interior. Marquette and Joliet floated down the Mississippi; Lewis and Clark poled up the Missouri for a first peek by Americans into the Louisiana Territory. Later these rivers became part of the mythology of westward expansion, rolling highways as pioneers floated their wagons down the Ohio on rafts to reach Missouri and then followed the Platte across Nebraska and Wyoming. And, of course, rivers provided early forms of power: Waterwheels drove millstones and otherwise furnished the energy for a developing culture.

Many charge that we repaid our debt to rivers by making them giant sewers to carry away the discharge of factories, the garbage of cities, the effluvia of an increasingly consumptive society. Today few contest the need to scrub the waters of America's rivers; indeed, that scrubbing and accompanying changes in practices have left many of America's rivers cleaner today than they were fifty years ago. Thus, the environmental issues regarding rivers today include not only the waterborne detritus of industry but also the use of the water itself.

This use sometimes pits state against state. As the Missouri River rises out of the breaks in Montana and courses east and south to its confluence with the Mississippi, it now passes through a series of locks and dams, behind which many lakes have grown for the entertainment of local residents. A battle has ensued between South Dakota and states downstream, including Missouri, over how much water should be released through the dams for the use of cities and industries and how much should be retained for the benefit of tourism in the upstream states. Other battles rage in the Northwest, where efforts are being made to limit the number of dams on streams in areas with a surplus of electricity.

Rivers, then, have changed drastically from the days when Mark Twain was still Samuel Clemens. For Twain the river was a constantly changing course. He would have agreed with the ancient Greek Heraclitus, who once observed that it is not possible to step twice into the same river. He meant, of course, that a river is moving water—moving, always moving, and therefore always new. And for the Greeks, as well as for Native Americans, a river was a living being. Not just the aquatic life below the surface but the river itself was alive, imbued with spirit. Much of today's environmental energy is directed at preserving that spirit.

The essays in this chapter deal with some of the complexities of controlling the rivers—debates on whether rivers should be dammed, on whether rivers have been tamed too much, on whether rivers and riparian life have purpose apart from that which humans assign. Several of the writers here approach rivers with a nearly Greek-like appreciation for the life-sustaining qualities a river symbolizes; others write of rivers we cannot know, rivers of a hundred years ago or more; still others approach the banks of these streams in no-nonsense prose borne along on the currents of events. Like Twain, get to know these rivers.

LOREN EISELEY

The Flow of the River

In the first chapter of The Immense Journey, *the book from which the next essay is taken, Loren Eiseley warns that the reader is about to encounter a "somewhat unconventional record of the prowlings of one mind which has sought to explore, to understand, and to enjoy the miracles of this world, both in and out of science." One of those miracles is the river.*

As an anthropologist and naturalist, Eiseley claimed to be "a man preoccupied with time." (That preoccupation seems to be common among people who reflect upon rivers, if not among us all.) Yet Eiseley, who died in 1977, was also a scientist, a university professor, president of the American Institute of Human Paleontology, and an author whose work appeared in numerous scientific journals and general-interest magazines.

The following essay may indeed be unconventional. As you read it, note the ways in which the anthropologist, the naturalist, and the man preoccupied with time merge and separate.

IF THERE IS MAGIC on this planet, it is contained in water. Its least stir even, as now in a rain pond on a flat roof opposite my office, is enough to bring me searching to the window. A wind ripple may be translating itself into life. I have a constant feeling that some time I may witness that momentous miracle on a city roof, see life veritably and suddenly boiling out of a heap of rusted pipes and old television aerials. I marvel at how suddenly a water beetle has come and is submarining there in a spatter of green algae. Thin vapors, rust, wet tar and sun are an alembic remarkably like the mind; they throw off odorous shadows that threaten to take real shape when no one is looking.

Once in a lifetime, perhaps, one escapes the actual confines of the flesh. Once in a lifetime, if one is lucky, one so merges with sunlight and air and running water that whole eons, the eons that mountains and deserts know, might pass in a single afternoon without discomfort. The mind has sunk away into its beginnings among old roots and the obscure tricklings and movings that stir inanimate things. Like the charmed fairy circle into which a man once stepped, and upon emergence learned that a whole century had passed in a single night, one can never quite define this secret; but it has something to do, I am sure, with common water.

Its substance reaches everywhere; it touches the past and prepares the future; it moves under the poles and wanders thinly in the heights of air. It can assume forms of exquisite perfection in a snowflake, or strip the living to a single shining bone cast up by the sea.

Many years ago, in the course of some scientific investigations in a remote western county, I experienced, by chance, precisely the sort of curious absorption by water—the extension of shape by osmosis—at which I have been hinting. You have probably never experienced in yourself the meandering roots of a whole watershed or felt your outstretched fingers touching, by some kind of clairvoyant extension, the brooks of snow-line glaciers at the same time that you were flowing toward the Gulf over the eroded debris of worn-down mountains. A poet, MacKnight Black, has spoken of being "limbed . . . with waters gripping pole and pole." He had the idea, all right, and it is obvious that these sensations are not unique, but they are hard to come by; and the sort of extension of the senses that people will accept when they put their ear against a sea shell, they will smile at in the confessions of a bookish professor. What makes it worse is the fact that because of a traumatic experience in childhood, I am not a swimmer, and am inclined to be timid before any large body of water. Perhaps it was just this, in a way, that contributed to my experience.

As it leaves the Rockies and moves downward over the high plains towards the Missouri, the Platte River is a curious stream. In the spring floods, on occasion, it can be a mile-wide roaring torrent of destruction, gulping farms and bridges. Normally, however, it is a rambling, dispersed series of streamlets flowing erratically over great sand and gravel fans that are, in part, the remnants of a mightier Ice Age stream bed. Quicksands and shifting islands haunt its waters. Over it the prairie suns beat mercilessly throughout the summer. The Platte, "a mile wide and an inch deep," is a refuge for any heat-weary pilgrim along its shores. This is particularly true on the high plains before its long march by the cities begins.

The reason that I came upon it when I did, breaking through a willow thicket and stumbling out through ankle-deep water to a dune in the shade, is of no concern to this narrative. On various purposes of science I have ranged over a good bit of that country on foot, and I know the kinds of bones that come gurgling up through the gravel pumps, and the arrowheads of shining chalcedony that occasionally spill out of water-loosened sand. On that day, however, the sight of sky and willows and the weaving net of water murmuring a little in the shallows on its way to the Gulf stirred me, parched as I was with miles of walking, with a new idea: I was going to float. I was going to undergo a tremendous adventure.

The notion came to me, I suppose, by degrees. I had shed my clothes and was floundering pleasantly in a hole among some reeds when a great desire to stretch out and go with this gently insistent water began to pluck at me. Now to this bronzed, bold, modern generation, the struggle I waged with timidity while standing there in knee-deep water can only seem farcical; yet actually for me it was not so. A near-drowning accident in childhood had scarred my reactions; in addition to the fact that I was a nonswimmer, this "inch-deep river" was treacherous with holes and quicksands. Death was not precisely infrequent along its wandering and illusory channels. Like all broad wastes of this kind, where neither water nor land quite prevails, its thickets were lonely and untraversed. A man in trouble would cry out in vain.

I thought of all this, standing quietly in the water, feeling the sand shifting away under my toes. Then I lay back in the floating position that left my face to the sky, and shoved off. The sky wheeled over me. For an instant, as I bobbed into the main channel, I had the sensation of sliding down the vast tilted face of the continent. It was then that I felt the cold needles of the alpine springs at my fingertips, and the warmth of the Gulf pulling me southward. Moving with me, leaving its taste upon my mouth and spouting under me in dancing springs of sand, was the immense body of the continent itself, flowing like the river was flowing, grain by grain, mountain by mountain, down to the sea. I was streaming over ancient sea beds thrust aloft where giant reptiles had once sported; I was wearing down the face of time and trundling cloud-wreathed ranges into oblivion. I touched my margins with the delicacy of a crayfish's antennae, and felt great fishes glide about their work.

I drifted by stranded timber cut by beaver in mountain fastnesses; I slid over shallows that had buried the broken axles of prairie schooners and the mired bones of mammoth. I was streaming alive through the hot and working ferment of the sun, or oozing secretively through shady thickets. I *was* water and the unspeakable alchemies that gestate and take shape in water, the slimy jellies that under the enormous magnification of the sun writhe and whip upward as great barbeled fish mouths, or sink indistinctly back into the murk out of which they arose. Turtle and fish and the pinpoint chirpings of individual frogs are all watery projections, concentrations—as man himself is a concentration—of that indescribable and liquid brew which is compounded in varying proportions of salt and sun and time. It has appearances, but at its heart lies water, and as I was finally edged gently against a sand bar and dropped like any log, I tottered as I rose. I knew once more the body's revolt against emergence into the harsh and unsupporting air, its reluctance to break contact with that mother element which still, at this late point in time, shelters and brings into being nine tenths of everything alive.

As for men, those myriad little detached ponds with their own swarming corpuscular life, what were they but a way that water has of going about beyond the reach of rivers? I, too, was a microcosm of pouring rivulets and floating driftwood gnawed by the mysterious animalcules of my own creation. I was three fourths water, rising and subsiding according to the hollow knocking in my veins: a minute pulse like the eternal pulse that lifts Himalayas and which, in the following systole, will carry them away.

Thoreau, peering at the emerald pickerel in Walden Pond, called them "animalized water" in one of his moments of strange insight. If he had been possessed of the geological knowledge so laboriously accumulated since his time, he might have gone further and amusedly detected in the planetary rumblings and eructations which so delighted him in the gross habits of certain frogs, signs of that dark interior stress which has reared sea bottoms up to mountainous heights. He might have developed an acute inner ear for the sound of the surf on Cretaceous beaches where now the wheat of Kansas rolls. In any case, he would have seen, as the long trail of life was unfolded by the fossil hunters, that his animalized water had changed its shapes eon by eon to the beating of the earth's dark millennial heart. In the swamps of the low continents, the amphibians had flourished and had their day; and as the long skyward swing—the isostatic response of the crust—had come about, the era of the cooling grasslands and mammalian life had come into being.

A few winters ago, clothed heavily against the weather, I wandered several miles along one of the tributaries of that same Platte I had floated down years before. The land was stark and ice-locked. The rivulets were frozen, and over the marshlands the willow thickets made such an array of vertical lines against the snow that tramping through them produced strange optical illusions and dizziness. On the edge of a frozen backwater, I stopped and rubbed my eyes. At my feet a raw prairie wind had swept the ice clean of snow. A peculiar green object caught my eye; there was no mistaking it.

Staring up at me with all his barbels spread pathetically, frozen solidly in the wind-ruffled ice, was a huge familiar face. It was one of those catfish of the twisting channels, those dwellers in the yellow murk, who had been about me and beneath me on the day of my great voyage. Whatever sunny dream had kept him paddling there while the mercury plummeted downward and that Cheshire smile froze slowly, it would be hard to say. Or perhaps he was trapped in a blocked channel and had simply kept swimming until the ice contracted around him. At any rate, there he would lie till the spring thaw.

At that moment I started to turn away, but something in the bleak, whiskered face reproached me, or perhaps it was the river calling to her children. I termed it science, however—a convenient rational phrase I reserve for such occasions—and decided that I would cut the fish out of the ice and take him home. I had no intention of eating him. I was merely struck by a sudden impulse to test the survival qualities of high-plains fishes, particularly fishes of this type who get themselves immured in oxygenless ponds or in cut-off oxbows buried in winter drifts. I blocked him out as gently as possible and dropped him, ice and all, into a collecting can in the car. Then we set out for home.

Unfortunately, the first stages of what was to prove a remarkable resurrection escaped me. Cold and tired after a long drive, I deposited the can with its melting water and ice in the basement. The accompanying corpse I anticipated I would either dispose of or dissect on the following day. A hurried glance had revealed no signs of life.

To my astonishment, however, upon descending into the basement several hours later, I heard stirrings in the receptacle and peered in. The ice had melted. A vast pouting mouth ringed with sensitive feelers confronted me, and the creature's gills labored slowly. A thin stream of silver bubbles rose to the surface and popped. A fishy eye gazed up at me protestingly.

"A tank," it said. This was no Walden pickerel. This was a yellow-green, mud-grubbing, evil-tempered inhabitant of floods and droughts and cyclones. It was the selective product of the high continent and the waters that pour across it. It had outlasted prairie blizzards that left cattle standing frozen upright in the drifts.

"I'll get the tank," I said respectfully.

He lived with me all that winter, and his departure was totally in keeping with his sturdy, independent character. In the spring a migratory impulse or perhaps sheer boredom struck him. Maybe, in some little lost corner of his brain, he felt, far off, the pouring of the mountain waters through the sandy coverts of the Platte. Anyhow, something called to him, and he went. One night when no one was about, he simply jumped out of his tank. I found him dead on the floor next morning. He had made his gamble like a man—or, I should say, a fish. In the proper place it would not have been a fool's gamble. Fishes in the drying shallows of intermittent prairie streams who feel their confinement and have the impulse to leap while there is yet time may regain the main channel and survive. A million ancestral years had gone into that jump, I thought as I looked at him, a million years of climbing through prairie sunflowers and twining in and out through the pillared legs of drinking mammoth.

"Some of your close relatives have been experimenting with air breathing," I remarked, apropos of nothing, as I gathered him up. "Suppose we meet again up there in the cottonwoods in a million years or so."

I missed him a little as I said it. He had for me the kind of lost archaic glory that comes from the water brotherhood. We were both projections out of that timeless ferment and locked as well in some greater unity that lay incalculably beyond us. In many a fin and reptile foot I have seen myself passing by—some part of myself, that is, some part that lies unrealized in the momentary shape I inhabit. People have occasionally written me harsh letters and castigated me for a lack of faith in man when I have ventured to speak of this matter in print. They distrust, it would seem, all shapes and thoughts but their own. They would bring God into the compass of a shopkeeper's understanding and confine Him to those limits, lest He proceed to some unimaginable and shocking act—create perhaps, as a casual afterthought, a being more beautiful than man. As for me, I believe nature capable of this, and having been part of the flow of the river, I feel no envy—any more than the frog envies the reptile or an ancestral ape should envy man.

Every spring in the wet meadows and ditches I hear a little shrilling chorus which sounds for all the world like an endlessly reiterated, "We're here, we're here, we're here." And so they are, as frogs, of course. Confident little fellows. I suspect that to some greater ear than ours, man's optimistic pronouncements about his role and destiny may make a similar little ringing sound that travels a small way out into the night. It is only its nearness that is offensive. From the heights of a mountain, or a marsh at evening, it blends, not too badly, with all the other sleepy voices that, in croaks or chirrups, are saying the same thing.

After a while the skilled listener can distinguish man's noise from the katydid's rhythmic assertion, allow for the offbeat of a rabbit's thumping, pick up the autumnal monotone of crickets, and find in all of them a grave pleasure without admitting any to a place of preeminence in his thoughts. It is when all these voices cease and the waters are still, when along the frozen river nothing cries, screams or howls, that the enormous mindlessness of space settles down upon the soul. Somewhere out in that waste of crushed ice and reflected stars, the black waters may be running, but they appear to be running without life toward a destiny in which the whole of space may be locked in some silvery winter of dispersed radiation.

It is then, when the wind comes straitly across the barren marshes and the snow rises and beats in endless waves against the traveler, that I remember best, by some trick of the imagination, my summer voyage on the river. I remember my green extensions, my catfish nuzzlings and minnow wrigglings, my gelatinous materializations out of the mother ooze. And as I walk on through the white smother, it is the magic of water that leaves me a final sign.

Men talk much of matter and energy, of the struggle for existence that molds the shape of life. These things exist, it is true; but more delicate, elusive, quicker than the fins in water, is that mysterious principle known as "organization," which leaves all other mysteries concerned with life stale and insignificant by comparison. For that without organization life does not persist is obvious. Yet this organization itself is not strictly the product of life, nor of selection. Like some dark and passing shadow within matter, it cups out the eyes' small windows or spaces the notes of a meadow lark's song in the interior of a mottled egg. That principle—I am beginning to suspect—was there before the living in the deeps of water.

The temperature has risen. The little stinging needles have given way to huge flakes floating in like white leaves blown from some great tree in open space. In the car, switching on the lights, I examine one intricate crystal on my sleeve before it melts. No utilitarian philosophy explains a snow crystal, no doctrine of use or disuse. Water has merely leapt out of vapor and thin nothingness in the night sky to array itself in form. There is no logical reason for the existence of a snowflake any more than there is for evolution. It is an apparition from that mysterious shadow world beyond nature, that final world which contains—if anything contains—the explanation of men and catfish and green leaves.

Content Considerations

1. What evidence does Eiseley give to show that water contains whatever magic is on this planet?
2. What kinds of journeys does the author undertake in this philosophic essay?
3. What do science and natural history have to do with Eiseley's journey?

The Writer's Strategies

1. In what ways might the title of this essay be interpreted?
2. Eiseley presents his ideas about the mysterious and connecting force of life as a series of anecdotes and personal experiences. What is the effect of this approach?

Writing Possibilities

1. Respond to Eiseley's essay, focusing on how you made sense of it and your reaction to its language and tone.
2. With a classmate, discuss the wildness of the river Eiseley writes about. Then write a paper exploring how important that wildness is to the discoveries Eiseley makes.
3. In an essay, consider the needs of cities, towns, and farms for water. Then weigh those needs against the possibility or desirability of leaving rivers natural. How would you assign priority?

CONGER BEASLEY, JR.

The Return of Beaver to the Missouri River

The Missouri River has a long history in American settlement; fur trappers worked it, and in 1804 explorers Meriwether Lewis and William Clark followed it west, filing reports with the United States government about the land that was part of the Louisiana Purchase of 1803. Before Europeans arrived, of course, various Native American tribes had lived on and near its banks, and other tribes traversed and used the river in their travels and wanderings.

The Missouri was wild then, so wild that a desire to tame it grew. Now a series of dams and reservoirs upriver attempt to contain and control its flow, usually with great success. Industry makes use of the river, barges carry goods on it, and it supplies water to cities and towns. Each week a plane from the Army Corps of Engineers flies low across it to observe its flow. The Missouri has become vital to the way the region operates, and those in charge remain vigilant.

Writer Conger Beasley, Jr., grew up in St. Joseph, Missouri, and now lives in Kansas City. His latest book, Sundancers and River Demons, which includes the following essay, is a collection of pieces about, as its subtitle says, landscape and ritual—two parts of culture that may be quite closely connected.

Before you begin reading, write a paragraph about controlling rivers so that the water may be harnessed for other uses; follow it with a paragraph about what is gained and what is lost through such control.

IN AN AUTOBIOGRAPHICAL VOLUME entitled *The River and I*, John Neihardt recounts the first time he ever looked upon the Missouri River. It was sometime in the late 1880s, and the place was Kansas City. The river was in full flood, a "yellow swirl that spread out into the wooded bottomlands," demolishing entire towns. "There was a dreadful fascination about it," Neihardt remembers, "the fascination of all huge and irresistable things. I had caught my first wee glimpse into the infinite. . . ."

Some seventy years later, in the spring of 1953, I stood on a bluff in St. Joseph, Missouri, and watched the last great flood of that unruly river ravage the bottomlands between my home town and the hills of distant Kansas. Augmented by several weeks of ferocious rains, tributaries in Iowa and Nebraska had disgorged an unprecedented volume of water into the Missouri, which quickly overflowed its banks. Levees crumbled, dikes collapsed, water swept across wheat and alfalfa fields, carrying houses, cattle, barns, and automobiles with it. From bluff to bluff between the two states, a distance of maybe five miles, the river was stippled with foamy whirlpools and entire trees. I remember watching the procession in stunned silence with my father and his friends. All my life (I was then twelve) I had heard of the river and watched it from passing cars and trains and even viewed it once or twice from airplanes, but I had never been on it in a boat or (God forbid) swum in it. That was unthinkable. The river was too capricious to attempt such a feat. There were boiling eddies that could devour the strongest swimmer, deadly snags and sawyers that could rip open the stoutest hull, animals with pointy teeth that could tear off a leg or arm. No, the river was a creature to observe from a distance and to cross as quickly as possible; it was not a place to linger in idle contemplation or recreational enjoyment. It was an unbridled monster in dire need of hobbling.

"That sure is a hell of a lot of water," remarked one of my father's friends.

"The airport's gone. Elwood's under water. A few feet more and it will wash over the Pony Express Bridge," said another.

"Yeah, but this won't keep up for long," declared a third. "Once they close off those dams up in Dakota, this ole river's gonna get trimmed down to size."

I think of that exchange now, thirty years later, whenever I launch my raft or canoe out on the river. Within the scope of a few decades it has changed in character and shape. It no longer is as wide as it once was, neither does it flood as torrentially. Periodically, it spills over its banks and inundates the lowlands, but it no longer rolls from bluff to bluff or sweeps through entire communities, stranding people in trees. Rarely does it bring the media rushing to its cresting banks. The dams up in South Dakota and the Army Corps of Engineers have taken care of that. Over the years the Corps has deepened the channel and made it more accessible to barge traffic. More recently, the Corps has lined the banks with a solid wall of riprap and installed wingdikes which jut out into the water at right angles; silt, building up behind these protrusions, progressively narrows the river's width. Gradually, the Corps has exerted more and more control over the river, reducing it in size to a tawny ribbon whose least impulse can be carefully monitored. In the process commercial fishing has become almost nonexistent, and the meandering oxbows—remnants of the river's earlier path—have dried up, drastically reducing the acreage of precious wetlands, prime nesting places for waterfowl.

"You know what that river has become?" a man said to me recently in a bar in Kansas City. "An irrigation ditch, that's what. A goddamn irrigation ditch!"

He had grown up in St. Joseph in the 1920s and '30s, and had fished on the river as a boy. Once, on his sixteenth birthday, he had swum the width, from Kansas back to the Missouri side. When he told me that, I gazed at him with speechless admiration. When I was growing up, swimming the river was the most daring thing a boy could do, more daring than stealing the old man's car for a joyride or crawling through sewer pipes under a cemetery or putting your hand on a girl's breast or even engaging in BB-gun wars.

Despite considerable changes that have severely modified its character, the river is still regarded with trepidation by most people who live near it. The reasons for this are mystifying. Recently, as I was tying my canoe on top of my car, my neighbor—an amiable man in his sixties, a veteran of the Battle of the Bulge—strolled over to help me adjust the ropes and tie the knots. "Where you headed?" he asked after pronouncing the boat secure.

"I'm going on the river."

"The Missouri!"

"Yep. A day float down from the mouth of the Platte."

He pulled carefully on his cigarette. "Well, you want to think twice before doing that, don't you?"

"Why?"

"It's dangerous. There're whirlpools that can easily upset a canoe the size of yours."

"Have you ever been on the river?" I asked.

"No. But I grew up around here, and I know when to stay away from a place that doesn't want me."

As I drove to the river I thought about what he had said. He had encouraged me to enjoy an outing on Smithville Lake, a reservoir located northeast of Kansas City, filled with power boats and water skiers and beer-drinking people swaddled with layers of fat. Their presence aside, there's something about still water that doesn't engage my imagination the way moving water does. A river flows from point to point and around the next bend; scenes unfold in slow procession with subtle variations. The sense of motion is invigorating.

My neighbor's remarks recalled the look of incredulity on my father's face the first time I told him I was going on the river. I might just as well have put a gun to my head and pulled the trigger, he declared, for all the chance I had of surviving.

"But you don't understand," I protested. "The river has changed since the time you took me up on the bluff to watch the flood. You can still die there, I grant you. But it's not the power it once was."

"You're crazy," he concluded with a shake of the head. "You're crazy to tempt fate that way."

When the reality alters, the rhetoric seems to harden into place. At least that's what I concluded after talking with my father and neighbor—two men of the same generation and similar backgrounds and experiences. The Missouri River—the lower portion of it at least, from Gavin's Point Dam in South Dakota to its juncture with the Mississippi—is but a slip of its former brawly self; nonetheless, the popular perception of it remains the same. The folklore of the river still evokes images of greedy whirlpools and menacing trees and aquatic carnivores. Elements of these images persist, though in sadly reduced form. Added to these fears, of course, is the relatively new one of pollution, though like many rivers in the United States the Missouri is less contaminated now than it was twenty-five years ago, primarily because the stockyards of Omaha, St. Joseph, and Kansas City no longer pump their refuse directly into it. But the rhetoric persists, almost as if people need to believe all the bad things they've heard. The river is still configured in the local imagination—and not just by people my father's age—as an excess in need of correcting.

The fact of the matter is that the Missouri has been "corrected"—overcorrected to a fault, I would say: dammed, diked, dredged, and drained to suit the needs of the dying barge traffic industry—so it will no longer flood valuable property along the banks; so it will no longer serve as a breeding ground for superfluous fish and wildlife. Certainly as a cultural and recreational resource it has been sadly ignored. In Kansas City, for example, there is virtually no access to the river within the city limits; there is no museum or park along the banks where the river can be viewed and appreciated. Memories of the devastating 1951 flood are still vivid here; while that kind of destruction will never occur again, does it really make sense to construct more wingdikes and drain more oxbows and lay down more riprap so that, within the city boundaries at least, the river will purl harmlessly as water through a sluice?

Enough water flows past Kansas City in a single day to satisfy its needs for an entire year. When I tell this to my river-running friends in New Mexico and the arid Southwest, they express envy and delight. But when they actually view the river and see the wingdikes with the sandbars filling in behind them and the miles of concrete chunks lining the banks, they shrug and turn away. The river isn't very interesting, they seem to say. It isn't very wild.

And yet parts of it still are. You have to search for them, but they are there. Great blue herons still poke for frogs along the banks. Kingfishers rattle noisily between the trees. Borne by sultry thermals, turkey vultures hover over the bottomlands, scouting for carrion. Fish erupt from the scuddy current in flashes of sun-dappled scales. And the river still churns along its ancient bed, down from the Dakotas, across the loamy, fertile midsection of the continent, to its fabled confluence with the Mississippi. Always, even in its present denatured form, there is a sense of movement, of process, of rhythm . . . of a metabolism older and wiser and more meaningful than anything yet invented by human ingenuity.

Historically, the Missouri River has defined one segment of the progressive western border of the American continent. It provided the pathway into the heart of the northern plains and brought trappers and explorers to the verge of the Rockies. In states like the two Dakotas it marks the boundary between one form of terrain and another. East of the river, the land is sectioned into small undulating farms with a distinct Midwestern feel; west of it, the grass diminishes in height, the range opens up to the horizon, and the sky arches endlessly like a yawning mouth. When the Teton Sioux first crossed the river in significant numbers in the mid-eighteenth century, their culture changed dramatically. For decades in southwest Minnesota they had been a woodland community, dwelling in deep pine forests, hunting and fishing on lakes, content with occasional forays onto the plains. Once they crossed the river their transformation into a fierce warrior society, the most respected of all Plains Indians, was assured. Armed with French rifles and mounted on Spanish horses, they created, through legendary heroes like Crazy Horse and Sitting Bull, a reputation for valor that endures to this day. Ahead of them lay the Badlands and the Black Hills and battlefields like the Rosebud and the Little Big Horn. Behind them, frothy and unpredictable—a Rubicon of the sensibility that forever distinguishes the western imagination from all others on the continent—flowed the massive, untidy, indefatigable Missouri.

Unquestionably, George Caleb Bingham was the premier artist of the Missouri River, if not the entire border region, and in the Metropolitan Museum of Art in New York City hangs one of his finest works, "Fur Traders Descending the Missouri." It depicts a man, probably of French extraction, and a boy, most likely a half-breed, sitting in a pirogue laden with furs. The time is early evening; roseate tints from the descending sun tinge the river's placid face and the trees in the background. The man smokes a pipe and dips his paddle in the water, more to steer than to accelerate the pirogue's speed. The boy leans against a hide-covered chest and stares dreamily into the artist's eyes. On the bow, tethered by a short rope, sits one of the most enigmatic figures in all of American painting . . . a dark, bristly, wolfish-looking animal with pointed ears and a glistening snout that appears to be looking down at its reflected image in the water—or is it staring into the artist's eyes?

Blake's tyger holds less portent for me than this creature. I like to think that, intentionally or not, Bingham captured in this curious figure the true feeling of wilderness that the Missouri River once held for explorers and adventurers. That feeling has been described accurately and at great length by observers from Lewis and Clark to John Neihardt and James Willard Schultz; but no where else for me in all the art and writing produced by the region does it exist so powerfully. Whenever I bemoan the loss of the river's freedom, I look at that painting and am content that Bingham at least was able to capture a portion of what it once was and to pass it on for others to savor. Whenever I paddle my canoe on the silty current, I imagine the animal sitting in the bow, staring back at me with all the irony and inscrutability that two hundred years of bitter history can produce.

The Midwest is a sadly misunderstood place, routinely dismissed by Californians and Atlantic seaboarders, scorned by Rocky Mountain enthusiasts, and grudgingly defended by its own inhabitants. In a culture that celebrates spectacular surfaces, such a quiet, unruffled landscape is easy to ridicule. "I don't like to go west of the Alleghenies," a lawyer in New York once told me when I was a student there. "Missouri, Kansas . . . places like that. It's the same old thing, over and over and over again."

But it's not. The rivers of southern Missouri differ from the rivers of eastern Kansas. The foliage along the banks, the soil composition, the fish and animals vary in subtle, yet significant, ways. A sensitivity to the nuances of topography sharpens the eye and instructs the mind in the difficult task of making distinctions between organic forms. There is a moral here. The way we perceive landscape can have a direct bearing upon the way we perceive society and the human beings who comprise it. Dismissing a landscape because it does not conform to preconceptions is a prejudice as galling as dismissing people because of the color of their skin or the beliefs they profess. It violates the biological urge toward multiplicity and diversity that energizes our planet. By adjusting the rhetoric of perception to the reality of the fact perceived—by making the two more consonant and therefore truthful—our sensibilities can be sharpened and refined, and wherever we go in the world, instead of adopting the prevailing stereotype, we can encounter the reality, the genuine forms, that reside underneath.

One evening, after floating all day down from Atchison, Kansas, a friend and I passed under the Leavenworth bridge just as the sun was about to set. Our destination was Parkville, Missouri, a small town a few miles upriver from Kansas City. The time was late summer; a full moon was due to appear in about an hour, and despite the obvious dangers of floating at night, we intended to do just that, guided by the moonlight and the phosphorescent markers on the channel buoys. Barge traffic had been light that day, but we needed to be wary of the occasional tree limb that bobbed just under the surface.

The moon, huge and full, came up over the trees on the east bank. The light spilled onto the leaves and spread in a wavering beam across the water. We watched with fascination as the sky and river seemed to swell under the eerie light. On the west bank, willows and cottonwoods stood out in bold relief; between them, dark and moist as the entrance of a cave, the shadows were alive with sounds

Suddenly, close by, there was a loud crash as if a rock had been chucked into the river, followed by another and another, echoing back and forth and far downstream. "What the hell was that?" my friend exclaimed; and I confess that at that moment images of river demons, passed on to me by another generation, surfaced in my brain. A moment later I saw a creature with a sleek head and a flat tail slip off the bank and disappear into the water. "Beaver," I muttered, almost in disbelief. One of the stories I had heard as a boy was that the beaver had been trapped out in these parts, along with the otter and mink, leaving only the muskrat, a durable species.

Additional explosions sounded up and down the channel, signaling our presence. "Beaver," I whispered, and suddenly I had a vision of the river as it once was—wide, tumultuous, shoally with islands, teeming with birds and fish and animals. If this were 1832, their pelts would fetch hard silver down in Westport or in the trading posts of Blacksnake Hills. But it wasn't; that era, with its magnificent vistas and murderous events, was over. The future stretched before us with the same chimerical uncertainty as the river's path in the moonlight. Tonight we were just drifting along, enjoying the sights, the steady current, the moist air that lapped against our cheeks. As if in acknowledgment, more beaver boomed their warning signal. We laughed and called out to them. This time, I thought, we'll share the river together.

Content Considerations

1. What is the significance of the beavers' return to the river?
2. How have the dams changed the character of the river?
3. Upon what events or conditions was the river's reputation based?

The Writer's Strategies

1. What is the author's attitude toward changes wrought upon the Missouri River?
2. How does Beasley build the notion of conflict, of battle between river and people?

Writing Possibilities

1. Through prewriting and discussion, rediscover a place you knew as a child; build this memory into an essay about what that place has become.
2. Write about the ways landscape or geography have formed or influenced your culture. Gather ideas from other people through interviews.
3. With a classmate, examine Beasley's idea that people may sometimes need to adjust "the rhetoric of perception to the reality of the fact perceived" (paragraph 30). Together, find examples of such incongruities and present them in a paper.

Henry David Thoreau

Concord River

Probably best known for Walden, *his paean to simplicity and natural living, and next for the themes of moral individualism and majority-of-one contained in "On Resistance to Civil Government," Thoreau very early showed evidence of his view regarding mankind and nature. Perhaps no line of his is more famous than: "In wilderness is the salvation of the world."*

In 1839 Thoreau and his brother John floated down a couple of local rivers in Massachusetts, and ten years later he recounted the experience in his first book, A Week on the Concord and Merrimack Rivers. *The following piece is the first chapter from that work.*

As you read, look at the writing as revealing not only the pace of life and the physical setting of Thoreau's New England but also a bit of Thoreau himself. See if you can understand something of the man and his life view through his comments about the Concord River and the interaction of humans with nature along its banks.

"Beneath low hills, in the broad interval
Through which at will our Indian rivulet
Winds mindful still of sannup and of squaw,
Whose pipe and arrow oft the plough unburies,
Here, in pine houses, built of new-fallen trees,
Supplanters of the tribe, the farmers dwell."

—Emerson

THE MUSKETAQUID, or Grass-ground River, though probably as old as the Nile or Euphrates, did not begin to have a place in civilized history, until the fame of its grassy meadows and its fish attracted settlers out of England in 1635, when it received the other but kindred name of CONCORD from the first plantation on its banks, which appears to have been commenced in a spirit of peace and harmony. It will be Grass-ground River as long as grass grows and water runs here; it will be Concord River only while men lead peaceable lives on its banks. To an extinct race it was grass-ground, where they hunted and fished, and it is still perennial grass-ground to Concord farmers, who own the Great Meadows, and get the hay from year to year. "One branch of it," according to the Historian of Concord, for I love to quote so good authority, "rises in the south part of Hopkinton, and another from a pond and a large cedar swamp in Westborough," and flowing between Hopkinton and Southborough, through Framingham, and between Sudbury and Wayland, where it is sometimes called Sudbury River, it enters Concord at the south part of the town, and after receiving the North or Assabeth River, which has its source a little further to the north and west, goes out at the northeast angle, and flowing between Bedford, and Carlisle, and through Billerica, empties into the Merrimack at Lowell. In Concord it is, in summer, from four to fifteen feet deep, and from one hundred to three hundred feet wide, but in the spring freshets, when it overflows its banks, it is in some places nearly a mile wide. Between Sudbury and Wayland the meadows acquire their greatest breadth, and when covered with water, they form a handsome chain of shallow vernal lakes, resorted to by numerous gulls and ducks. Just above Sherman's Bridge, between these towns, is the largest expanse, and when the wind blows freshly in a raw March day, heaving up the surface into dark and sober billows or regular swells, skirted as it is in the distance with alder swamps and smoke-like maples, it looks like a smaller Lake Huron, and is very pleasant and exciting for a landsman to row or sail over. The farm-houses along the Sudbury shore, which rises gently to a considerable height, command fine water prospects at this season. The shore is more flat on the Wayland side, and this town is the greatest loser by the flood. Its farmers tell me that thousands of acres are flooded now, since the dams have been erected, where they remember to have seen the white honeysuckle or clover growing once, and they could go dry with shoes only in summer. Now there is nothing but blue-joint and sedge and cut-grass there, standing in water all the year round. For a long time, they made the most of the driest season to get their hay, working sometimes till nine o'clock at night, sedulously paring with their scythes in the twilight round the hummocks left by the ice; but now it is not worth the getting, when they can come at it, and they look sadly round to their wood-lots and upland as a last resource.

It is worth the while to make a voyage up this stream, if you go no farther than Sudbury, only to see how much country there is in the rear of us; great hills, and a hundred brooks, and farm-houses, and barns, and hay-stacks, you never saw before, and men everywhere, Sudbury, that is *Southborough* men, and Wayland, and Nine-Acre-Corner men, and Bound Rock, where four towns bound on a rock in the river, Lincoln, Wayland, Sudbury, Concord. Many waves are there agitated by the wind, keeping nature fresh, the spray blowing in your face, reeds and rushes waving; ducks by the hundred, all uneasy in the surf, in the raw wind, just ready to rise, and now going off with a clatter and a whistling, like riggers straight for Labrador, flying against the stiff gale with reefed wings, or else circling round first, with all their paddles briskly moving, just over the surf, to reconnoitre you before they leave these parts; gulls wheeling overhead, muskrats swimming for dear life, wet and cold, with no fire to warm them by that you know of; their labored homes rising here and there like hay-stacks; and countless mice and moles and winged titmice along the sunny, windy shore; cranberries tossed on the waves and heaving up on the beach, their little red skiffs beating about among the alders;—such healthy natural tumult as proves the last day is not yet at hand. And there stand all around the alders, and birches, and oaks, and maples full of glee and sap, holding in their buds until the waters subside. You shall perhaps run aground on Cranberry Island, only some spires of last year's pipegrass above water, to show where the danger is, and get as good a freezing there as anywhere on the North-west Coast. I never voyaged so far in all my life. You shall see men you never heard of before, whose names you don't know, going away down through the meadows with long ducking guns, with water-tight boots, wading through the fowl-meadow grass, on bleak, wintry, distant shores, with guns at half cock, and they shall see teal, blue-winged, green-winged, shelldrakes, whistlers, black ducks, ospreys, and many other wild and noble sights before night, such as they who sit in parlors never dream of. You shall see rude and sturdy, experienced and wise men, keeping their castles, or teaming up their summer's wood, or chopping alone in the woods, men fuller of talk and rare adventure in the sun and wind and rain, than a chestnut is of meat; who were out not only in '75 and 1812, but have been out every day of their lives; greater men than Homer, or Chaucer, or Shakspeare, only they never got time to say so; they never took to the way of writing. Look at their fields, and imagine what they might write, if ever they should put pen to paper. Or what have they not written on the face of the earth already, clearing, and burning, and scratching, and harrowing, and plowing, and subsoiling, in and in, and out and out, and over and over, again and again, erasing what they had already written for want of parchment.

As yesterday and the historical ages are past, as the work of today is present, so some flitting perspectives, and demi-experiences of the life that is in nature are in time veritably future, or rather outside to time, perennial, young, divine, in the wind and rain which never die.

Concord River is remarkable for the gentleness of its current, which is scarcely perceptible, and some have referred to its influence the proverbial moderation of the inhabitants of Concord, as exhibited in the Revolution, and on later occasions. It has been proposed that the town should adopt for its coat of arms a field verdant, with the Concord circling nine times round. I have read that a descent of an eighth of an inch in a mile is sufficient to produce a flow. Our river has, probably, very near the smallest allowance. The story is current, at any rate, though I believe that strict history will not bear it out, that the only bridge ever carried away on the main branch, within the limits of the town, was driven up stream by the wind. But wherever it makes a sudden bend it is shallower and swifter, and asserts its title to be called a river. Compared with the other tributaries of the Merrimack, it appears to have been properly named Musketaquid, or Meadow River, by the Indians. For the most part, it creeps through broad meadows, adorned with scattered oaks, where the cranberry is found in abundance, covering the ground like a mossbed. A row of sunken dwarf willows borders the stream on one or both sides, while at a greater distance the meadow is skirted with maples, alders, and other fluviatile trees, overrun with the grape vine, which bears fruit in its season, purple, red, white, and other grapes. Still further from the stream, on the edge of the firm land, are seen the gray and white dwellings of the inhabitants. According to the valuation of 1831, there were in Concord two thousand one hundred and eleven acres, or about one-seventh of the whole territory, in meadow; this standing next in the list after pasturage and unimproved lands, and, judging from the returns of previous years, the meadow is not reclaimed so fast as the woods are cleared.

The sluggish artery of the Concord meadows steals thus unobserved through the town, without a murmur or a pulse-beat, its general course from south-west to north-east, and its length about fifty miles; a huge volume of matter, ceaselessly rolling through the plains and valleys of the substantial earth, with the moccasined tread of an Indian warrior, making haste from the high places of the earth to its ancient reservoir. The murmurs of many a famous river on the other side of the globe reach even to us here, as to more distant dwellers on its banks; many a poet's stream floating the helms and shields of heroes on its bosom. The Xanthus or Scamander is not a mere dry channel and bed of a mountain torrent, but fed by the ever-flowing springs of fame;—

"And thou Simois, that as an arrowe, clere
Through Troy rennest, aie downward to the sea;"—

and I trust that I may be allowed to associate our muddy but much abused Concord River with the most famous in history.

"Sure there are poets which did never dream
Upon Parnassus, nor did taste the stream

Of Helicon; we therefore may suppose
Those made not poets, but the poets those."

The Mississippi, the Ganges, and the Nile, those journeying atoms from the Rocky Mountains, the Himmaleh, and Mountains of the Moon, have a kind of personal importance in the annals of the world. The heavens are not yet drained over their sources, but the Mountains of the Moon still send their annual tribute to the Pasha without fail, as they did to the Pharaohs, though he must collect the rest of his revenue at the point of the sword. Rivers must have been the guides which conducted the footsteps of the first travellers. They are the constant lure, when they flow by our doors, to distant enterprise and adventure, and, by a natural impulse, the dwellers on their banks will at length accompany their currents to the lowlands of the globe, or explore at their invitation the interior of continents. They are the natural highways of all nations, not only levelling the ground, and removing obstacles from the path of the traveller, quenching his thirst, and bearing him on their bosoms, but conducting him through the most interesting scenery, the most populous portions of the globe, and where the animal and vegetable kingdoms attain their greatest perfection.

I had often stood on the banks of the Concord, watching the lapse of the current, an emblem of all progress, following the same law with the system, with time, and all that is made; the weeds at the bottom gently bending down the stream, shaken by the watery wind, still planted where their seeds had sunk, but ere long to die and go down likewise; the shining pebbles, not yet anxious to better their condition, the chips and weeds, and occasional logs and stems of trees, that floated past, fulfilling their fate, were objects of singular interest to me, and at last I resolved to launch myself on its bosom, and float wither it would bear me.

Content Considerations

1. What is the nature of the Concord River?
2. Thoreau writes about a river trip in late summer of 1839. How had the river changed by that time from what he knew of it through history? What caused the changes?
3. What is Thoreau's attitude toward the men he meets?

The Writer's Strategies

1. Near the end of paragraph 2 Thoreau compares working the earth to writing. How does the analogy work? What attitude does it express?

2. Thoreau begins by talking about the river's history, then discusses its geographical location. From what other perspectives does he discuss the river?

Writing Possibilities

1. In an essay, show how the epigraph (the verse that appears before the piece) relates to the ideas and themes Thoreau presents.
2. Play with writing an essay similar to Thoreau's—find an interesting outside place and observe. Take notes if you wish. Then, in writing, describe what you have seen and suggest what it could mean.
3. Using Thoreau's essay and your knowledge and experience, focus an essay on the importance of rivers in the growth of civilization or, if you choose, their importance in an individual's life.

JOHN MUIR

The River Floods

John Muir (1838–1914) was one of America's most active conservationists. He was the first president of the Sierra Club, he fought to make Yosemite a national park, and he lost a bitter battle to save the Hetch Hetchy Valley area from being flooded to form a reservoir.

The Sierra range captivated him; he found in its wildness—not only its landscape and animal life but its storms and floods, its clouds and thunder— the essence of spirituality. Muir's observation of the area led him to propose that the Yosemite Valley had been formed by glaciers, not, as other scientists had insisted, by a collapse of the earth. His keen eye and early study of geology, botany, and chemistry served him well in his lifelong attempt to understand the causes of things, to find reasons, to create possibilities.

He also had wide experience of the world. He worked in a factory, in a mill, as a sheepherder; he walked from Indiana to Florida, visited Cuba, and then continued to San Francisco. At thirty, he first saw Yosemite, and there he lived, wandering and writing but emerging to fight for wilderness and its conservation.

In the following essay Muir experiences a flood in the mountains. Characteristically, he does not try to flee, but instead moves closer in order to see, feel, hear, and touch—and then put into words—the storm he has tasted. Before you read his account, write for a few minutes about where you would go for an experience such as this and whether you would want to.

THE SIERRA RIVERS are flooded every spring by the melting of the snow as regularly as the famous old Nile. They begin to rise in May, and in June high-water mark is reached. But because the melting does not go on rapidly over all the fountains, high and low, simultaneously, and the melted snow is not reinforced at this time of year by rain, the spring floods are seldom very violent or destructive. The thousand falls, however, and the cascades in the cañons are then in full bloom, and sing songs from one end of the range to the other. Of course the snow on the lower tributaries of the rivers is first melted, then that on the higher fountains most exposed to sunshine, and about a month later the cooler, shadowy fountains send down their treasures, thus allowing the main trunk streams nearly six weeks to get their waters hurried through the foot-hills and across the lowlands to the sea. Therefore very violent spring floods are avoided, and will be as long as the shading, restraining forests last. The rivers of the north half of the range are still less subject to sudden floods, because their upper fountains in great part lie protected from the changes of the weather beneath thick folds of lava, just as many of the rivers of Alaska lie beneath folds of ice, coming to the light farther down the range in large springs, while those of the high Sierra lie on the surface of solid granite, exposed to every change of temperature. More than ninety per cent of the water derived from the snow and ice of Mount Shasta is at once absorbed and drained away beneath the porous lava folds of the mountain, where mumbling and groping in the dark they at length find larger fissures and tunnel-like caves from which they emerge, filtered and cool, in the form of large springs, some of them so large they give birth to rivers that set out on their journeys beneath the sun without any visible intermediate period of childhood. Thus the Shasta River issues from a large lake-like spring in Shasta Valley, and about two thirds of the volume of the McCloud River gushes forth suddenly from the face of a lava bluff in a roaring spring seventy-five yards wide.

These spring rivers of the north are of course shorter than those of the south whose tributaries extend up to the tops of the mountains. Fall River, an important tributary of the Pitt or Upper Sacramento, is only about ten miles long, and is all falls, cascades, and springs from its head to its confluence with the Pitt. Bountiful springs, charmingly embowered, issue from the rocks at one end of it, a snowy fall a hundred and eighty feet high thunders at the other, and a rush of crystal rapids sing and dance between. Of course such streams are but little affected by the weather. Sheltered from evaporation their flow is nearly as full in the autumn as in the time of general spring floods. While those of the high Sierra diminish to less than the hundredth part of their springtime prime, shallowing in autumn to a series of silent pools among the rocks and hollows of their channels, connected by feeble, creeping threads of water, like the sluggish sentences of a tired writer, connected by a drizzle of "ands" and "buts." Strange to say, the greatest floods occur in the winter, when one would suppose all the wild waters would be muffled and chained in frost and snow. The same long, all-day storms of the so-called Rainy Season in California, that give rain to the lowlands, give dry frosty snow to the mountains. But at rare intervals warm rains and warm winds invade the mountains and push back the snow line from 2000 feet to 8000, or even higher, and then come the big floods.

I was usually driven down out of the High Sierra about the end of November, but the winter of 1874 and 1875 was so warm and calm that I was tempted to seek general views of the geology and topography of the basin of Feather River in January. And I had just completed a hasty survey of the region, and made my way down to winter quarters, when one of the grandest flood-storms that I ever saw broke on the mountains. I was then in the edge of the main forest belt at a small foot-hill town called Knoxville, on the divide between the waters of the Feather and Yuba rivers. The cause of this notable flood was simply a sudden and copious fall of warm wind and rain on the basins of these rivers at a time when they contained a considerable quantity of snow. The rain was so heavy and long-sustained that it was, of itself, sufficient to make a good wild flood, while the snow which the warm wind and rain melted on the upper and middle regions of the basins was sufficient to make another flood equal to that of the rain. Now these two distinct harvests of flood waters were gathered simultaneously and poured out on the plain in one magnificent avalanche. The basins of the Yuba and Feather, like many others of the Sierra, are admirably adapted to the growth of floods of this kind. Their many tributaries radiate far and wide, comprehending extensive areas, and the tributaries are steeply inclined, while the trunks are comparatively level. While the flood-storm was in progress the thermometer at Knoxville ranged between 44° and 50°; and when warm wind and warm rain fall simultaneously on snow contained in basins like these, both the rain and that portion of the snow which the rain and wind melt are at first sponged up and held back until the combined mass becomes sludge, which at length, suddenly dissolving, slips and descends all together to the trunk channel; and since the deeper the stream the faster it flows, the flooded portion of the current above overtakes the slower foot-hill portion below it, and all sweeping forward together with a high, overcurling front, debouches on the open plain with a violence and suddenness that at first seem wholly unaccountable. The destructiveness of the lower portion of this particular flood was somewhat augmented by mining gravel in the river channels, and by levees which gave way after having at first restrained and held back the accumulating waters. These exaggerating conditions did not, however, greatly influence the general result, the main effect having been caused by the rare combination of flood factors indicated above. It is a pity that but few people meet and enjoy storms so noble as this in their homes in the mountains, for, spending themselves in the open levels of the plains, they are likely to be remembered more by the bridges and houses they carry away than by their beauty or the thousand blessings they bring to the fields and gardens of Nature.

On the morning of the flood, January 19th, all the Feather and Yuba landscapes were covered with running water, muddy torrents filled every gulch and ravine, and the sky was thick with rain. The pines had long been sleeping in sunshine; they were now awake, roaring and waving with the beating storm, and the winds sweeping along the curves of hill and dale, streaming through the woods, surging and gurgling on the tops of rocky ridges, made the wildest of wild storm melody.

It was easy to see that only a small part of the rain reached the ground in the form of drops. Most of it was thrashed into dusty spray like that into which small waterfalls are divided when they dash on shelving rocks. Never have I seen water coming from the sky in denser or more passionate streams. The wind chased the spray forward in choking drifts, and compelled me again and again to seek shelter in the dell copses and back of large trees to rest and catch my breath. Wherever I went, on ridges or in hollows, enthusiastic water still flashed and gurgled about my ankles, recalling a wild winter flood in Yosemite when a hundred waterfalls came booming and chanting together and filled the grand valley with a sea-like roar.

After drifting an hour or two in the lower woods, I set out for the summit of a hill 900 feet high, with a view to getting as near the heart of the storm as possible. In order to reach it I had to cross Dry Creek, a tributary of the Yuba that goes crawling along the base of the hill on the northwest. It was now a booming river as large as the Tuolumne at ordinary stages, its current brown with mining-mud washed down from many a "claim," and mottled with sluice-boxes, fence-rails, and logs that had long lain above its reach. A slim foot-bridge stretched across it, now scarcely above the swollen current. Here I was glad to linger, gazing and listening, while the storm was in its richest mood—the gray rain-flood above, the brown river-flood beneath. The language of the river was scarcely less enchanting than that of the wind and rain; the sublime overboom of the main bouncing, exulting current, the swash and gurgle of the eddies, the keen dash and clash of heavy waves breaking against rocks, and the smooth, downy hush of shallow currents feeling their way through the willow thickets of the margin. And amid all this varied throng of sounds I heard the smothered bumping and rumbling of boulders on the bottom as they were shoving and rolling forward against one another in a wild rush, after having lain still for probably 100 years or more.

The glad creek rose high above its banks and wandered from its channel out over many a briery sand-flat meadow. Alders and willows waist-deep were bearing up against the current with nervous trembling gestures, as if afraid of being carried away, while supple branches bending confidingly, dipped lightly and rose again, as if stroking the wild waters in play. Leaving the bridge and passing on through the storm-thrashed woods, all the ground seemed to be moving. Pine-tassels, flakes of bark, soil, leaves, and broken branches were being swept forward, and many a rock-fragment, weathered from exposed ledges, was now receiving its first rounding and polishing in the wild streams of the storm. On they rushed through every gulch and hollow, leaping, gliding, working with a will, and rejoicing like living creatures.

Nor was the flood confined to the ground. Every tree had a water system of its own spreading far and wide like miniature Amazons and Mississippis.

Toward midday, cloud, wind, and rain reached their highest development. The storm was in full bloom, and formed, from my commanding outlook on the hilltop, one of the most glorious views I ever beheld. As far as the eye could reach, above, beneath, around, wind-driven rain filled the air like one vast waterfall. Detached clouds swept imposingly up the valley, as if they were endowed with independent motion and had special work to do in replenishing the mountain wells, now rising above the pine-tops, now descending into their midst, fondling their arrowy spires and soothing every branch and leaf with gentleness in the midst of all the savage sound and motion. Others keeping near the ground glided behind separate groves, and brought them forward into relief with admirable distinctness; or, passing in front, eclipsed whole groves in succession, pine after pine melting in their gray fringes and bursting forth again seemingly clearer than before.

The forms of storms are in great part measured, and controlled by the topography of the regions where they rise and over which they pass. When, therefore, we attempt to study them from the valleys, or from gaps and openings of the forest, we are confounded by a multitude of separate and apparently antagonistic impressions. The bottom of the storm is broken up into innumerable waves and currents that surge against the hillsides like sea-waves against a shore, and these, reacting on the nether surface of the storm, erode immense cavernous hollows and cañons, and sweep forward the resulting detritus in long trains, like the moraines of glaciers. But, as we ascend, these partial, confusing effects disappear and the phenomena are beheld united and harmonious.

The longer I gazed into the storm, the more plainly visible it became. The drifting cloud detritus gave it a kind of visible body, which explained many perplexing phenomena, and published its movements in plain terms, while the texture of the falling mass of rain rounded it out and rendered it more complete. Because raindrops differ in size they fall at different velocities and overtake and clash against one another, producing mist and spray. They also, of course, yield unequal compliance to the force of the wind, which gives rise to a still greater degree of interference, and passionate gusts sweep off clouds of spray from the groves like that torn from wave-tops in a gale. All these factors of irregularity in density, color, and texture of the general rain mass tend to make it the more appreciable and telling. It is then seen as one grand flood rushing over bank and brae, bending the pines like weeds, curving this way and that, whirling in huge eddies in hollows and dells, while the main current pours grandly over all, like ocean currents over the landscapes that lie hidden at the bottom of the sea.

I watched the gestures of the pines while the storm was at its height, and it was easy to see that they were not distressed. Several large Sugar Pines stood near the thicket in which I was sheltered, bowing solemnly and tossing their long arms as if interpreting the very words of the storm while accepting its wildest onsets with passionate exhilaration. The lions were feeding. Those who have observed sunflowers feasting on sunshine during the golden days of Indian summer know that none of their gestures express thankfulness. Their celestial food is too heartily given, too heartily taken to leave room for thanks. The pines were evidently accepting the benefactions of the storm in the same whole-souled manner; and when I looked down among the budding hazels, and still lower to the young violets and fern-tufts on the rocks, I noticed the same divine methods of giving and taking, and the same exquisite adaptations of what seems an outbreak of violent and uncontrollable force to the purposes of beautiful and delicate life. Calms like sleep come upon landscapes, just as they do on people and trees, and storms awaken them in the same way. In the dry midsummer of the lower portion of the range the withered hills and valleys seem to lie as empty and expressionless as dead shells on a shore. Even the highest mountains may be found occasionally dull and uncommunicative as if in some way they had lost countenance and shrunk to less than half their real stature. But when the lightnings crash and echo in the cañons, and the clouds come down wreathing and crowning their bald snowy heads, every feature beams with expression and they rise again in all their imposing majesty.

Storms are fine speakers, and tell all they know, but their voices of lightning, torrent, and rushing wind are much less numerous than the nameless still, small voices too low for human ears; and because we are poor listeners we fail to catch much that is fairly within reach. Our best rains are heard mostly on roofs, and winds in chimneys; and when by choice or compulsion we are pushed into the heart of a storm, the confusion made by cumbersome equipments and nervous haste and mean fear, prevent our hearing any other than the loudest expressions. Yet we may draw enjoyment from storm sounds that are beyond hearing, and storm movements we cannot see. The sublime whirl of planets around their suns is as silent as raindrops oozing in the dark among the roots of plants. In this great storm, as in every other, there were tones and gestures inexpressibly gentle manifested in the midst of what is called violence and fury, but easily recognized by all who look and listen for them. The rain brought out the colors of the woods with delightful freshness, the rich brown of the bark of the trees and the fallen burs and leaves and dead ferns; the grays of rocks and lichens; the light purple of swelling buds, and the warm yellow greens of the libocedrus and mosses. The air was steaming with delightful fragrance, not rising and wafting past in separate masses, but diffused through all the atmosphere. Pine woods are always fragrant, but most so in spring when the young tassels are opening and in warm weather when the various gums and balsams are softened by the sun. The wind was now chafing their innumerable needles and the warm rain was steeping them. Monardella grows here in large beds in the openings, and there is plenty of laurel in dells and manzanita on the hillsides, and the rosy, fragrant chamœbatia carpets the ground almost everywhere. These, with the gums and balsams of the woods, form the main local fragrance-fountains of the storm. The ascending clouds of aroma wind-rolled and rain-washed became pure like light and traveled with the wind as part of it. Toward the middle of the afternoon the main flood cloud lifted along its western border revealing a beautiful section of the Sacramento Valley some twenty or thirty miles away, brilliantly sun-lighted and glistering with rain-sheets as if paved with silver. Soon afterward a jagged bluff-like cloud with a sheer face appeared over the valley of the Yuba, dark-colored and roughened with numerous furrows like some huge lava-table. The blue Coast Range was seen stretching along the sky like a beveled wall, and the somber, craggy Marysville Buttes rose impressively out of the flooded plain like islands out of the sea. Then the rain began to abate and I sauntered down through the dripping bushes reveling in the universal vigor and freshness that inspired all the life about me. How clean and unworn and immortal the woods seemed to be!—the lofty cedars in full bloom laden with golden pollen and their washed plumes shining; the pines rocking gently and settling back into rest, and the evening sunbeams spangling on the broad leaves of the madroños, their tracery of yellow boughs relieved against dusky thickets of Chestnut Oak; liverworts, lycopodiums, ferns were exulting in glorious revival, and every moss that had ever lived seemed to be coming crowding back from the dead to clothe each trunk and stone in living green. The steaming ground seemed fairly to throb and tingle with life; smilax, fritillaria, saxifrage, and young violets were pushing up as if already conscious of the summer glory, and innumerable green and yellow buds were peeping and smiling everywhere.

As for the birds and squirrels, not a wing or tail of them was to be seen while the storm was blowing. Squirrels dislike wet weather more than cats do; therefore they were at home rocking in their dry nests. The birds were hiding in the dells out of the wind, some of the strongest of them pecking at acorns and manzanita berries, but most were perched on low twigs, their breast feathers puffed out and keeping one another company through the hard time as best they could.

When I arrived at the village about sundown, the good people bestirred themselves, pitying my bedraggled condition as if I were some benumbed castaway snatched from the sea, while I, in turn, warm with excitement and reeking like the ground, pitied them for being dry and defrauded of all the glory that Nature had spread round about them that day. 15

Content Considerations

1. How do forests prevent flooding?
2. What differences did Muir find in viewing the storm from the lower woods and from the summit?
3. What conditions lead to flooding in the Sierras, according to Muir?

The Writer's Strategies

1. Muir uses various literary devices in this essay, especially onomatopoeia and personification. Using examples for support, explain how his use of these devices affects your reading.
2. What is the tone of this essay? How is it expressed?

Writing Possibilities

1. Compare the kinds of debris and detritus swept downstream in the flood Muir describes to the kinds likely to be swept away during a flood in the city. Which seems more dangerous to you?
2. Write an essay explaining the new information you were able to glean from Muir's essay.
3. In an essay, consider both the good and the bad that floods do, then lead into your position on how much river control is appropriate.

John McPhee

Los Angeles Against the Mountains

The piece that follows is excerpted from The Control of Nature *(1989), in which professor and writer John McPhee looks at civilization's attempts to make nature accommodate the life it constructs. This particular piece looks at flood control in Los Angeles or, actually, how the city attempts to direct and contain the water and debris set loose during heavy rains.*

Like most cities, Los Angeles has an intense interest in protecting its inhabitants from harm and their possessions from destruction. It also has an interest in providing water both for use within its boundaries and for sale elsewhere. The cost is high, and success is not guaranteed.

McPhee begins this piece with a story about the Genofile family, whose home filled with mud, rocks, and water during a hard rain. Before you read the story, write for a few minutes about what people do to try to control nature and how much you think they should do.

In Los Angeles versus the San Gabriel Mountains, it is not always clear which side is losing. For example, the Genofiles, Bob and Jackie, can claim to have lost and won. They live on an acre of ground so high that they look across their pool and past the trunks of big pines at an aerial view over Glendale and across Los Angeles to the Pacific bays. The setting, in cool dry air, is serene and Mediterranean. It has not been everlastingly serene.

On a February night some years ago, the Genofiles were awakened by a crash of thunder—lightning striking the mountain front. Ordinarily, in their quiet neighborhood, only the creek beside them was likely to make much sound, dropping steeply out of Shields Canyon on its way to the Los Angeles River. The creek, like every component of all the river systems across the city from mountains to ocean, had not been left to nature. Its banks were concrete. Its bed was concrete. When boulders were running there, they sounded like a rolling freight. On a night like this, the boulders should have been running. The creek should have been a torrent. Its unnatural sound was unnaturally absent. There was, and had been, a lot of rain.

The Genofiles had two teen-age children, whose rooms were on the uphill side of the one-story house. The window in Scott's room looked straight up Pine Cone Road, a cul-de-sac, which, with hundreds like it, defined the northern limit of the city, the confrontation of the urban and the wild. Los Angeles is overmatched on one side by the Pacific Ocean and on the other by very high mountains. With respect to these principal boundaries, Los Angeles is done sprawling. The San Gabriels, in their state of tectonic youth, are rising as rapidly as any range on earth. Their loose inimical slopes flout the tolerance of the angle of repose. Rising straight up out of the megalopolis, they stand ten thousand feet above the nearby sea, and they are not kidding with this city. Shedding, spalling, self-destructing, they are disintegrating at a rate that is also among the fastest in the world. The phalanxed communities of Los Angeles have pushed themselves hard against these mountains, an aggression that requires a deep defense budget to contend with the results. Kimberlee Genofile called to her mother, who joined her in Scott's room as they looked up the street. From its high turnaround, Pine Cone Road plunges downhill like a ski run, bending left and then right and then left and then right in steep christiania turns for half a mile above a three-hundred-foot straightaway that aims directly at the Genofiles' house. Not far below the turnaround, Shields Creek passes under the street, and there a kink in its concrete profile had been plugged by a six-foot boulder. Hence the silence of the creek. The water was now spreading over the street. It descended in heavy sheets. As the young Genofiles and their mother glimpsed it in the all but total darkness, the scene was suddenly illuminated by a blue electrical flash. In the blue light they saw a massive blackness, moving. It was not a landslide, not a mudslide, not a rock avalanche; nor by any means was it the front of a conventional flood. In Jackie's words, "It was just one big black thing coming at us, rolling, rolling with a lot of water in front of it, pushing the water, this big black thing. It was just one big black hill coming toward us."

In geology, it would be known as a debris flow. Debris flows amass in stream valleys and more or less resemble fresh concrete. They consist of water mixed with a good deal of solid material, most of which is above sand size. Some of it is Chevrolet size. Boulders bigger than cars ride long distances in debris flows. Boulders grouped like fish eggs pour downhill in debris flows. The dark material coming toward the Genofiles was not only full of boulders; it was so full of automobiles it was like bread dough mixed with raisins. On its way down Pine Cone Road, it plucked up cars from driveways and the street. When it crashed into the Genofiles' house, the shattering of safety glass made terrific explosive sounds. A door burst open. Mud and boulders poured into the hall. We're going to go, Jackie thought. Oh, my God, what a hell of a way for the four of us to die together.

The parents' bedroom was on the far side of the house. Bob Genofile was in there kicking through white satin draperies at the panelled glass, smashing it to provide an outlet for water, when the three others ran in to join him. The walls of the house neither moved nor shook. As a general contractor, Bob had built dams, department stores, hospitals, six schools, seven churches, and this house. It was made of concrete block with steel reinforcement, sixteen inches on center. His wife had said it was stronger than any dam in California. His crew had called it "the fort." In those days, twenty years before, the Genofiles' acre was close by the edge of the mountain brush, but a developer had come along since then and knocked down thousands of trees and put Pine Cone Road up the slope. Now Bob Genofile was thinking, I hope the roof holds. I hope the roof is strong enough to hold. Debris was flowing over it. He told Scott to shut the bedroom door. No sooner was the door closed than it was battered down and fell into the room. Mud, rock, water poured in. It pushed everybody against the far wall. "Jump on the bed," Bob said. The bed began to rise. Kneeling on it—on a gold velvet spread—they could soon press their palms against the ceiling. The bed also moved toward the glass wall. The two teen-agers got off, to try to control the motion, and were pinned between the bed's brass railing and the wall. Boulders went up against the railing, pressed it into their legs, and held them fast. Bob dived into the muck to try to move the boulders, but he failed. The debris flow, entering through windows as well as doors, continued to rise. Escape was still possible for the parents but not for the children. The parents looked at each other and did not stir. Each reached for and held one of the children. Their mother felt suddenly resigned, sure that her son and daughter would die and she and her husband would quickly follow. The house became buried to the eaves. Boulders sat on the roof. Thirteen automobiles were packed around the building, including five in the pool. A din of rocks kept banging against them. The stuck horn of a buried car was blaring. The family in the darkness in their fixed tableau watched one another by the light of a directional signal, endlessly blinking. The house had filled up in six minutes, and the mud stopped rising near the children's chins.

Stories like that do not always have such happy endings. A man went outside to pick up his newspaper one morning, heard a sound, turned, and died of a heart attack as he saw his house crushed to pieces with his wife and two children inside. People have been buried alive in their beds. But such cases are infrequent. Debris flows generally are much less destructive of life than of property. People get out of the way.

If they try to escape by automobile, they have made an obvious but imperfect choice. Norman Reid backed his Pontiac into the street one January morning and was caught from behind by rock porridge. It embedded the car to the chrome strips. Fifty years of archival news photographs show cars of every vintage standing like hippos in chunky muck. The upper halves of their headlights peep above the surface. The late Roland Case Ross, an emeritus professor at California State University, told me of a day in the early thirties when he watched a couple rushing to escape by car. She got in first. While her husband was going around to get in his side, she got out and ran into the house for more silverware. When the car at last putt-putted downhill, a wall of debris was nudging the bumper. The debris stayed on the vehicle's heels all the way to Foothill Boulevard, where the car turned left.

Foothill Boulevard was U.S. Route 66—the western end of the rainbow. Through Glendora, Azusa, Pasadena, it paralleled the mountain front. It strung the metropolitan border towns. And it brought in emigrants to fill them up. The real-estate line of maximum advance now averages more than a mile above Foothill, but Foothill receives its share of rocks. A debris flow that passed through the Monrovia Nursery went on to Foothill and beyond. With its twenty million plants in twelve hundred varieties, Monrovia was the foremost container nursery in the world, and in its recovery has remained so. The debris flow went through the place picking up pots and cans. It got into a greenhouse two hundred feet long and smashed out the southern wall, taking bougainvillea and hibiscus with it. Arby's, below Foothill, blamed the nursery for damages, citing the hibiscus that had come with the rocks. Arby's sought compensation, but no one was buying beef that thin.

In the same storm, large tree trunks rode in the debris like javelins and broke through the sides of houses. Automobiles went in through picture windows. A debris flow hit the gym at Azusa Pacific College and knocked a large hole in the upslope wall. In the words of Cliff Hamlow, the basketball coach, "If we'd had students in there, it would have killed them. Someone said it sounded like the roar of a jet engine. It filled the gym up with mud, and with boulders two and three feet in diameter. It went out through the south doors and spread all over the football field and track. Chain-link fencing was sheared off—like it had been cut with a welder. The place looked like a war zone." Azusa Pacific College wins national championships in track, but Coach Hamlow's basketball team (12–18) can't get the boulders out of its game.

When a debris flow went through the Verdugo Hills Cemetery,

which is up a couple of switchbacks on the mountain front, two of the central figures there, resting under impressive stones, were "Hiram F. Hatch, 1st Lieut. 6th Mich. Inf., December 24, 1843–October 12, 1922," and "Henry J. Hatch, Brigadier General, United States Army, April 28, 1869–December 31, 1931." The two Hatches held the hill while many of their comrades slid below. In all, thirty-five coffins came out of the cemetery and took off for lower ground. They went down Hillrose Street and were scattered over half a mile. One came to rest in the parking lot of a supermarket. Many were reburied by debris and, in various people's yards, were not immediately found. Three turned up in one yard. Don Sulots, who had moved into the fallout path two months before, said, "It sounded like thunder. By the time I made it to the front door and got it open, the muck was already three feet high. It's quite a way to start off life in a new home—mud, rocks, and bodies all around."

Most people along the mountain front are about as mindful of debris flows as those corpses were. Here today, gone tomorrow. Those who worry build barricades. They build things called deflection walls—a practice that raises legal antennae and, when the caroming debris breaks into the home of a neighbor, probes the wisdom of Robert Frost. At least one family has experienced so many debris flows coming through their back yard that they long ago installed overhead doors in the rear end of their built-in garage. To guide the flows, they put deflection walls in their back yard. Now when the boulders come they open both ends of their garage, and the debris goes through to the street.

Between Harrow Canyon and Englewild Canyon, a private street called Glencoe Heights teased the mountain front. Came a time of unprecedented rain, and the neighborhood grew ever more fearful—became in fact so infused with catastrophic anticipation that it sought the drastic sort of action that only a bulldozer could provide. A fire had swept the mountainsides, leaving them vulnerable, dark, and bare. Expecting floods of mud and rock, people had piled sandbags and built heavy wooden walls. Their anxiety was continuous for many months. "This threat is on your mind all the time," Gary Lukehart said. "Every time you leave the house, you stop and put up another sandbag, and you just hope everything will be all right when you get back." Lukehart was accustomed to losing in Los Angeles. In the 1957 Rose Bowl, he was Oregon State's quarterback. A private street could not call upon city or county for the use of heavy equipment, so in the dead of night, as steady rain was falling, a call was put in to John McCafferty—bulldozer for hire. McCafferty had a closeup knowledge of the dynamics of debris flows: he had worked the mountain front from San Dimas to

Sierra Madre, which to him is Sarah Modri. ("In those canyons at night, you could hear them big boulders comin'. They sounded like thunder.") He arrived at Glencoe Heights within the hour and set about turning the middle of the street into the Grand Canal of Venice. His Cat was actually not a simple dozer but a 955 loader on tracks, with a two-and-a-quarter-yard bucket seven feet wide. Cutting water mains, gas mains, and sewers, he made a ditch that eventually extended five hundred feet and was deep enough to take in three thousand tons of debris. After working for five hours, he happened to be by John Caufield's place ("It had quit rainin', it looked like the worst was over") when Caufield came out and said, "Mac, you sure have saved my bacon."

McCafferty continues, "All of a sudden, we looked up at the mountains—it's not too far from his house to the mountains, maybe a hundred and fifty feet—and we could just see it all comin'. It seemed the whole mountain had come loose. It flowed like cement." In the ditch, he put the Cat in reverse and backed away from the oncoming debris. He backed three hundred feet. He went up one side of the ditch and was about halfway out of it when the mud and boulders caught the Cat and covered it over the hood. In the cab, the mud pushed against McCafferty's legs. At the same time, debris broke into Caufield's house through the front door and the dining-room window, and in five minutes filled it to the eaves.

Other houses were destroyed as well. A garage left the neighborhood with a car in it. One house was buried twice. (After McCafferty dug it out, it was covered again.) His ditch, however, was effective, and saved many places on slightly higher ground, among them Gary Lukehart's and the home of John Marcellino, the chief executive officer of Mackinac Island Fudge. McCafferty was promised a lifetime supply of fudge. He was on the scene for several days, and in one span worked twenty-four hours without a break. The people of the street brought him chocolate milkshakes. He had left his lowbed parked around the corner. When at last he returned to it and prepared to go home, he discovered that a cop had given him a ticket.

A metropolis that exists in a semidesert, imports water three hundred miles, has inveterate flash floods, is at the grinding edges of two tectonic plates, and has a microclimate tenacious of noxious oxides will have its priorities among the aspects of its environment that it attempts to control. For example, Los Angeles makes money catching water. In a few days in 1983, it caught twenty-eight million dollars' worth of water. In one period of twenty-four hours, however, the ocean hit the city with twenty-foot waves, a tornado made its own freeway, debris flows poured from the San Gabriel front, and an earthquake shook the region. Nature's invoice was forty million dollars. Later, twenty million more was spent dealing with the mountain debris.

There were those who would be quick—and correct—in saying that were it not for the alert unflinching manner and imaginative strategies by which Los Angeles outwits the mountains, nature's invoices at such times would run into the billions. The rear-guard defenses are spread throughout the city and include more than two thousand miles of underground conduits and concrete-lined open stream channels—a web of engineering that does not so much reinforce as replace the natural river systems. The front line of battle is where the people meet the mountains—up the steep slopes where the subdivisions stop and the brush begins.

Strung out along the San Gabriel front are at least a hundred and twenty bowl-shaped excavations that resemble football stadiums and are often as large. Years ago, when a big storm left back yards and boulevards five feet deep in scree, one neighborhood came through amazingly unscathed, because it happened to surround a gravel pit that had filled up instead. A tungsten filament went on somewhere above Los Angeles. The county began digging pits to catch debris. They were quarries, in a sense, but exceedingly bizarre quarries, in that the rock was meant to come to them. They are known as debris basins. Blocked at their downstream ends with earthfill or concrete constructions, they are also known as debris dams. With clean spillways and empty reservoirs, they stand ready to capture rivers of boulders—these deep dry craters, lying close above the properties they protect. In the overflowing abundance of urban nomenclature, the individual names of such basins are obscure, until a day when they appear in a headline in the Los Angeles *Times:* Harrow, Englewild, Zachau, Dunsmuir, Shields, Big Dalton, Hog, Hook East, Hook West, Limekiln, Starfall, Sawpit, Santa Anita. For fifty miles, they mark the wild boundary like bulbs beside a mirror. Behind chain links, their idle ovate forms more than suggest defense. They are separated, on the average, by seven hundred yards. In aggregate, they are worth hundreds of millions of dollars. All this to keep the mountains from falling on Johnny Carson.

The principal agency that developed the debris basins was the hopefully named Los Angeles County Flood Control District, known familiarly through the region as Flood Control, and even more intimately as Flood. ("When I was at Flood, one of our dams filled with debris overnight," a former employee remarked to me. "If any more rain came, we were going to have to evacuate the whole of Pasadena.") There has been a semantic readjustment, obviously intended to acknowledge that when a flood pours out of the mountains it might be half rock. The debris basins are now in the charge of the newly titled Sedimentation Section of the Hydraulic Division of the Los Angeles County Department of Public Works. People still call it Flood. By whatever name the agency is called, its essential tactic remains unaltered. This was summarized for me in a few words by an engineer named Donald Nichols, who pointed out that eight million people live below the mountains on the urban coastal plain, within an area large enough to accommodate Philadelphia, Detroit, Chicago, St. Louis, Boston, and New York. He said, "To make the area inhabitable, you had to put in lined channels on the plain and halt the debris at the front. If you don't take it out at the front, it will come out in the plain, filling up channels. A filled channel won't carry diddly-boo."

To stabilize mountain streambeds and stop descending rocks even before they reach the debris basins, numerous crib structures (barriers made of concrete slats) have been emplaced in high canyons—the idea being to convert plunging streams into boulder staircases, and hypothetically cause erosion to work against itself. Farther into the mountains, a dozen dams of some magnitude were built in the nineteen-twenties and thirties to control floods and conserve water. Because they are in the San Gabriels, they inadvertently trap large volumes of debris. One of them—the San Gabriel Dam, in the San Gabriel River—was actually built as a debris-control structure. Its reservoir, which is regularly cleaned out, contained, just then, twenty million tons of mountain.

The San Gabriel River, the Los Angeles River, and the Big Tujunga (Bigta Hung-ga) are the principal streams that enter the urban plain, where a channel that filled with rock wouldn't carry diddly-boo. Three colossal debris basins—as different in style as in magnitude from those on the mountain front—have been constructed on the plain to greet these rivers. Where the San Gabriel goes past Azusa on its way to Alamitos Bay, the Army Corps of Engineers completed in the late nineteen-forties a dam ninety-two feet high and twenty-four thousand feet wide—this to stop a river that is often dry, and trickles most of the year. Santa Fe Dam, as it is called, gives up at a glance its own story, for it is made of boulders that are shaped like potatoes and are generally the size of watermelons. They imply a large volume of water flowing with high energy. They are stream-propelled, stream-rounded boulders, and the San Gabriel is the stream. In Santa Fe Basin, behind the dam, the dry bed of the San Gabriel is half a mile wide. The boulder-strewn basin in its entirety is four times as wide as that. It occupies eighteen hundred acres in all, nearly three square miles, of what would be prime real estate were it not for the recurrent arrival of rocks. The scene could have been radioed home from Mars, whose cobbly face is in part the result of debris flows dating to a time when Mars had surface water.

The equally vast Sepulveda Basin is where Los Angeles receives and restrains the Los Angeles River. In Sepulveda Basin are three golf courses, which lend ample support to the widespread notion that everything in Los Angeles is disposable. Advancing this national prejudice even further, debris flows, mudslides, and related phenomena have "provided literary minds with a ready-made metaphor of the alleged moral decay of Los Angeles." The words belong to Reyner Banham, late professor of the history of architecture at University College, London, whose passionate love of Los Angeles left him without visible peers. The decay was only "alleged," he said. Of such nonsense he was having none. With his "Los Angeles: The Architecture of Four Ecologies," Banham had become to this deprecated, defamed, traduced, and disparaged metropolis what Pericles was to Athens. Banham knew why the basins were there and what the people were defending. While all those neurasthenic literary minds are cowering somewhere in ethical crawl space, the quality of Los Angeles life rises up the mountain front. There is air there. Cool is the evening under the crumbling peaks. Cool descending air. Clean air. Air with a view. "The financial and topographical contours correspond almost exactly," Banham said. Among those "narrow, tortuous residential roads serving precipitous house-plots that often back up directly on unimproved wilderness" is "the fat life of the delectable mountains."

People of Gardena, Inglewood, and Watts no less than Azusa and Altadena pay for the defense of the mountain front, the rationale being that debris trapped near its source will not move down and choke the channels of the inner city, causing urban floods. The political City of Los Angeles—in its vague and tentacular configuration—actually abuts the San Gabriels for twenty miles or so, in much the way that it extends to touch the ocean in widely separated places like Venice, San Pedro, and Pacific Palisades. Los Angeles County reaches across the mountains and far into the Mojave Desert. The words "Los Angeles" as generally used here refer neither to the political city nor to the county but to the multinamed urban integrity that has a street in it seventy miles long (Sepulveda Boulevard) and, from the Pacific Ocean at least to Pomona, moves north against the mountains as a comprehensive town.

The debris basins vary greatly in size—not, of course, in relation to the populations they defend but in relation to the watersheds and washes above them in the mountains. For the most part, they are associated with small catchments, and the excavated basins are commensurately modest, with capacities under a hundred thousand cubic yards. In a typical empty reservoir—whatever its over-all dimensions may be—stands a columnar tower that resembles a campanile. Full of holes, it is known as a perforated riser. As the basin fills with a thick-flowing slurry of water, mud, and rock, the water goes into the tower and is drawn off below. The county calls this water harvesting.

Like the freeways, the debris-control system ordinarily functions but occasionally jams. When the Genofiles' swimming pool filled with cars, debris flows descended into other neighborhoods along that part of the front. One hit a culvert, plugged the culvert, crossed a road in a bouldery wave, flattened fences, filled a debris basin, went over the spillway, and spread among houses lying below, shoving them off their foundations. The debris basins have caught as much as six hundred thousand cubic yards in one storm. Over time, they have trapped some twenty million tons of mud and rock. Inevitably, sometimes something gets away.

At Devils Gate—just above the Rose Bowl, in Pasadena—a dam was built in 1920 with control of water its only objective. Yet its reservoir, with a surface of more than a hundred acres, has filled to the brim with four million tons of rock, gravel, and sand. A private operator has set up a sand-and-gravel quarry in the reservoir. Almost exactly, he takes out what the mountains put in. As one engineer has described it, "he pays Flood, and Flood makes out like a champ."

Content Considerations

1. Explain how a debris flow is created.
2. What defenses against nature has Los Angeles erected? How effective are they?
3. What part does economics play in the problem this essay presents?

The Writer's Strategies

1. Who does the author believe to be winning—Los Angeles or the mountains? How can you tell?
2. The story of the Genofiles holds this piece together. How does it do so?

Writing Possibilities

1. Using the information McPhee provides, construct a first-person narrative on watching a debris flow. Or, if you have witnessed such an occurrence yourself, tell your own story.
2. In a paper, discuss the ways a natural debris flow might differ from one created by human practices.

3. Working with a classmate, construct a plan to lessen the destruction a debris flow can cause. In your plan, note the defenses McPhee has explained; argue against or in support of them.

Annie Dillard

The Present

Annie Dillard's Pilgrim at Tinker Creek, *a nonfiction book awarded the Pulitzer Prize in 1974, presents the author's observations of the Blue Ridge area of Virginia and the reflections and musings those observations provoked. Since 1974 she has written several other books, including* The Writing Life *(1989), in which she wonders at writing, at what prompts it and how it is sustained. She says, "People love pretty much the same things best. A writer looking for subjects inquires not after what he loves best, but after what he alone loves at all."*

Dillard has, in a sense, done this. Her inquiries rise from places and memories, and her writing of them sparks recognition in her readers. In this selection from Pilgrim at Tinker Creek, *she brings to bear a variety of personal experiences and thoughts as she contemplates, among other notions, how water heals. Before you read her essay, write about the ways in which your own memories and experiences affect how you see the world and how you tend to respond to the particulars of your surroundings.*

WHAT ELSE IS GOING ON right this minute while ground water creeps under my feet? The galaxy is careening in a slow, muffled widening. If a million solar systems are born every hour, then surely hundreds burst into being as I shift my weight to the other elbow. The sun's surface is now exploding; other stars implode and vanish, heavy and black, out of sight. Meteorites are arcing to earth invisibly all day long. On the planet the winds are blowing: the polar easterlies, the westerlies, the northeast and southeast trades. Somewhere, someone under full sail is becalmed, in the horse latitudes, in the doldrums; in the northland, a trapper is maddened, crazed, by the eerie scent of the chinook, the snoweater, a wind that can melt two feet of snow in a day. The pampero blows, and the tramontane, and the Boro, sirocco, levanter, mistral. Lick a finger: feel the now.

Spring is seeping north, towards me and away from me, at sixteen miles a day. Caribou straggle across the tundra from the spruce-fir forests of the south, first the pregnant does, hurried, then the old and unmated does, then suddenly a massing of bucks, and finally the diseased and injured, one by one. Somewhere, people in airplanes are watching the sun set and peering down at clustered houselights, stricken. In the montana in Peru, on the rain-forested slopes of the Andes, a woman kneels in a dust clearing before a dark shelter of overlapping broadleaves; between her breasts hangs a cross of smooth sticks she peeled with her teeth and lashed with twistings of vine. Along estuary banks of tidal rivers all over the world, snails in black clusters like currants are gliding up and down the stems of reed and sedge, migrating every moment with the dip and swing of tides. Behind me, Tinker Mountain, and to my left, Dead Man Mountain, are eroding one thousandth of an inch a year.

The tomcat that used to wake me is dead; he was long since grist for an earthworm's casting, and is now the clear sap of a Pittsburgh sycamore, or the honeydew of aphids sucked from that sycamore's high twigs and sprayed in sticky drops on a stranger's car. A steer across the road stumbles into the creek to drink; he blinks, he laps; a floating leaf in the current catches against his hock and wrenches away. The giant water bug I saw is dead, long dead, and its moist gut and rigid casing are both, like the empty skin of the frog it sucked, dissolved, spread, still spreading right now, in the steer's capillaries, in the windblown smatter of clouds overhead, in the Sargasso Sea. The mockingbird that dropped furled from a roof . . . but this is no time to count my dead. That is nightwork. The dead are staring, underground, their sleeping heels in the air.

The sharks I saw are roving up and down the coast. If the sharks cease roving, if they still their twist and rest for a moment, they die. They need new water pushed into their gills; they need dance. Somewhere east of me, on another continent, it is sunset, and starlings in breathtaking bands are winding high in the sky to their evening roost. Under the water just around the bend downstream, the coot feels with its foot in the creek, rolling its round red eyes. In the house a spider slumbers at her wheel like a spinster curled in a corner all day long. The mantis egg cases are tied to the mock-orange hedge; within each case, within each egg, cells elongate, narrow, and split; cells bubble and curve inward, align, harden or hollow or stretch. The Polyphemus moth, its wings crushed to its back, crawls down the driveway, crawls down the driveway, crawls. . . . The snake whose skin I tossed away, whose homemade, personal skin is now tangled at the county dump—that snake in the woods by the quarry stirs now, quickens now, prodded under the leafmold by sunlight, by the probing root of May apple, the bud of bloodroot. And where are you now?

I stand. All the blood in my body crashes to my feet and instantly heaves to my head, so I blind and blush, as a tree blasts into leaf spouting water hurled up from roots. What happens to me? I stand before the sycamore dazed; I gaze at its giant trunk.

Big trees stir memories. You stand in their dimness, where the very light is blue, staring unfocused at the thickest part of the trunk as though it were a long, dim tunnel—: the Squirrel Hill tunnel. You're gone. The egg-shaped patch of light at the end of the blackened tunnel swells and looms; the sing of tire tread over brick reaches an ear-splitting crescendo; the light breaks over the hood, smack, and full on your face. You have achieved the past.

Eskimo shamans bound with sealskin thongs on the igloo floor used to leave their bodies, their skins, and swim "muscle-naked" like a flensed seal through the rock of continents, in order to placate an old woman who lived on the sea floor and sent or withheld game. When he fulfilled this excruciating mission, the Eskimo shaman would awake, returned to his skin exhausted from the dark ardors of flailing peeled through rock, and find himself in a lighted igloo, at a sort of party, among dear faces.

In the same way, having bored through a sycamore trunk and tunneled beneath a Pennsylvania mountain, I blink, awed by the yellow light, and find myself in a shady side of town, in a stripped dining room, dancing, years ago. There is a din of trumpets, upbeat and indistinct, like some movie score for a love scene played on a city balcony; there is an immeasurably distant light glowing from half-remembered faces. . . . I stir. The heave of my shoulders returns me to the present, to the tree, the sycamore, and I yank myself away, shove off and moving, seeking live water.

Live water heals memories. I look up the creek and here it comes, the future, being borne aloft as on a winding succession of laden trays. You may wake and look from the window and breathe the real air, and say, with satisfaction or with longing, "This is it." But if you look up the creek, if you look up the creek in any weather, your spirit fills, and you are saying, with an exulting rise of the lungs, "Here it comes!"

Here it comes. In the far distance I can see the concrete bridge where the road crosses the creek. Under that bridge and beyond it the water is flat and silent, blued by distance and stilled by depth. It is so much sky, a fallen shred caught in the cleft of banks. But it pours. The channel here is straight as an arrow; grace itself is an archer. Between the dangling wands of bankside willows, beneath the overarching limbs of tulip, walnut, and Osage orange, I see the creek pour down. It spills toward me streaming over a series of sandstone tiers, down, and down, and down. I feel as though I stand at the foot of an infinitely high staircase, down which some exuberant spirit is flinging tennis ball after tennis ball, eternally, and the one thing I want in the world is a tennis ball.

There must be something wrong with a creekside person who, all things being equal, chooses to face downstream. It's like fouling your own nest. For this and a leather couch they pay fifty dollars an hour? Tinker Creek doesn't back up, pushed up its own craw, from the Roanoke River; it flows down, easing, from the northern, unseen side of Tinker Mountain. "Gravity, to Copernicus, is the nostalgia of things to become spheres." This is a curious, tugged version of the great chain of being. Ease is the way of perfection, letting fall. But, as in the classic version of the great chain, the pure trickle that leaks from the unfathomable heart of Tinker Mountain, this Tinker Creek, widens, taking shape and cleaving banks, weighted with the live and intricate impurities of time, as it descends to me, to where I happen to find myself, in this intermediate spot, halfway between here and there. Look upstream. Just simply turn around; have you no will? The future is a spirit, or a distillation of *the* spirit, heading my way. It is north. The future is the light on the water; it comes, mediated, only on the skin of the real and present creek. My eyes can stand no brighter light than this; nor can they see without it, if only the undersides of leaves.

Trees are tough. They last, taproot and bark, and we soften at their feet. "For we are strangers before thee, and sojourners, as were all our fathers: our days on the earth are as a shadow, and there is none abiding." We can't take the lightning, the scourge of high places and rare airs. But we can take the light, the reflected light that shines up the valleys on creeks. Trees stir memories; live water heals them. The creek is the mediator, benevolent, impartial, subsuming my shabbiest evils and dissolving them, transforming them into live moles, and shiners, and sycamore leaves. It is a place even my faithlessness hasn't offended; it still flashes for me, now and tomorrow, that intricate, innocent face. It waters an undeserving world, saturating cells with lodes of light. I stand by the creek over rock under trees.

It is sheer coincidence that my hunk of the creek is strewn with boulders. I never merited this grace, that when I face upstream I scent the virgin breath of mountains, I feel a spray of mist on my cheeks and lips, I hear a ceaseless splash and susurrus, a sound of water not merely poured smoothly down air to fill a steady pool, but tumbling live about, over, under, around, between, through an intricate speckling of rock. It is sheer coincidence that upstream from me the creek's bed is ridged in horizontal croppings of sandstone. I never merited this grace, that when I face upstream I see the light on the water careening towards me, inevitably, freely, down a graded series of terraces like the balanced winged platforms on an infinite, inexhaustible font. "Ho, if you are thirsty, come down to the water; ho, if you are hungry, come and sit and eat." This is the present, at last. I can pat the puppy any time I want. This is the now, this flickering, broken light, this air that the wind of the future presses down my throat, pumping me buoyant and giddy with praise.

My God, I look at the creek. It is the answer to Merton's prayer, "Give us time!" It never stops. If I seek the senses and skill of children, the information of a thousand books, the innocence of puppies, even the insights of my own city past, I do so only, solely, and entirely that I might look well at the creek. You don't run down the present, pursue it with baited hooks and nets. You wait for it, empty-handed, and you are filled. You'll have fish left over. The creek is the one great giver. It is, by definition, Christmas, the incarnation. This old rock planet gets the present for a present on its birthday every day.

Here is the word from a subatomic physicist: "Everything that has already happened is particles, everything in the future is waves." Let me twist his meaning. Here it comes. The particles are broken; the waves are translucent, laving, roiling with beauty like sharks. The present is the wave that explodes over my head, flinging the air with particles at the height of its breathless unroll; it is the live water and light that bears from undisclosed sources the freshest news, renewed and renewing, world without end.

Content Considerations

1. How does Dillard explain the cycles of life, the oneness of all living things?
2. What meanings of "present" fit the way that word is used in this essay?
3. Paraphrase Dillard's last paragraph. How does its meaning fit the essay?

The Writer's Strategies

1. The metaphor of the river as time holds this piece together. How does the metaphor work?
2. At times this essay is almost a catalog—a listing of kinds of wind, of animals, of geography, of people in particular places who do particular things. How does this listing reinforce or illustrate Dillard's ideas?

Writing Possibilities

1. Big trees stir memories for Dillard. What stirs your memories? Write about it and the memories it provokes.
2. Natural history writing usually goes beyond observation and fact to reflection and contemplation. Write about why you think this happens.
3. Suppose that Tinker Creek were controlled—its flow dammed and its water diverted. Write about the changes to its ecosystem you think would occur.

Constance Elizabeth Hunt

Creating an Endangered Ecosystems Act

Quite often the response to an environmental problem occurs long after the problem has been first noticed. The problem is noticed, then studied, then studied more; solutions are proposed, studied, studied, studied more. Interested parties—agencies, companies, individuals—enter the circle of debate, offering their perspectives, objections, and expectations. While all this is occurring, the problem often becomes more serious.

Hasty action may sometimes be worse than late action or none at all, but according to many environmentalists, the best way to handle environmental problems is by preventing them. Prevention takes an enormous amount of individual and collective energy, but is likely to save both energy and money in the long run.

The following proposal centers on action that would protect riparian habitats from development and dam projects that endanger them. Such a proposal would, author Constance Elizabeth Hunt argues, prevent or partially solve several environmental problems. Before you read her proposal, included in her book, Down by the River *(with Verne Huser)*, discuss what environmental concerns might benefit from such an act.

THE FEDERAL GOVERNMENT is nothing more than an instrument of the public will. When it is guided by the voices of the people, the government is forced to weigh the requests of one interest group against those of another and to incorporate equanimity into its decisions and actions. When the government hears only a few voices, however, such as those of paid lobbyists representing an irrigation district or a commercial barge company, its response to those voices is unbalanced and, as a result, irreplaceable resources are destroyed.

One possible solution to the rapid loss of landscape and biological diversity in the United States is the establishment of an endangered ecosystems act. Current statutes are inadequate to protect natural communities threatened with extirpation. The National Environmental Policy Act (NEPA) creates the opportunity for interested parties to comment on federal programs but contains no provisions for prohibiting programs that are environmentally unsound. Section 404 of the Clean Water Act is limited in scope and is administered by the Army Corps of

Engineers, an agency that functions primarily as a federal engineering firm. The Endangered Species Act, the nation's toothiest environmental law, protects habitat only if it is essential to the survival of an endangered species. This stopgap legislation promotes the backward process of protecting habitat after a species has reached the brink of extinction. The decline in populations of most species in modern America is related to loss of habitat; therefore, a species' becoming endangered is an indicator that its habitat has been seriously impaired by abuse or encroachment. If rare *habitats* were protected by federal statute, the number of species requiring protection under the Endangered Species Act in order to survive would be much smaller.

When a federal action is halted by implementation of the Endangered Species Act, frequently it is the endangered species' habitat that is endangered by the proposed action. When habitat loss is the true problem posed by the planned action, the action too often proceeds without being stripped of its potential menace to the environment. Before 1981, for example, changes in the Colorado River's aquatic habitat brought the Colorado squawfish, humpback chub, and bonytail chub to the brink of extinction. The Fish and Wildlife Service refused to allow any further federal development that would significantly add to the depletion of water from the Colorado River system until it had sufficient information to be certain that the water losses would not wipe out the fish. By 1981, however, pressure from western water interests was great enough to change the service's position. These water brokers argued that the Service should not hold up projects when it couldn't demonstrate that the projects would harm the endangered species. For the next 4 years, therefore, the Service allowed water development projects to proceed while it attempted to gather information on the projects' impacts and the ecological needs of the fish. Instead of instituting a more natural water management regime to restore aquatic habitat conditions, researchers yanked many of the fish from the river and bred them in captivity. Meanwhile, fish habitat in the Colorado continued to decline, riparian ecology was harmed as well, and Congress considered plans for more water development on the river.

The situation is similar on the Platte. Critical habitat has been designated for the endangered whooping crane, but the species' recovery depends to a large extent on a captive breeding program. This breeding effort, which began in 1967, has met with success; but meanwhile, populations of two more bird species that depend on essentially the same Platte River riparian habitat, the spotted sandpiper and interior least tern, have declined to the point where protection under the Endangered Species Act is warranted.

The Fish and Wildlife Service's jeopardy opinion for Cliff Dam was a move toward recognizing that habitat, rather than an individual species or population, was the resource endangered by the dam. By proposing habitat-oriented alternatives, the service offered to trade an active bald eagle nesting site for a riparian community and flows to maintain the aquatic ecosystem.

As a species nears extinction, the costs involved in recovering the species increase. Where mere restrictions on development in a certain habitat type may be sufficient to protect the species early in its decline, the expensive facilities and personnel required for feeding, breeding, and studying a chronically endangered wildlife species offer only dim hope for its recovery later on. While the funds and energies of the responsible federal agencies are expended on trying to resuscitate the dying species, the species' habitat in the wild may continue to decline, bringing still more species to the brink of extinction. If the endangered species is successfully reared in captivity, biologists may proudly march out of their labs with a potentially growing population of the species only to find that no habitat exists to support it.

Early protection of potentially endangered species is not the only rationale for endangered ecosystems legislation. Many ecosystems harbor unique associations of plant and animal species. Although these species may exist in large numbers in other associations in different regions of the country and therefore are not candidates for the endangered species list, the associations are entities in themselves and should be preserved. The population decline of Arizona Bell's vireos, for example, is an indicator of the deteriorating health of riparian habitat on the lower Colorado River. Although the species is not in danger of extinction, the habitat is. The sturgeon and striped bass fisheries on the Apalachicola River are similar indicators of ecosystem illness. Is the existence of these fishes in other waterways sufficient justification for the death of the Apalachicola's natural ecology?

The same valuative arguments for preserving individual species, such as their importance in providing the world with future genetic resources and their values as sources of scientific knowledge, as portions of our children's rightful heritage, and even as God's creations with an inherent right to exist, apply to the preservation of ecosystems as well. These values cannot be economically quantified, so when land and water use decisions are made under pressure from participants in the national marketplace, such as irrigation equipment companies or constructors of marinas, the nonmonetary values of species and ecosystems often are not considered.

Cities are expanding, and agriculture is claiming more land. Ecosystems are being squeezed and cut into parcels incapable of supporting free-ranging animals. As the acreage of natural habitats decreases, the numbers of animals whose home range needs are satisfied by the remaining available habitat also decreases. Predators at the top of the food chain, such as mountain lions and wolves, require habitats ranging in size from 36 to 114,000 square miles. These animals often are the first link in the biological food web to be lost from an area, as the endangered status of the wolf and Florida panther (a mountain lion subspecies) indicates. The loss of predators from a region often triggers booms in populations of prey species, such as deer, which quickly outgrow the capacity of the ecosystem to feed them and often painfully starve to death in huge numbers.

Habitat shrinkage causes or contributes to the extinctions of wildlife species. The last known North American population of the ivory-billed woodpecker, a bird specifically adapted to life in mature bottomland hardwood forests, disappeared in 1948 when the 120-square-mile Singer tract in Louisiana was cut for soybean cultivation. Leadership at the federal level is necessary if we wish to preserve the natural heritage of the United States. The following pages contain a proposed legislative program to accomplish this goal.

Research

An endangered ecosystem act would appropriate federal funds to supply matching grants to the states for the purpose of surveying and designating ecosystems as endangered (in danger of complete extirpation) and threatened (close to becoming endangered). The surveying and designation work would be done on a state-by-state basis by committees consisting of representatives from federal resource management agencies, state conservation agencies, the research community, and any interested private groups, such as conservation organizations, livestock growers, and timber companies. The committees would address issues such as the biological richness of proposed ecosystem types; the relative scarcity of these types on a state, regional, and national basis; the minimum size of unit necessary to preserve the majority of wildlife species dependent on that ecosystem; and the land and water uses that are compatible or incompatible with the preservation of the ecosystem. The size and location of the ecosystem and any possible connections of natural areas by corridors would be identified in the planning stage. Once the most sensitive ecosystem types were identified, the committees would develop maps illustrating their locations throughout the state. The committees would also suggest land use guidelines to be applied to threatened and endangered ecosystems.

PROTECTION

Committee maps depicting the location and approximate extent of the endangered and threatened ecosystems would serve as guidelines for the protection of those ecosystems. For threatened ecosystems, federal resource management agencies would develop local guidelines restricting land and water uses that the committees found incompatible with ecosystem preservation. Such guidelines for threatened riparian habitats, for example, would create buffer zones where logging would not be allowed, would restrict grazing, and would prohibit the use of all-terrain cycles.

Endangered ecosystems on federal land would be managed by guidelines more restrictive of land and water use than those for threatened ecosystems. The guidelines would make preservation of the ecosystem's self-sustaining biological integrity the primary management goal. In endangered riparian ecosystems, for example, no structural alteration of streams would be allowed. The legislation would allocate funds to federal agencies for restoration and creation of ecosystem types that have been identified as endangered on public lands and would provide grants to the states for restoration and creation of these types on state and private lands. Further authorization would permit appropriations of federal funds to obtain conservation easements on endangered ecosystems on private lands.

The endangered ecosystems legislation could actually save money for the Treasury. Restricting development on floodplains could save billions of dollars every year in flood relief payments. Additional incentives to take agricultural land out of production would reduce the demand for crop deficiency payments. Restrictions on structural alterations of rivers would save initial outlays for construction materials and labor as well as funds used to maintain the alterations and correct erosion problems downstream from the structures.

Prohibitions on logging in threatened and endangered ecosystem types such as riparian canyons and stands of old-growth timber would save funds that the Forest Service would otherwise spend on roads and timber landings. Finally, the Fish and Wildlife Service would save some of the funds it would otherwise spend in recovering endangered species.

The United States, a country wealthy in both knowledge and natural resources, is already a world leader in the field of conservation. To maintain this lead, our society must be able to adapt to changing conditions. Our nation has outgrown its frontier infancy, when its youthful urge to expand into and harness nature was more beneficial to the present generation and less threatening to future generations. In the sunset of the twentieth century, a social policy comprising careful stewardship of our natural resources and meticulous planning and orderly growth of our communities is more appropriate to our maturing society.

Riparian ecosystems are perhaps the best places to "test-drive" a new American conservation ethic. These ecosystems provide water quality and flood mitigation benefits to society. Preserved in their natural condition, riparian corridors can provide recreational opportunities and aesthetic pleasure to communities. Linked with larger parcels of protected natural areas, riparian zones can hold together vital wildlife habitats, allowing dispersal of animals and decreasing the genetic bottlenecks that can occur in isolated wildlife populations, while maintaining migratory corridors and seasonal shelter. If one could adequately weigh the copious benefits that could be reaped through conservation and restoration of riparian habitats, the benefit-cost analysis game would no longer be fixed. Nature would be the clear winner in a fair game. 17

Content Considerations

1. What is the purpose of an endangered ecosystems act?
2. How does controlling a river affect its ecosystems?
3. According to the author, what economic benefits would be gained by protecting entire ecosystems?

The Writer's Strategies

1. How is this piece organized? How does its organization suit its purpose?
2. What tone does the author achieve? How?

Writing Possibilities

1. Summarize and respond to Hunt's proposal, focusing on the parts that seem especially strong or weak to you.
2. Comment on the idea that government is "an instrument of the public will" (paragraph 1). Do you agree? Support your position with examples.
3. Propose an environmental law you would like your legislature to adopt. Explain the need for it, its consequences, and how it would be established and enforced.

Additional Writing and Research

Connections

1. Both McPhee and Muir describe mountain flooding. Compare their accounts—content, attitudes, and conclusions.
2. Determine your own position regarding how to balance what nature does and what people do. Use any four of the readings for support and reference.
3. The essays of Beasley, Eiseley, and Thoreau all recount river journeys. What does each author discover?
4. The wilderness of rivers plays a part in the pieces by Muir, Hunt, and Thoreau. Using these three, discuss the value their authors find in rivers and wilderness.
5. Dillard and Eiseley link time with rivers. Examine what each has to say about this connection, then add your own perspective.
6. Using any four of the pieces in this chapter for support, construct your own argument for what you believe is the proper amount of control of rivers.

Researches

1. Investigate a river in your area, using local sources to determine the amount and kind of control it receives and the reasons for imposing such controls.
2. Read about the Army Corps of Engineers and interview someone at its nearest office. What is the agency's history? Its purpose? Its current mission?
3. Discover how controlling a river affects wildlife and habitat. Try to locate instances in which control has had little environmental impact.
4. Become familiar with the purposes of river control and how those purposes are achieved—that is, what kinds of structures are designed for which kinds of perceived problems.
5. Find out about your area's flood control plan; also research the area's history of flooding. (Old newspapers may be especially helpful for this, as may interviews with longtime residents.) Evaluate how well the plan seems to suit the problem.
6. Locate current arguments over the water of one river—arguments about how much water is released, who gets to use it, and how it is to be used. Focus on how such arguments are solved.

4

Developing the Deserts

FOR MANY AMERICANS the vast deserts of the Southwest evoke the stereotypes of late-night westerns: John Wayne riding through wind-sculpted rock and saguaro cactus, Indians watching silently from ridgetops, and prospectors crawling over blistering sand in search of the mother lode. Few have actually experienced the desert beyond driving through it in an air-conditioned car or truck, and for them the images might be of earth-toned rock, desiccated plants, and heat shimmering off the highway. Those who climb from their cars to walk or camp or hike report other images, images of richness, variety, difference. They see colors and hear sounds lost to more fleeting visitors. But most travelers, lingering or not, find something frightening about the desert, about its inhospitable surface crawling with creatures that harm.

Early European explorers in America found little redeeming value in its desert areas, labeling much of the Southwest "uninhabitable." Mormons, however, in the mid-nineteenth century, adopted the Utah desert as their promised land and, through irrigation and river control, caused it to bloom. Adding water to these arid regions changed the way they were valued by non-Native Americans. The regions became places to farm and ranch, to build cities, to develop housing and industry. No longer were desert areas equated with privation and death.

Adding water to the desert also changed the plant and animal life that had adapted to aridity. It changed nature in rivers, lakes, and streams. It still does. Water allows people to live on ranches and farms in desert regions, in small towns and communities, in the created oases of metropolitan areas. But as desert populations increased, views diverged, problems multiplied, and clashes of beliefs and practices rose exponentially. The American deserts may be settled, but desert issues are not.

For years few questioned the greening of the desert, the flight of humans and industry to the Southwest, or the miracle of diverting water hundreds of miles away. Now, with a greater understanding of what is happening to underground water supplies, to the rivers of the west, and to the deserts themselves, serious questions are being raised by naturalists, scientists, environmentalists, and those who simply love the desert.

The issues are not simple ones. Perhaps we should not attempt to build cities in a desert if doing so changes forever the character of the region and affects the lives of its plants and animals. Perhaps it is not even possible to build cities that last; long-term continued development may become a mirage as water supplies dwindle. Or perhaps protecting desert areas from continued development violates the rights of those whose families have lived in the region for generations.

The following essays and articles put forth a variety of arguments for and against further development in the deserts—development of resources, of the land itself, of recreational areas. Some will argue that deserts cannot support the life that demands suburbs with green lawns, golf courses, and high-rises. They argue there is not enough water to continue indefinitely and any water policy failing to recognize that truth is not a policy but a scheme. Or a scam.

There are other views. You will meet a rancher whose family has raised cattle on leased public land for five generations and whose livelihood and home have become threatened by a proposed national park. You will read the comments of off-road enthusiasts and of mining interests, both of whom see the desert as land to be used for human enjoyment and profit. You will tour the backroads and the crossroads of the Southwest and come face to face with what one writer calls the "indomitable goofiness" of desert life.

Developing the desert seems to contain a great irony—people appear to be innately afraid of deserts, yet no region in the United States is growing as fast as the Sunbelt. The names we have bestowed on the driest regions (Arizona means "arid zone") may themselves be messages indicating our fear of these desiccated landscapes: Death Valley, Jornado del Muerto, Camino del Diablo. The names register our horror of such inhospitable climes, our basic human need for water wherever we live or travel.

Formerly the haven of artists and outlaws, deserts have for decades attracted retirees, industry and business moguls, suburbanites, and tourists. Yet nearly all of the immigrants to the desert regions seem to want the desert on human terms, a playground or farm or homesite with plenty of water and air-conditioned comfort. Deserts are for looking at, not for living in; they are for transforming, not for maintaining.

But, of course, nature still has a say in any attempt at transformation. Parts of the West, including California, have experienced years of drought, further reducing available water supplies and forcing people to consider whether the region in which they have chosen to live will sustain the style of life they practice.

What exactly is a desert? A trackless wasteland of wind-rippled sand? A macrocosmic bell jar of cactus and sandstone? A dead zone without a drop of water, with no living creatures worth mentioning? Or is it something more, a unique bio-niche that sustains meaningful plant and animal life? As you read, create your own definition, your own understanding of "desertness" and how such regions may be viewed and treated.

MARY AUSTIN

Land of Little Rain

Mary Austin (1868–1934) was a prolific writer—thirty-five books and hundreds of essays, poems, and reviews; she specialized in the southwest and its people, particularly the Native Americans of the desert and their relationship to the land. Austin admired the descriptiveness of Native American names; "land of little rain" was much better than "desert" because it did not carry the baggage of predisposition.

The following piece is the first chapter of Austin's 1903 classic of the same title. In it she describes the land that most writers had found empty and forbidding; she relishes the harshly beautiful environment. Here Austin frequently refers to practices of Native Americans in the region, practices that allow them to exist in harmony with the land.

As you read this, consider ways in which we use deserts today and how development has changed the desert from what Austin describes.

EAST AWAY FROM THE SIERRAS, south from Panamint and Amargosa, east and south many an uncounted mile, is the Country of Lost Borders.

Ute, Paiute, Mojave, and Shoshone inhabit its frontiers, and as far into the heart of it as a man dare go. Not the law, but the land sets the limit. Desert is the name it wears upon the maps, but the Indian's is the better word. Desert is a loose term to indicate land that supports no man; whether the land can be bitted and broken to that purpose is not proven. Void of life it never is, however dry the air and villainous the soil.

This is the nature of that country. There are hills, rounded, blunt, burned, squeezed up out of chaos, chrome and vermilion painted, aspiring to the snow-line. Between the hills lie high level-looking plains full of intolerable sun glare, or narrow valleys drowned in a blue haze. The hill surface is streaked with ash drift and black, unweathered lava flows. After rains water accumulates in the hollows of small closed valleys, and, evaporating, leaves hard dry levels of pure desertness that get the local name of dry lakes. Where the mountains are steep and the rains heavy, the pool is never quite dry, but dark and bitter, rimmed about with the efflorescence of alkaline deposits. A thin crust of it lies along the marsh over the vegetating area, which has neither beauty nor freshness. In the broad wastes open to the wind the sand drifts in hummocks about the stubby shrubs, and between them the soil shows saline traces. The sculpture of the hills here is more wind than water work, though the quick storms do sometimes scar them past many a year's redeeming. In all the Western desert edges there are essays in miniature of the famed, terrible Grand Cañon, to which, if you keep on long enough in this country, you will come at last.

Since this is a hill country one expects to find springs, but not to depend upon them; for when found they are often brackish and unwholesome, or maddening, slow dribbles in a thirsty soil. Here you find the hot sink of Death Valley, or high rolling districts where the air has always a tang of frost. Here are the long heavy winds and breathless calms on the tilted mesas where dust devils dance, whirling up into a wide, pale sky. Here you have no rain when all the earth cries for it, or quick downpours called cloud-bursts for violence. A land of lost rivers, with little in it to love; yet a land that once visited must be come back to inevitably. If it were not so there would be little told of it.

This is the country of three seasons. From June on to November it lies hot, still, and unbearable, sick with violent unrelieving storms; then on until April, chill, quiescent, drinking its scant rain and scanter snows; from April to the hot season again, blossoming, radiant, and seductive. These months are only approximate; later or earlier the rain-laden wind may drift up the water gate of the Colorado from the Gulf, and the land sets its seasons by the rain.

The desert floras shame us with their cheerful adaptations to the seasonal limitations. Their whole duty is to flower and fruit, and they do it hardly, or with tropical luxuriance, as the rain admits. It is recorded in the report of the Death Valley expedition that after a year of abundant rains, on the Colorado desert was found a specimen of Amaranthus ten feet high. A year later the same species in the same place matured in the drought at four inches. One hopes the land may breed like qualities in her human offspring, not tritely to "try," but to do. Seldom does the desert herb attain the full stature of the type. Extreme aridity and extreme altitude have the same dwarfing effect, so that we find in the high Sierras and in Death Valley related species in miniature that reach a comely growth in mean temperatures. Very fertile are the desert plants in expedients to prevent evaporation, turning their foliage edgewise toward the sun, growing silky hairs, exuding viscid gum. The wind, which has a long sweep, harries and helps them. It rolls up dunes about the stocky stems, encompassing and protective, and above the dunes, which may be, as with the mesquite, three times as high as a man, the blossoming twigs flourish and bear fruit.

There are many areas in the desert where drinkable water lies within a few feet of the surface, indicated by the mesquite and the bunch grass *(Sporobolus airoides).* It is this nearness of unimagined help that makes the tragedy of desert deaths. It is related that the final breakdown of that hapless party that gave Death Valley its forbidding name occurred in a locality where shallow wells would have saved them. But how were they to know that? Properly equipped it is possible to go safely across that ghastly sink, yet every year it takes its toll of death, and yet men find there sun-dried mummies, of whom no trace or recollection is preserved. To underestimate one's thirst, to pass a given landmark to the right or left, to find a dry spring where one looked for running water—there is no help for any of these things.

Along springs and sunken watercourses one is surprised to find such water-loving plants as grow widely in moist ground, but the true desert breeds its own kind, each in its particular habitat. The angle of the slope, the frontage of a hill, the structure of the soil determines the plant. South-looking hills are nearly bare, and the lower tree-line higher here by a thousand feet. Cañons running east and west will have one wall naked and one clothed. Around dry lakes and marshes the herbage preserves a set and orderly arrangement. Most species have well-defined areas of growth, the best index the voiceless land can give the traveler of his whereabouts.

If you have any doubt about it, know that the desert begins with the creosote. This immortal shrub spreads down into Death Valley and up to the lower timber-line, odorous and medicinal as you might guess from the name, wandlike, with shining fretted foliage. Its vivid green is grateful to the eye in a wilderness of gray and greenish white shrubs. In the spring it exudes a resinous gum which the Indians of those parts know how to use with pulverized rock for cementing arrow points to shafts. Trust Indians not to miss any virtues of the plant world!

Nothing the desert produces expresses it better than the unhappy growth of the tree yucca. Tormented, thin forests of it stalk drearily in the high mesas, particularly in that triangular slip that fans out eastward from the meeting of the Sierras and coastwise hills where the first swings across the southern end of the San Joaquin Valley. The yucca bristles with bayonet-pointed leaves, dull green, growing shaggy with age, tipped with panicles of fetid, greenish bloom. After death, which is slow, the ghostly hollow network of its woody skeleton, with hardly power to rot, makes the moonlight fearful. Before the yucca has come to flower, while yet its bloom is a creamy cone-shaped bud of the size of a small cabbage, full of sugary sap, the Indians twist it deftly out of its fence of daggers and roast it for their own delectation. So it is that in those parts where man inhabits one sees young plants of *Yucca arborensis* infrequently. Other yuccas, cacti, low herbs, a thousand sorts, one finds journeying east from the coastwise hills. There is neither poverty of soil nor species to account for the sparseness of desert growth, but simply that each plant requires more room. So much earth must be preëmpted to extract so much moisture. The real struggle for existence, the real brain of the plant, is underground; above there is room for a rounded perfect growth. In Death Valley, reputed the very core of desolation, are nearly two hundred identified species.

Above the lower tree-line, which is also the snow-line, mapped out abruptly by the sun, one finds spreading growth of piñon, juniper, branched nearly to the ground, lilac and sage, and scattering white pines.

There is no special preponderance of self-fertilized or wind-fertilized plants, but everywhere the demand for and evidence of insect life. Now where there are seeds and insects there will be birds and small mammals, and where these are, will come the slinking, sharp-toothed kind that prey on them. Go as far as you dare in the heart of a lonely land, you cannot go so far that life and death are not before you. Painted lizards slip in and out of rock crevices, and pant on the white hot sands. Birds, hummingbirds even, nest in the cactus scrub; woodpeckers befriend the demoniac yuccas; out of the stark, treeless waste rings the music of the night-singing mockingbird. If it be summer and the sun well down, there will be a burrowing owl to call. Strange, furry, tricksy things dart across the open places, or sit motionless in the conning towers of the creosote. The poet may have "named all the birds without a gun," but not the fairy-footed, ground-inhabiting, furtive, small folk of the rainless regions. They are too many and too swift; how many you would not believe without seeing the footprint tracings in the sand. They are nearly all night workers, finding the days too hot and white. In mid-desert where there are no cattle, there are no birds of carrion, but if you go far in that direction the chances are that you will find yourself shadowed by their tilted wings. Nothing so large as a man can move unspied upon in that country, and they know well how the land deals with strangers. There are hints to be had here of the way in which a land forces new habits on its dwellers. The quick increase of suns at the end of spring sometimes overtakes birds in their nesting and effects a reversal of the ordinary manner of incubation. It becomes necessary to keep eggs cool rather than warm. One hot, stifling spring in the Little Antelope I had occasion to pass and repass frequently the nest of a pair of meadowlarks, located unhappily in the shelter of a very slender weed.

I never caught them sitting except near night, but at midday they stood, or drooped above it, half fainting with pitifully parted bills, between their treasure and the sun. Sometimes both of them together with wings spread and half lifted continued a spot of shade in a temperature that constrained me at last in a fellow feeling to spare them a bit of canvas for permanent shelter. There was a fence in that country shutting in a cattle range, and along its fifteen miles of posts one could be sure of finding a bird or two in every strip of shadow; sometimes the sparrow and the hawk, with wings trailed and beaks parted, drooping in the white truce of noon.

If one is inclined to wonder at first how so many dwellers came to be in the loneliest land that ever came out of God's hands, what they do there and why stay, one does not wonder so much after having lived there. None other than this long brown land lays such a hold on the affections. The rainbow hills, the tender bluish mists, the luminous radiance of the spring, have the lotus charm. They trick the sense of time, so that once inhabiting there you always mean to go away without quite realizing that you have not done it. Men who have lived there, miners and cattle-men, will tell you this, not so fluently, but emphatically, cursing the land and going back to it. For one thing there is the divinest, cleanest air to be breathed anywhere in God's world. Some day the world will understand that, and the little oases on the windy tops of hills will harbor for healing its ailing, house-weary broods. There is promise there of great wealth in ores and earths, which is no wealth by reason of being so far removed from water and workable conditions, but men are bewitched by it and tempted to try the impossible.

You should hear Salty Williams tell how he used to drive eighteen and twenty-mule teams from the borax marsh to Mojave, ninety miles, with the trail wagon full of water barrels. Hot days the mules would go so mad for drink that the clank of the water bucket set them into an uproar of hideous, maimed noises, and a tangle of harness chains, while Salty would sit on the high seat with the sun glare heavy in his eyes, dealing out curses of pacification in a level, uninterested voice until the clamor fell off from sheer exhaustion. There was a line of shallow graves along that road; they used to count on dropping a man or two of every new gang of coolies brought out in the hot season. But when he lost his swamper, smitten without warning at the noon halt, Salty quit his job; he said it was "too durn hot." The swamper he buried by the way with stones upon him to keep the coyotes from digging him up, and seven years later I read the penciled lines on the pine headboard, still bright and unweathered.

But before that, driving up on the Mojave stage, I met Salty again crossing Indian Wells, his face from the high seat, tanned and ruddy as a harvest moon, looming through the golden dust above his eighteen mules. The land had called him.

The palpable sense of mystery in the desert air breeds fables, chiefly of lost treasure. Somewhere within its stark borders, if one believes report, is a hill strewn with nuggets; one seamed with virgin silver; an old clayey water-bed where Indians scooped up earth to make cooking pots and shaped them reeking with grains of pure gold. Old miners drifting about the desert edges, weathered into the semblance of the tawny hills, will tell you tales like these convincingly. After a little sojourn in that land you will believe them on their own account. It is a question whether it is not better to be bitten by the little horned snake of the desert that goes sidewise and strikes without coiling, than by the tradition of a lost mine.

And yet—and yet—is it not perhaps to satisfy expectation that one falls into the tragic key in writing of desertness? The more you wish of it the more you get, and in the mean time lose much of pleasantness. In that country which begins at the foot of the east slope of the Sierras and spreads out by less and less lofty hill ranges toward the Great Basin, it is possible to live with great zest, to have red blood and delicate joys, to pass and repass about one's daily performance an area that would make an Atlantic seaboard State, and that with no peril, and, according to our way of thought, no particular difficulty. At any rate, it was not people who went into the desert merely to write it up who invented the fabled Hassaympa, of whose waters, if any drink, they can no more see fact as naked fact, but all radiant with the color of romance. I, who must have drunk of it in my twice seven years' wanderings, am assured that it is worth while.

For all the toll the desert takes of a man it gives compensations, deep breaths, deep sleep, and the communion of the stars. It comes upon one with new force in the pauses of the night that the Chaldeans were a desert-bred people. It is hard to escape the sense of mastery as the stars move in the wide clear heavens to risings and settings unobscured. They look large and near and palpitant; as if they moved on some stately service not needful to declare. Wheeling to their stations in the sky, they make the poor world-fret of no account. Of no account you who lie out there watching, nor the lean coyote that stands off in the scrub from you and howls and howls.

Content Considerations

1. What comparisons does Austin make between how Native Americans and non-Native Americans live in the desert?
2. What possibilities for developing the desert does Austin envision?
3. What does this essay say about social attitudes and conditions at the turn of the century?

The Writer's Strategies

1. How does Austin make known her attitude toward and emotion for the desert?
2. How does Austin use her observations of desert conditions to make the reader aware of desert life?

Writing Possibilities

1. Write an essay discussing how plant and animal life has adapted to desert conditions.
2. Respond to Austin's declaration, "Not the law but the land sets the limit" (paragraph 2). Include in your response some comment on how her declaration connects to desert development.
3. In her preface to *Land of Little Rain* (1903), Austin says, "The earth is no wanton to give up all her best to every comer, but keeps a sweet, separate intimacy for each." Write about what this quotation suggests regarding how people relate to the land.

JOHN ALCOCK

A Treasure of Complexities

In this essay, a piece that prefaced a series of desert photographs in Wilderness *magazine, author and professor of zoology John Alcock recreates a sense of the rich diversity of the American desert regions. Each is different, he argues, and various sections of each desert give life to animals and plants that do not and perhaps cannot exist elsewhere.*

At issue here is biological diversity—a term used to denote all species of life on earth in their varied and diverse forms and the need to maintain them. Development of a desert, a coast, a wetland, a prairie, or any other geographic entity necessarily affects the plant and animal life of that entity. One form of life may disappear, causing the disappearance of others in the ecosystem whose life functions depended on it.

Biological diversity can be harmed in many ways—pollution of air, land, and water; introduction of nonnative species; the human practices of logging, building, mining, manufacturing, enjoying recreation. It can also be protected in many ways, most of which require limited human contact. Before you read this essay, list as many kinds of desert plants and animals as you can; as you read, make a note of those the author mentions that you had never heard of.

I watch—quietly, but not quietly enough—as a mule deer drinks at the edge of Burro Creek, then jerks its head up in alarm, water dripping from its muzzle. The deer turns and leaps in one fluid motion, disappearing into the maze of mesquites that border this stretch of stream. It is spring in central Arizona and the mesquite bosque is life itself. Fresh green leaves of the year cover every tree. Long yellow catkins dangle from branch tips like overgrown woolly bear caterpillars. The catkins swarm with minute yellow bees that are dwarfed by an occasional honey bee forager. Lucy's warblers sing with territorial determination, having converted the bosque into a warbler subdivision. Their songs rise and fall from every point along the stream even though the pale silver-grey singers remain hidden within the canopy of the trees. A phainopepla flutters toward a clump of mistletoe growing parasitically from the limb of a mesquite. The white wing patches of the black phainopepla flicker in the filtered light of the trees. The mistletoe still has a few small pink berries left for the bird to harvest. In the bottom of a deep pool three fat round-tail chubs over a foot long skulk in the shade created by a rock wall and a cottonwood bordering the stream.

The chunky fish are on the threatened species list for the state of Arizona and in decline throughout their range, but they are doing well for the moment in Burro Creek. . . .

BURRO CREEK is a kind of paradigm: just one of thousands of examples that could be drawn from the western United States to make the point that deserts are far more complex than the stereotyped image of a bleak, grey wasteland pimpled with sand dunes and creased with waterless arroyos—places that hardly anybody lives in or cares about.

Much of the desert terrain in North America falls under the control of the Bureau of Land Management. The BLM manages millions of acres in the Sonoran Desert of southern Arizona, the Mojave Desert of California and Nevada, the Chihuahuan Desert of west Texas and New Mexico, and the Great Basin Desert that embraces most of Nevada and portions of Utah, Idaho, Oregon, and California. These regions are difficult and hostile places for humans but are home to a surprisingly diverse array of organisms, including many that are found nowhere else.

Each one of our western deserts has its own distinctive features and characteristic assemblages of plants and animals. The Sonoran Desert boasts the most diverse collection of cacti of the North American deserts. Here and only here can you find the giant saguaros and the columnar organ pipe cactus, for example. The Mojave Desert has far fewer cacti and desert trees than the Sonoran but it has its own specialties and the unique flavor conferred on it by its Joshua trees, huge furry yuccas that dominate the landscape of the Mojave in the same wonderful way that saguaros claim dominion over the Sonoran desert. The Great Basin is the most northern of the American deserts and the coldest. Its serene plains are covered by sagebrush and saltbrush—and by snow in the winter. The Chihuahuan Desert also features drought-tolerant shrubs and succulents, like sotols, yuccas, agaves, and acacias, which mingle with desert grasses in the plains and plateaus of New Mexico.

Although it is possible to characterize each desert by the presence of a few key plant species, these thumbnail sketches do not begin to do justice to the diversity of life that exists within their boundaries. Even in one small corner of a western desert, there can be astonishing differences in the plants and animals found within a few acres. Burro Creek offers a case in point. In the narrow riparian corridor running along the stream, mesquite bosques, cottonwoods, and willows shelter and feed Lucy's warblers, phainopeplas, and common black-hawks, the last a candidate for the threatened species list in Arizona. But move several hundred yards away from the watercourse and you are in a dramatically different world. Here, the mesquites give way to little creosote bushes with their spindly limbs and minimalist leaves. Thorny acacias pull at trousers and pluck at sleeves. In a narrow dry wash, sand lies in eddies and curls that do not move. On the boulder-strewn hillsides, agaves point their needle-tipped leaves to the sky and the twisted green limbs of foothill palo verdes bask in the penetrating sunlight. It is a place that grunting white-collared peccaries find amenable.

Travel north along Burro Creek to the upper watershed and you enter another very different habitat, a block of semidesert grassland, with its own special congregation of plant and animal occupants. This spartan prairie supports a population of pronghorn antelope, the Burro Creek part of a Chihuahuan subspecies which has been reduced to a fraction of its original numbers. Semidesert grasslands like these are dominated by species of grass that can manage on just 10 to 15 inches of rain a year. Although native species of grass have largely been replaced in central Arizona by introduced species that cope better with cattle grazing, a few native inhabitants such as Arizona cottontop, black grama, and sideoats grama grass persist in places. The semidesert grasslands on Burro Mesa run off into the distance largely unblemished by scruffy mesquites and juniper thickets. Tufts of black grama grass sway erratically in time to the gusts of wind that rush over the plateau, and prairie falcons and Say's phoebe grace the sky.

This is beautiful country and much of its beauty and value lies in the variety of species that the land sustains. A great deal has been said recently about the importance of biological diversity by biologists and conservationists who have been made keenly aware that an appreciation of diversity for diversity's sake is not universal. Many Americans are emphatic that the value of living things and natural habitats is proportional to their potential contribution to economic gain or utility for off-road vehicle recreation.

Admirers of biological diversity have worked hard to counter this utilitarian view. By maintaining as many species as possible, they say, we protect organisms that might eventually prove to have some special, but now unknown, benefit for mankind (a cancer cure is a favorite for persons employing this argument). Or, they maintain, given the subtle interactions among species, the loss of even one apparently insignificant species may have unexpected (and disastrous) ramifications that might ultimately affect human welfare, while studies of desert plants, they say, might help to expand future desert agriculture.

All these claims are probably correct, but the biologists that I know present them without much enthusiasm, probably because they are based on the utilitarian premise that the universe revolves around human beings and their economic goals. Many of us, however, value biological diversity simply because of the pure aesthetic appeal of natural differences. I just like the *idea* of the Death Valley Mormon tea, for instance, a plant not terribly different from other Mormon teas but not exactly the same either. Or what about the balloon vine, a species found in the United States only in the Coyote Mountains of Arizona? I have never seen the plant but it gives me pleasure to know that it is out there somewhere. Each plant and animal species has its unique features and unique puzzles to solve. There are volumes of genetic information contained in the tissues that make up a balloon vine or a Joshua tree or a desert tortoise. The community of living things found along a stream in the Sonoran Desert offers a continual array of surprises and questions that could never be asked in the absence of these species. A world from which mesquite bosques had been removed would be a monochrome world deprived of Lucy's warblers and the *Perdita* bees that depend on mesquite pollen. It would be a world from which a whole raft of delicious complexities and untold stories would have disappeared, a world in which the possibilities for curiosity had been diminished.

And this is where the BLM and the concept of wilderness areas have their part to play. Burro Creek and the Coyote Mountains of Arizona are in the BLM empire as are the Henry Mountains and the Dirty Devil River canyons of Utah, the King Range and the New York Mountains of California, the Owyhee Plateau of Idaho, the Black Rock Desert of Nevada, the Centennial Mountains of Montana, and thousands of other sites, not all of them desert, but all of them irreplaceable. As part of the Federal Land Policy and Management Act of 1976, Congress requested that the BLM consider all roadless tracts in excess of 5,000 acres to find those that offered outstanding opportunities for solitude or primitive recreation.

The BLM's ultimate charge was to recommend to the Congress via the President and Secretary of the Interior those lands that it deemed suitable additions for the National Wilderness Preservation System. In Arizona, for example, the BLM first identified several million acres as possible wilderness study sites, including two blocks along Burro Creek. Then, after environmental impact studies had been completed, the BLM informed Congress that in their view only about 1 million acres actually deserved wilderness status—and only one 20,000-acre block along Burro Creek made the final grade.

Intense resistance to placing more than a small fraction of BLM lands in wilderness has greeted all attempts to enlarge the agency's recommendations in the West. Ranchers, miners, off-road recreationists and other elements of the anti-wilderness crowd have fought to restrict the amount of BLM land that would be placed in designated wilderness. Wilderness proponents, for their part, have been just as determined to expand wilderness proposals, and legislation to that effect is now pending in Arizona, California, and Utah, while conservationists in Oregon, Nevada, and New Mexico are sharpening their own knives for battle.

The fight is both predictable and necessary. The simple fact is that the BLM manages desert lands that contain a vast array of plant and animal species that would benefit greatly from wilderness designation. Plants like the Panamint liveforever, Stephen's beardtongue, Eureka dune-grass, and the Mimbres figwort. Animals like peregrine and prairie falcons, Sanborn's long-nosed bats, fringe-toed lizards, and the Yuma clapper rail. Many plants and animals in the BLM's care are scarce or endangered, highly localized in their distribution, and sensitive to human activities.

Thus, whether it appreciates it or not, the BLM has the opportunity to be a kind of zoological and botanical librarian, shepherding a legion of unusual desert-competent organisms into the twenty-first century so that their stories can be read in the future by persons able and willing to decipher them. Or perhaps we should be more emphatic. The BLM has the responsibility to be a guardian of the wealth of biological diversity that exists in all the lands that it manages for all Americans.

Burro Creek, Arizona. January 1990. While politicians, stockmen, miners, ORV enthusiasts, and wilderness advocates argue over what should and should not be included in the Arizona wilderness bill, I visit this small wilderness world again. The mesquite bosques are brown and leafless, looking for all the world like a dead forest. It is really only on hold, I know, still waiting for winter rains. If the rains do come, returning Lucy's warblers will find the mesquites decked out in green spring canopies, offering shelter, food, and safe song perches galore. The drought that has hit the Southwest has reduced Burro Creek in some places to shallow, scattered pools maintained by beavers, an unlikely desert mammal, but one thoroughly at home in the creek. The beaver dams may be only a foot or two high but they create habitat utilized by many desert visitors. A flock of American wigeons explodes from a sheltered cover and a great blue heron lumbers after them. A black phoebe accompanies me upstream, leaping exuberantly into the air from rocks in the stream after small midges dancing in the sunshine. A castanet click of its beak announces each attempt to snare a midge. From its perch on boulders set among a sparse forest of saguaros, a bald eagle leaps into space and calmly strokes away well above the creek bed. As it passes over a perfectly still pool, the mirror image of the eagle travels across the water's surface, wingbeat matching wingbeat for a few glorious moments.

Watching, I reflect that the very real possibility of losing some elements of the diverse natural world is as aesthetically repellent to some of us as would be the disappearance of all Picassos or Pollocks or Renoirs. To be told that the art of Manet would suffice for the loss of all Monets would hardly satisfy lovers of art.

"It's amazing what happens to a place if you leave it alone," Paul Theroux wrote in The O-Zone. "It just goes its own way. It stays alive. It grows. It gets better."

Maybe, I think, as the eagle flies, maybe it is time to leave it alone.

Content Considerations

1. What is Alcock trying to persuade the reader to believe or do?
2. What arguments support the need to maintain biological diversity?
3. What part does the Federal Land Policy and Management Act of 1976 play in the conservation of wilderness areas?

The Writer's Strategies

1. Review the plants and animals Alcock describes or mentions. What purposes could be served by including so many in the article?
2. The author uses an analogy that compares deserts to works of art. How does this analogy work?

Writing Possibilities

1. Construct an argument that ranchers, miners, or off-road recreationists threaten or do not threaten desert life. From among your classmates, find someone who disagrees with you; exchange papers, discuss your opposing viewpoints, then revise your argument so that it addresses the opposition's points.
2. Compare and contrast Alcock's position on desert regions with your own position. In what ways do you agree with Alcock? In what ways do you disagree?
3. Explore the idea that people need some amount of natural, untouched beauty in the land where they live.

Gary Paul Nabhan

An Overture

Author and botanist Gary Paul Nabhan has written extensively about the southwestern portion of Arizona that is the traditional home of the Papago Indians. In his work he studies native agriculture and wild plants, and he is cofounder of Native Seeds/SEARCH, an organization dedicated to such study and conservation.

In his writing Nabhan often reflects upon how people live in and of the land with a harmony evident in their practices. Here he addresses the question of how water relates to desert development, as do many of the writers included in this chapter, and examines the possibilities of living as the earth seems to dictate. His essay is the first chapter of The Desert Smells Like Rain, *published in 1982, the first of his several books. In 1990 Nabhan received a grant from the MacArthur Foundation.*

As you read "An Overture," compare the lives of the Papago with those of residents of desert cities—Tucson or Las Vegas, for example. To what extent is it possible to transplant a culture from one environment to another?

> With many dust storms, with many lightnings, with
> many thunders, with many rainbows, it started to go.
> From within wet mountains, more clouds came out
> and joined it.
>
> —Joseph Pancho, *Mockingbird Speech*

LAST SATURDAY BEFORE DUSK, the summer's 114-degree heat broke to 79 within an hour. A fury of wind whipped up, pelting houses with dust, debris, and gravel. Then a scatter of rain came, as a froth of purplish clouds charged across the skies. As the last of the sun's light dissipated, we could see Baboquivari Peak silhouetted on a red horizon, lightning dancing around its head.

The rains came that night—they changed the world.

Crusty dry since April, the desert floor softened under the rain's dance. Near the rain-pocked surface, hundreds of thousands of wild sprouts of bloodroot amaranth are popping off their seedcoats and diving toward light. Barren places will soon be shrouded in a veil of green.

Desert arroyos are running again, muddy water swirling after a head of suds, dung, and detritus. Where sheetfloods pool, buried animals awake, or new broods hatch. At dawn, dark egg-shaped clouds of flying ants hover over ground, excited in the early morning light.

In newly filled waterholes, spadefoot toads suddenly congregate. The males bellow. They seek out mates, then latch onto them with their special nuptial pads. The females spew out egg masses into the hot murky water. For two nights, the toad ponds are wild with chanting while the Western spadefoot's burnt-peanut-like smell looms thick in the air.

A yellow mud turtle crawls out of the drenched bottom of an old adobe borrow pit where he had been buried through the hot dry spell. He plods a hundred yards over to a floodwater reservoir and dives in. He has no memory of how many days it's been since his last swim, but the pull of the water—*that* is somehow familiar.

This is the time when the Papago Indians of the Sonoran Desert celebrate the coming of the rainy season moons, the *Jujkiabig Mamsad,* and the beginning of a new year.

Fields lying fallow since the harvest of the winter crop are now ready for another planting. If sown within a month after summer solstice, they can produce a crop quick enough for harvest by the Feast of San Francisco, October 4.

When I went by the Madrugada home in Little Tucson on Monday, the family was eagerly talking about planting the flash-flood field again. At the end of June, Julian wasn't even sure if he would plant this year—no rain yet, too hot to prepare the field, and hardly any water left in their *charco* catchment basin.

Now, a fortnight later, the pond is nearly filled up to the brim. Runoff has fed into it through four small washes. Sheetfloods have swept across the field surface. Julian imagines big yellow squash blossoms in his field, just another month or so away. It makes his mouth water.

Once I asked a Papago youngster what the desert smelled like to him. He answered with little hesitation:

"The desert smells like rain."

His reply is a contradiction in the minds of most people. How could the desert smell like rain, when deserts are, by definition, places which lack substantial rainfall?

The boy's response was a sort of Papago shorthand. Hearing Papago can be like tasting a delicious fruit, while sensing that the taste comes from a tree with roots too deep to fathom.

The question had triggered a scent—creosote bushes after a storm—

their aromatic oils released by the rains. His nose remembered being out in the desert, overtaken: *the desert smells like rain.*

Most outsiders are struck by the apparent absence of rain in deserts, feeling that such places lack something vital. Papago, on the other hand, are intrigued by the unpredictability rather than the paucity of rainfall—theirs is a dynamic, lively world, responsive to stormy forces that may come at any time.

A Sonoran Desert village may receive five inches of rain one year and fifteen the next. A single storm may dump an inch and a half in the matter of an hour on one field and entirely skip another a few miles away. Dry spells lasting four months may be broken by a single torrential cloudburst, then resume again for several more months. Unseasonal storms, and droughts during the customary rainy seasons, are frequent enough to reduce patterns to chaos.

The Papago have become so finely tuned to this unpredictability that it shapes the way they speak of rain. It has also ingrained itself deeply in the structure of their language.

Linguist William Pilcher has observed that the Papago discuss events in terms of their probability of occurrence, avoiding any assumption that an event will happen for sure:

> . . . it is my impression that the Papago abhor the idea of making definite statements. I am still in doubt as to how close a rain storm must be before one may properly say *t'o tju:* (It is going to rain on us), rather than *tki'o tju:ks* (something like: It looks like it may be going to rain on us).

Since few Papago are willing to confirm that something will happen until it does, an element of surprise becomes part of almost everything. Nothing is ever really cut and dried. When rains do come, they're a gift, a windfall, a lucky break.

Elderly Papago have explained to me that rain is more than just water. There are different ways that water comes to living things, and what it brings with it affects how things grow.

Remedio Cruz was once explaining to me why he plants the old White Sonora variety of wheat when he does. He had waited for some early January rains to gently moisten his field before he planted.

"That Pap'go wheat—it's good to plant just in January or early February. It grows good on just the *rain*water from the sky. It would not do good with water from the *ground*, so that's why we plant it when those soft winter rains come to take care of it."

In the late 1950s, a Sonoran Desert ecologist tried to simulate the gentle winter rains in an attempt to make the desert bloom. Lloyd Tevis used untreated groundwater from a well, sprayed up through a sprinkler, to encourage wildflower germination on an apparently lifeless patch of desert. While Tevis did trigger germination of one kind of desert wildflower with a little less than two inches of fake rain, none germinated with less than an inch. In general, production of other wildflowers required more than three or four inches of fake rain.

Tevis was then surprised to see what happened when less than an inch of real rain fell on his experimental site in January. He noticed in the previously sparse vegetation "a tremendous emergence of seedlings. Real rain demonstrated an extraordinary superiority over the artificial variety to bring about a high rate of germination." With one particular kind of desert wildflower, seedlings were fifty-six times more numerous after nearly an inch of real rain than they were after the more intense artificial watering.

The stimulating power of rain in the desert is simply more than moisture. Be it the nutrients released in a rainstorm, or the physical force of the water, there are other releasing mechanisms associated with rainwater. But even if someone worked up a better simulation of rain using *fortified* groundwater, would it be very useful in making the desert bloom?

Doubtful. Remedio himself wonders about the value of groundwater pumping for farming, for water is something he *sings* rather than pumps into his field. Every summer, Remedio and a few elderly companions sing to bring the waters from the earth and sky to meet each other. Remedio senses that only with this meeting will his summer beans, corn, and squash grow. A field relying solely on groundwater would not have what it takes. He has heard that well water has some kind of "medicine" (chemical) in it that is no good for crops. In addition, he believes that groundwater pumping as much as twenty miles away adversely affects the availability of moisture to his field.

I joined in a study with other scientists to compare the nutritive value of tepary beans grown in Papago flashflood fields with those grown in modern Anglo-American-style groundwater-irrigated fields nearby. The protein content of the teparies grown in the traditional flashflood environments tended to be higher than that of the same tepary bean varieties grown with water pumped from the ground. Production appeared to be more efficient in the Papago fields—more food energy was gained with less energy in labor and fuel spent. No wonder—it is a way of agriculture that has fine-tuned itself to local conditions over generations.

There they are, Julian and Remedio—growing food in a desert too harsh for most kinds of agriculture—using cues that few of us would ever notice. Their sense of how the desert works comes from decades of day-to-day observations. These perceptions have been filtered through a cultural tradition that has been refined, honed, and handed down over centuries of living in arid places.

If others wish to adapt to the Sonoran Desert's peculiarities, this ancient knowledge can serve as a guide. Yet the best guide will tell you: there are certain things you must learn on your own. The desert is unpredictable, enigmatic. One minute you will be smelling dust. The next, the desert can smell just like rain.

Content Considerations

1. What seems to be the main idea of the essay?
2. Two scientific studies receive considerable attention in this piece. Why? How do they relate to the point of the essay?
3. To what practices does the author attribute the Papagos' success in agriculture?

The Writer's Strategies

1. Nabhan begins this piece with a description of a storm. How does that description prepare you for the rest of the essay?
2. What examples does the author give to illustrate how an "element of surprise becomes a part of almost everything" in the Papago culture? How do these examples help you make meaning of the essay?

Writing Possibilities

1. Using the examples and illustrations Nabhan provides, write about how the Papago culture differs from your own or another culture.
2. While observing an area or relying upon your memory of it, search for the "cues" the area offers about its use and possibilities. Write about them.
3. With a group of classmates, discuss the "things you must learn on your own" about the region where you live. Write a paper focusing on these things.

CHARLIE HAAS

Desert Sojourn

People do, of course, live in deserts. Some desert cities are well-known: Los Angeles, Phoenix, Las Vegas. Hundreds of thousands of people live in these oases created by water transported mile after mile, by electricity produced in power plants built on dammed rivers. But many obscure little towns and cities exist in the desert just as they do across America. Settled because water projects made irrigation possible or because a mine began to produce, these towns and cities still exist, some growing, some dying. Many of them have changed character in their struggle to keep desert forces at bay.

Writer Charlie Haas tours some of them in a road trip that begins in San Diego, California, and ends just west of Tucson, Arizona. He wanders off the highways and back into towns, spends nights in motels and at least one on the desert floor, pondering as he goes just who is in charge of life in the desert.

Before you read the following essay, published in Esquire in 1987, make a list of what you expect the author to describe, what stories you expect him to tell. As you read, keep score: Who is winning, desert or civilization?

THE WEST COAST, as you may have suspected, is a sort of hoax. You live out there, seeing everything built up and domesticated, steak houses and skate rentals right down to the ocean, and you think, *That's it. This is as far west as you can go, and this is tamed, so everything's tamed, from east to unwild west.*

But there's a gap in that reasoning, and that gap is America's southwestern desert, the weird open spaces, where gaps of reasoning come easily to man and nature alike. Take the front page of this morning's *Arizona Daily Star*, with two salient facts above the fold. One, when police came to break up a fight at a Short Stop store in Tucson yesterday, frisky onlookers responded by winging debris at the cops, and the situation soon escalated, requiring twenty-eight police cars, one helicopter, and a number of arrests to restore order. Two, the high temperature in Tucson yesterday was 107 degrees.

The *Star* draws no correlation here, but to your correspondent, squinting at the paper on an Arizona morning, the two stories are as one. Feeling this heat, you can just *see* those gaps arising in the reasoning of the average convenience-store customer, the average cop, or, really, anyone. It's probably just *fine* to build cities in this climate—a sporty idea, really—as long as someone remembers to keep a few of those choppers cued up.

Driving east from San Diego, we're up to four thousand feet before the bottom drops out. Halfway down Montezuma Grade, the green drains from the landscape, replaced by the red-and-brown rock of a gaping canyon, its floor crossed by dirt roads, its mouth opening into flat, hazy vastness.

At the bottom is the Anza-Borrego Desert, California's largest state park, which occupies much of San Diego County and spills into Imperial as well. The roads through the sandy landscape gleam with water mirages, and animals dart in front of the car: jackrabbits; fat chuckwalla lizards and iguanas; roadrunners ridiculously like the one in the Warner Bros. cartoons; and antelope squirrels, who look like two-story chipmunks, doubling their white tails over their bodies to reflect the heat.

The desert is beautiful but not pretty. In the flashbulb sunlight of morning, there are no clouds or shadows, and there's nothing graceful, like the smooth-hilled forests of most big parks, or rectilinear, like city skylines: everything is jagged. The Laguna Mountains crack the distant sky, and the sand at our feet is a coarse rummage of alluvial rock. The ocotillo plant, with its bare, angling spindles, looks like lightning trying to jump from the ground.

Pat Flanagan, a biologist who offers "rent-a-naturalist" tours of the desert, loves it here. She will cheerfully tell you all the weird-wildlife facts you want (agave century plants are pollinated by bats, which may be a start toward explaining tequila; the loggerhead shrike, a predatory songbird, impales snakes to death on its hooked bill, then resumes its sweet warbling; the kangaroo rat, which doesn't drink water, gets moisture from dry seeds and has a highly evolved kidney to produce urine up to five times as thick as ours), and she can show you thousand-year-old Indian petroglyphs and three-million-year-old fossilized camel footprints. But her greater skill is in teaching people *how* to look at the desert, where much of the life and color is low to the ground, camouflaged, or coded into the sand and stone.

"Winter is the best time to come," Flanagan says, as we set off on an exploratory drive. "The light is the most special then, and the weather is perfect." Today, in June, the air has a power-station crackle—beetles buzzing, hummingbirds dive-bombing, four big ravens flying in a spooky straight line over a stagecoach road that connected St. Louis to San

Francisco in the 1850s. In a smoke tree, clouds of bees float over deep-blue flowers and brilliant-orange dye glands.

Flanagan gives us small, plastic, hand-held magnifying lenses. In brush that looks like a gray blur from the road, the lenses find lush pink pollen baskets or blossoms like tiny artichokes covered with purple fur. We look closely at these tiny amazements and then stare up at the big amazement of the empty sky and western-movie peaks, and there's nothing between the big and little wonders but the heat, an amazement in itself: you don't walk in it, you wade.

People here keep reminding you that the heat is dry, which sounds like a plus until you encounter the Delayed Sweat Factor. On a roasting late morning, we climb up Palm Canyon, hoping to see the bighorn sheep that come to water at a palm oasis at the top of the trail. We gasp as we walk, and drink water like crazy, but we don't seem to be sweating even a little. The sheep show their endangered, Edward R. Murrow-ish faces, we hike back down, and my wife and I head for our hotel, La Casa del Zorro, to check in. Our room isn't ready, so we decide to have lunch, and as soon as we sit down in the mildly formal dining room, *whoosh*—the sweat starts showering out of us like thousands of tiny time-release pills that go to work when you need them least. We've been sweating all along, of course, but the dry heat has been instantly evaporating it, and now we're indoors and it's coming out in torrents, as nice families sit under authentic stagecoach paintings and eat artfully arranged chicken salads, and the nervously smiling people who work here seem to be trying to decide whether to pour the iced tea into our glasses or over our heads.

As it turns out, a resort such as La Casa del Zorro is a good place to hole up in the desert afternoon—soft drinks half-frozen to slush in the vending machines, and a congenial pool scene where people loll in the water, drinking and talking a lot, not swimming much. A comfortable room, with calming southwestern decor and a shelf of pop potboilers to read, goes for a cut summer rate that would get you a poor motel room in L.A. Of course, when you go outside and the heat hits you, you laugh out loud and go straight into the pool or back inside, where the all-weather cable-TV channel reports a high since midnight of 110 degrees.

On another night we sleep out in the desert, in a flat spot where the mountains are miles out in all directions. Soon after sunset, a full moon comes roaring up, as brighter-than-life as the sun is here. I wake in the middle of the night, sit up in that flat dish of moonlight, and realize that what woke me is a distant coyote—not the expected howl, but an intricate song in a weird voice, half dog, half rooster. As I stumble around in the morning, a ball of jumping cholla cactus attaches its several spines to my arm. My Swiss army knife gets them out okay, though only a desert-style gap in reasoning could have caused me to choose the wood-saw blade as the surgical instrument. I leave the Anza-Borrego with a new respect for our colorful friend, the viciously underhanded international criminal, cholla cactus.

Respect is the only choice, because nature is in charge here. The desert, with its white earth, ghost trees, and space-case wildlife, has an appeal, but the appeal has an edge—the feeling that people have made only a dent here and that the critters could have the place back any time they decide the dent has gotten ugly enough, which could be now—as in the local story of the condo development where the people have pollinated the date palms by hand, only to have coyotes come and wait under the trees for the fruits to fall off into their jaws. If nothing else, a few days out here, surrounded by nature's freak adaptation and predatory blitzkrieg, give you a new perspective on the desert city of Las Vegas—a notion that all that crazed architecture, sexual display, desperate gambling, and Mario Puzo behavior may only be an attempt by human beings to *keep up*.

"You're not hunting all the time. You regulate your body temperature by moving in and out of the sun. You can lead a *slow* life."

That's Pat Flanagan talking about desert reptiles, but with any luck at all she could mean you. When you're wound up enough to need relaxation forced on you, the desert can be just the thing. Especially the driving. On a long desert drive—one with stops for sodas at every gas station, and radio stations where the records are C & W but the deejay patter is Navaho—the featureless landscape can turn your mind loose into cowboy-picture or long-haul-trucking fantasies in just a few elastic hours.

At a place like Salton City, east of Borrego, you can participate in a public fantasy as well. In the early 1900s, an irrigation project bringing Colorado River water to the Imperial Valley got out of control, and the saltwater Salton Sea was formed more than two hundred feet below sea level. Wonderfully, the local people decided that, now that they had a sea, they would have a beach town, complete with streets called Sea View and Salt Air, in the middle of the desert. The sparser sections of town—isolated cracker-box houses on salt-white ground, with the odd bare paloverde tree or cactus garden—do a wonderful impression of Laguna Beach on the morning after nuclear war.

But just down the road is a marina, where people bring motor-boats hitched to RVs and scoot around on the flat, pungent water. At a friendly store/restaurant, fishing tackle and live mudsuckers are for sale alongside a bulletin board of Polaroids showing people with their catches of the hardy corbina and leaflets discounting the purported selenium danger. A more densely populated street nearby is under the bracing spell of what-the-hell retirement-zone decor—wing-spinning wooden ducks, observation decks atop mobile homes, and a restaurant called the Tiki, with wooden gods in the sandlot and thatch over the counter.

If Salton City doesn't thrill you with humanity's indomitable goofiness, something is seriously wrong, but you do get another chance. Up the highway, in date-growing Oasis and Indio, you plunge straight from the beach fantasy into a reverie of North Africa—gorgeous old signs with Rudolph Valentino lettering advertise such local attractions as El Morocco Motel, under the abutting sunbursts of the palms' high fronds. The Indio Date Shop has delicious fresh-date milk shakes and postcards of the Palm Springs homes of Red Skelton, Lucille Ball, and Frank Sinatra—a perfect send-off for the drive east into Arizona.

Part of the desert's odd fun is the juxtaposition of the awesome scenery with a human culture of deep corniness. In the tourist-strip town of Quartzsite on I-10, a sun-bleached bar is surrounded by nautical pilings and called the Quartzsite Yacht Club. In Tucson, where the waitresses patrol with iced-tea pitchers instead of coffeepots, the souvenir hats say ARIZONA—HOTTER THAN HELL. People slip their jokes into conversation in flat monotones: at a gas station, I hum along with a radio blasting Thunderclap Newman's "Something in the Air," and the guy who takes my money says, in a pinched southwestern twang, "Okay, that's five-twenny-seven out of six don't sing along with that you'll show your age senny-three's your change have a nice one." It may be that, when it's hot enough to make your eyeballs and eardrums hurt, timing takes too much time—the fewer pauses for effect, the faster you're back in the shade.

Off Highway 19, just outside Tucson, is the 190-year-old Mission San Xavier del Bac, an ornate amalgam of terra-cotta friezes, iron balustrades, and chalk-white walls surrounded by scrubby flatness and subsistence living. The inner walls of the church are covered with harsh, primitive religious paintings—the standard Christian subject matter and poses, but the style a brutal world away from the Renaissance. In an alcove near the front, the reclining figure of St. Francis Xavier lies under a pale green blanket to which people pin photographs, First Communion pins, and hospital ID bracelets. Taped hymns play on the PA as an old Indian woman shuffles past the saint, rattling a prayer under her breath.

Outside in the sun, some Indian women and little girls have an Indian Fry Bread stand made of tree limbs and very old plywood; they pat out flats of dough and fry them in a kettle of oil over a wooden fire that also heats the beefy red chili and bean fillings (best meal of the trip, three dollars). In places like the Mission, you feel that you've arrived in the Southwest you carry in your head—a free-fire corridor of Catholic converts, pagan holdouts, Native Americans, Mexican-Americans, crabby retired Americans, UFO buffs, missionaries, hermits, the possessed, the dispossessed, and the permanently cooled out. It's an exciting feeling, though part of the excitement takes the form of wanting to get *right* back in the car.

In Bisbee, a former copper-mining town in the hills of Tucson, the air is cooler, the landscape a little softer. It looks like the town in the westerns: old pioneer-style houses and solid civic buildings, whitewash and brick, under cooling trees. This is one of those western places where history, tourism, and a quiet hippie insurgence run together—the art galleries, "art" galleries, and pleasant bed-and-breakfasts sit beside the Cochise Private Industry Council storefront and the county courthouse.

In the afternoon, gold light gleams on the houses that climb the narrow switchback streets up the hill, and Bisbee seems like the capital of the lizard-slow life. A thirtyish couple sit on the porch of their place at the foot of Quality Hill, watching their two dogs run around the yard. He's skinny, with a beard and a baseball cap that says BEING SEXY IS A HARD JOB BUT SOMEBODY'S GOT TO DO IT; she's in a blue terry top and shorts, smoking cigarettes and poking her toes in the dirt.

"Everybody's easygoing here," he says. "Nobody's in a hurry. You just do what you have to do and get it done when you can. Your biggest decision might be which one of four or five bars to go to at night."

"There's no McDonald's or anything," she says. "That's what I love about it: the total lack of convenience."

"When you come through that tunnel into town," he says, "that's a time warp. It's still the Sixties here. A lot of hippies."

"But, now, the hippies can be fine people," she says. "Some of them are the finest people I've met."

Wait, I thought *these* were the hippies. The hippies must be *very* relaxed.

As relaxed, perhaps, as the grinning gentleman who floats down the street toward me as I leave their house—his feet about four inches off the ground, his T-shirt green enough to pass for lizard skin, with printing on it reading JUST SAY NO.

At dawn on our last day here, I drive out to a section of the Saguaro National Monument just west of Tucson. The huge saguaro cacti tower over mesquite, creosote bush, and paloverde. From the first short climb of the Hugh Norris Trail, the view is amazing: hundreds of green-skinned cacti receding into the valley, like a dinosaur-world version of a redwood forest.

The trail follows the ridge for eleven miles, but that's for another season. Some fall or winter when it's rainy at home, we'll come back to the last, best hotel we stayed in, the Arizona Inn—a splendid 1930s resort with beautiful high-ceilinged reading and dining rooms and a courtyard ramada shading cushioned wooden chairs. We'll hike the whole trail and further investigate Tucson's famous Mexican food (the green corn tamales and mammoth guacamole cheese crisps at El Terrero were a valuable start in this important area) and try to find out why a local bar is called DIRTBAG'S—A PART OF GROWING UP.

I think we'll have a great time, and I think you would, too. But I'd be remiss if I didn't tell you that one of the best parts of a desert visit is coming home, like all those people who stuck it out on the westward stagecoach looking around at the gentle landscape overstocked with green, and saying, "Oh. Yeah. Let us settle *here*."

Content Considerations

1. What does the author mean when he writes that the biologist Pat Flanagan teaches people *how* to look at the desert?
2. Summarize the ways temperature affects human behavior, according to the author.
3. How does nature remain in charge despite human attempts to develop the desert, according to Haas?

The Writer's Strategies

1. What purposes does humor serve in this piece?
2. How do you think the author actually feels about the desert? In what ways does he reveal this?

Writing Possibilities

1. With several classmates, discuss the essential character of the land and people in your region. Together or individually, write an article that presents this region to others.
2. Summarize the author's conclusions about desert life. In what ways do you agree or disagree with these conclusions?

JOHN NICHOLS

Meeting on the Mesa

Author John Nichols published his first novel, The Sterile Cuckoo, *in 1965. Since then he has published several works of nonfiction as well as novels, including the New Mexico trilogy—*The Milagro Beanfield War, The Magic Journey, *and* The Nirvana Blues.

In this essay from On the Mesa *Nichols tells of a group of farmers and ranchers who meet to discuss their problems with the land and government policies. Central to this meeting is the idea that government policies on how the land is used and how development rights are granted have changed the character of the land where the men live and work. The men remember better times when a less developed land was used in ways consistent with its character.*

Before you read the next few pages, write brief sketches of the kinds of people you expect to meet in Nichols's essay. As you read, try to connect descriptions and qualities with the sketches you wrote.

LOS CORDOVAS is a small, largely agricultural settlement five miles southwest of Taos. Near the heart of the still-intact old plaza is a community center. Toward it, on a bright Sunday afternoon, I steer my truck, bounding uncomfortably over the potholed dirt road. Beside me is Dr. Paul Sears, a noted author, professor, and conservationist who must be in his nineties. Between his legs he holds a cane; on his lap rests a dated cross-section of a sagebrush trunk.

In the community center parking lot eight men are waiting. I help Dr. Sears out and shake hands with the men. Most are cattle or sheep ranchers who graze their animals on the mesa. After the amenities, we all enter the building and get down to business.

Andres Martínez chairs the gathering. A former itinerant sheep shearer, Navajoland storekeeper, and Taos dairyman, Andres is now retired, but he keeps his octogenarian's body and nimble brain active in local affairs.

Another octogenarian is Bernabé Chavez, who grew up over in Carson, studied at the old stone schoolhouse, and in his youth tended sheep all around Tres Orejas.

Other weathered participants are in their fifties and sixties. One of them, Pacomio Mondragón runs over seven hundred sheep on the mesa; several times I have visited his spring lambing camps just south of Tres Orejas.

Manuel Martínez keeps his sheep on the mesa east of Tres Orejas and south of the Carson post office in the Petaca Arroyo.

Delfino Valerio, a Ranchos de Taos cattleman, has corrals and several stock ponds on the mesa.

These men, and a few others I know less personally, are concerned about their future on the mesa. Most have some reservations about how the Forest Service and the BLM govern their lives west of the gorge. About wild river sections or transmission lines, they know little. What they do understand is that each year it's more difficult for them to eke out a living in that arid country.

Mostly, the talk is in Spanish; I scribble notes, pretending to understand everything (when in reality I only catch about every fifth word). Andres and I will carry their observations back to our organization. No high passions or noisy rantings disrupt the meeting. These patient men keep their anger and frustrations under control. Yet the calm concern in their voices is sad to hear. Everybody remembers a time when things were better. Especially over fifty years ago when the mesa was a vast grassland the old timers love to recall. Their memories are not in default: according to Dr. Sears, his dated sagebrush trunk proves that the plateau was once rich in forage grasses.

"We had thousands of animals living on that grass," one old man says quietly. "Then came the Floresta and the BLM. They said the land was overgrazed; they made us get off. That's when the sagebrush took over, when they started to 'manage' the land."

Another man backs him up. "It's like they do with the forest. I remember when the first planes came and sprayed the trees with insecticide near my ranch. My cows got sick. In one year I lost eleven head. When I opened them up their blood was clotted, like plasma; it was just like glue. But those insecticides didn't do any good. Look at the forest today—from Mora to Santa Fe the trees are dying of budworm. They control the forest so much, they stop all the fires, they eliminate most of the grazing until—heck—they manage the health right out of the forest. Today the forest is so inbred that the trees and all the animals have lost their hardiness. They have become diseased and weak and sickly."

Starting in the twenties, they remember, the government began reducing permit allotments almost yearly, diminishing the herds around Taos. Until recently such policies were vigorously applied to Taos County, and today few animals remain on government land. And the surviving ranchers are afraid the Reagan Congress will jack up the grazing fees, effectively eliminating small operators like themselves from leasing public land.

According to one old timer: "Right now, whether you're rich or poor, whether you're big or small, you pay the same grazing fee. That's discriminatory. Whether the quality of grazing is good or bad, you pay the same. And that hurts people like us, in an area where the grazing is not too good anyway. In the old days, at least grazing fees went up and down according to the quality of pasturage and the price of wool and lambs. But nowadays the fees arbitrarily seem to go higher and higher. They don't want us over there anymore. They just want us to close up shop and go home." 13

Another problem they discuss involves Floresta policies requiring sheep permitees to move their camps every few days, even in midwinter when snow covers the ground and little grazing damage can occur. 14

"Each time I have to move my camp it hurts my animals," says a plump rancher. "They get warm in one place, then have to move and bed down in cold snow. I'm not allowed to build a corral to protect the sheep. I'm trucking in hay, but they have to move anyway. I could plow off snow in each new area, but the Floresta won't let me bring in a tractor. They say they're trying to protect the mesa. But what about protecting *me?*" 15

By contrast, apparently the government will concede almost anything to the electric company, no matter how destructive to grazing, wildlife, and the mesa. And of course a big timber company may practically level entire forests, and a mine can lop off the tops of mountains (as Molycorp has done in Questa) with hardly a question of raising the government's ire. 16

"El dinero habla," laughs one old geezer. "Money talks." 17

Our talk shifts to another subject. While tourist roads along the Wild River are paved and dotted with modern camping areas that include bathrooms, cement barbecue pits, and metal picnic cabanas, roads to mesa lambing camps seem to be deliberately neglected. A recreational vehicle can cruise effortlessly to most any scenic overlook of the Rio Grande Gorge, but the tank trucks of sheep people must navigate along nearly impassible dirt arteries two or three times weekly, bringing water to their animals. 18

"Or how about this?" a rancher asks. "If the Government is so friendly to the electric company, and to the gas exploration people, why can't it drill at least a few water wells in the lambing areas? Then we wouldn't have to wreck our trucks on those terrible roads." 19

A senior statesman rises: "Years ago we had a well between Ojo Caliente and Carson. Thirty or forty families survived off that well. I 20 myself hauled water for nine hundred sheep. We didn't have any pumps, we did it all by hand. We broke our backs filling the barrels on the wagons. There would always be seven or eight wagons lined up all day long. Those wells are caved in now. But why can't the government redrill them, and make the water available to the ranchers?"

For two hours the dialogue continues. I watch the faces. Chastisement of the government is meted out with irony and a sense of humour. At heart, most of the men are fatalistic. None carry an illusion that their way of life will survive much beyond themselves. When they die, most likely it will be over.

Several of the men are twice my age: I have known them for fifteen years. I hate to consider a Taos Valley (or my own life) without them. I think they stand for a strong history, a viable culture, and a healthy balance we can ill afford to lose around here (around anywhere).

And if their perceptions of life are reduced to quaint myth and nostalgic memory, this valley may succumb at last to the kind of cynical rootlessness that defines and cripples so much of America.

Meeting over, we saunter outside and mill around under a cloudless sky. The sun brilliantly bakes the parking lot. We all tarry, leaning against fenders, laughing, liking each other, making small talk. Sunshine permeates right to the bone; within moments we are laved in a golden glow. Glancing around, I notice how much everyone appreciates the warmth, especially our oldest fogies, who keep talking, reluctant to leave, tenaciously hanging on to the moment.

"You know something?" says one of the octogenarians, "I just can't get enough of this damn sunshine. When I die, I wish they could fill up my grave with sunshine."

Content Considerations

1. What is at issue at the meeting Nichols writes about?
2. In what ways does Dr. Paul Sears contribute to the meeting?
3. According to the elders at the meeting, what events or conditions led to the deterioration of the plateau from a rich grassland to a scrubby mesa?

The Writer's Strategies

1. The structure of this essay begins with background, leads the reader to the meeting, outlines the issues, then presents a series of comments made by participants before moving to a general reflection of the issues. What do the participants' comments add to the essay? Why?

2. In what ways does the author reveal his attitudes toward the men at the meeting, toward the government's policies, and toward tradition?

Writing Possibilities

1. Explore Nichols's notion of "the kind of cynical rootlessness that defines and cripples so much of America" (paragraph 23). What does he mean? In writing, support or attack his notion. Use your own examples and illustrations.
2. Attend a community or school meeting or recall one you have attended. Write an account of the meeting following Nichols's pattern of organization.
3. Working with a classmate, construct a biography of one of the meeting participants. You may use information the author provides, or you may create other information.

PETER STEINHART

The Water Profiteers

Water is a requirement not only for initial development but for continuing development. As cities grow, as cities first begin, more water must be found and made available. But it looks as if no new water sources exist; thus, water problems center on how to allocate what water there is. For one place to receive water, apparently another must lose it—a deal that seems not only unfair but almost criminal to many people in many places.

But others see in the reallocation of water perfectly fair entrepreneurial opportunities, opportunities for water ranching, speculating on water rights, hoarding those rights. The problem is political as well as economic: Cities and counties often buy rights in distant areas to ensure their own continued development. That practice means buying ranches and farms that are then put out of production, thus decreasing the chances for development where those farms and ranches are located.

In the following article, reprinted from Audubon, *writer Peter Steinhart begins with a scene in the San Luis Valley in the south-central portion of Colorado, but discusses other areas wrestling with the same problems. Before you read, write a paragraph or two about the ways in which development depends upon water.*

COLORADO'S SAN LUIS VALLEY is almost as flat as a pan, eighty miles wide and a hundred long. The Sangre de Cristo Mountains rise steeply from the eastern edge of the valley, and the San Juans form its western wall. The Rio Grande trickles out of the San Juans, meanders past Alamosa, and wanders through the sage and sand and rabbit brush into New Mexico. Clouds form over the mountain rims every summer afternoon, but the sky directly over the valley remains as blue as an oyster's dream. It rains only seven inches a year on the valley floor, a pluvial stinginess that qualifies the place as a desert. But the mountains ringing the valley have for millions of years collected rainfall and snowmelt which percolate into the ground and run down into the valley's deep alluvial gravels. Here and there it bubbles up on the valley floor in springs and artesian wells. Farmers have sunk 3,000 wells and drawn from the earth prosperous grain and potato farms. There are 1,800 large pivot-irrigation systems in the valley, one of the densest concentrations in the world. Under the San Luis Valley may be two billion acre-feet of groundwater, an aquifer as rich and wet as the Ogallala Aquifer which for a century has irrigated Nebraska, Kansas, and Oklahoma.

Coloradans have long fought over such water. In dry years wells and irrigation ditches dry up; a farmer who sees his crop die in the field may deepen his well or siphon more water than he is entitled to from the ditch, and his neighbor may shoot him or call the sheriff. In the 1950s the Park County water commissioner was found dead at the bottom of a well, and for the next four years no one could be persuaded to serve as water commissioner. In the early 1980s a speculator named John Houston laid claim to all of Colorado's groundwater, and was stopped in his scheme only by a state supreme court ruling. About the same time a Texas company tried to buy up land in the San Luis Valley, planning to export the water to Texas.

On December 31, 1986, American Water Development Incorporated (AWDI), the new owners of the 155,000-acre Baca Ranch in the San Luis Valley, applied for rights to drill ninety-seven new wells and pump as much as 200,000 acre-feet of water a year. (An acre-foot is the amount of water it would take to flood an acre of land a foot deep—roughly the amount of water in a municipal swimming pool.) That water, quite clearly, would not be used to irrigate lands in the San Luis Valley. It would be exported over Poncha Pass to support future urban growth in the Front Range cities of Denver and Colorado Springs.

The idea of such a development defies a deeply ingrained belief that water, like air, is everybody's birthright—that it is impulsive but free and dependable. Despite the reality that water and profit are the same word in the West, despite the fact that everybody pays for water, our culture grew up in wetter places; in our minds, our water and someone else's money don't mix. Says Michael Entz, who grows wheat, barley, and potatoes on seven pivot-irrigation circles near the Baca Ranch, "You have somebody coming here to export your resource and try to turn a big profit. It's yours and somebody is trying to take it from you."

Mike Entz's father and grandfather farmed the valley before him. In a good year the groundwater he pumps from wells accounts for about thirty percent of his costs. Over the years the water table in his area has dropped from five feet below the surface of the ground to eighty feet below, and each additional foot his pumps must lift the water costs more. The deeper the water table drops, the more likely the water will contain salts and minerals that are toxic to crops. And there are two aquifers, a shallow one and a deeper one, separated by a layer of blue clay. The deeper one is believed to feed the Rio Grande. When wells draw down the lower aquifer, the State of Colorado restricts pumping in the San Luis Valley to meet its obligations under the Rio Grande

Compact, an interstate agreement to deliver water through the Rio Grande to New Mexico and Texas.

Because the state engineer, who is the director of the Colorado Division of Water Resources, believes all the water is allotted, farmers cannot get permission to deepen their wells. So, if AWDI exports water from the valley, the farmers fear they will get poorer-quality water, pay more for it, and possibly be left without water at all. A study done at the request of the San Luis Valley Water Conservancy District estimated that the cost of water from deeper wells would be $50 to $80 an acre-foot in the shallow aquifer, $210 to $370 an acre-foot in the deeper aquifer, but concluded: "Realistically, the highest cost a valley farmer can pay is about twenty dollars an acre-foot." Says a farmer in the H & R Supply Store in Center, "It will put us out of business."

It could also have dire consequences for the natural environment. That water once fostered broad wetlands and waterfowl populations. Today there are two national wildlife refuges, four Bureau of Land Management wildlife areas, and a handful of state wildlife areas in the valley, and all of them might be affected by the AWDI withdrawals. Mel Nail, former manager of the Alamosa and Monte Vista national wildlife refuges, fears that AWDI withdrawals will leave less water in the Rio Grande, and so reduce the amount of water available to the refuges. Says Nail, "I'm sure it will just shut off the water to BLM's Blanca Wildlife Habitat Management Area."

AWDI didn't file for rights to the water until the Bureau of Reclamation had condemned 38,000 acres of the Baca Ranch for wells and pipelines under the Closed Basin Project, a scheme which pumps shallow aquifer water into the Rio Grande to meet Colorado's interstate compact obligations. BuRec claims 80 percent of the water is salvaged from plants which suck it from the shallow water table and transpire it into the air. By pumping the top two feet off the aquifer, the bureau expected to meet the needs of farmers, the compact, and wildlife. But once the bureau began to pump Baca Ranch groundwater, AWDI feared others would invoke the western water doctrine that water not being put to beneficial use can be appropriated, and make claims on its water. So AWDI filed for its own rights. At first AWDI hoped to use the water in the valley to grow barley and perhaps support a brewery. But that proved uneconomical. "The only economical way to get rid of this water is to sell it," says Steve Vandiver, the district engineer. And having been freshly disappointed in its desires to build Two Forks Dam, the Denver metropolitan area "will have to look at other places for a water supply," Vandiver says. "They're not shy in seeking water supplies wherever they can get them."

The San Luis Valley is not a wealthy place. Says Vandiver, "As a whole, the San Luis Valley is starving to death." Its southern counties are among the poorest in the nation. "There isn't anything to keep people here but agriculture, and agriculture is on the ropes. The kids are going to Denver and becoming computer operators." In the face of the valley's poverty, the rhetoric of AWDI has not been reassuring.

Baca Ranch general manager Buddy Whitlock, says, "If the farmers are using 1.2 million acre-feet of water and this valley is still poor, that's not really the maximum beneficial use of the water." He tells of a local banker who confided to him, "It simply does not make any sense to put three acre-feet of water that's worth two thousand to seven thousand dollars an acre-foot on an acre of pasture to grow one hundred fifty pounds of beef."

It might be easier to envision this as a war between open-handed farmers in the valley and vault-paled investors from Wall Street. But in this newly urbanizing West, it gets harder and harder to tell who the good guys are. AWDI is a consortium of eastern and western corporate investors. But the board of directors includes Richard Lamm, formerly the governor of Colorado, and William Ruckelshaus, formerly administrator of the Environmental Protection Agency. The chairman of the board, until recently, was Maurice Strong, formerly head of the United Nations Environment Programme.

For his part, Strong declared that AWDI would work to "guarantee that nobody is damaged, within the limits of our supply." If farmers had to deepen their wells, he said, "we would compensate them for additional costs." He said that "the wetlands must be protected." He declared, "If we cannot pass the environmental tests and protect the valley interests, I would not want to continue with the project."

But the thirst for water in the West is insistent and growing. "If we walked away tomorrow," Strong conceded, "there would be ten others lining up to do it." Valley farmers do not trust AWDI. They and environmental agencies are fighting AWDI's application for water rights. Currently challenging AWDI are the Rio Grande Water Conservation District, the Colorado state engineer, the Colorado Division of Wildlife, the Bureau of Reclamation, the Bureau of Land Management, the U.S. Fish and Wildlife Service, the National Park Service, and a host of water districts and municipalities. Altogether more than thirty groups have gone to court against AWDI.

What is going on in the San Luis Valley is going on here and there all over the arid Southwest, where a growing urban population has arrived after farmers plugged up all the good dam sites and after the federal government stopped cheerfully building irrigation projects to subsidize farmers. In the West, water distribution is based upon appropriative rights. The first one using the water gets it, and his claims are defended by the courts from subsequent claims by later arrivals. The oldest rights are senior in times of drought. Rarely is anything left in the streambed for fish or wildlife. Since midcentury, the cities have more than doubled in population, and they are expected to add nearly fifty percent to their current numbers by the year 2020. But there is no water lying around for easy appropriation.

"The era of finding a new water hole is largely gone," says former Arizona Governor Bruce Babbitt. "There's no federal money. The Department of the Interior is brain dead. And as a result of that we will have the opportunity to take the future in our hands." 15

We have entered what Santa Fe water-policy consultant Steve Shupe calls "an era of reallocation." Marketing water that has already been allocated is increasingly being viewed as a substitute for new government-subsidized dams. Irrigation districts have long traded water among themselves. In California's San Joaquin Valley, for example, the 9,500-acre Broadview Water District recycles agricultural drainwater and sells some of the fresh-water it buys from the Bureau of Reclamation to neighboring irrigation districts. What is new in this age of reallocation is that urban centers are vying with agricultural areas for the water. And since urban users are willing to pay much more for water than farmers can pay, the cities are winning. 16

Arizona uses twice the water it receives in surface runoff; groundwater is the chief water supply for the state and it is mined at rates that promise ultimate exhaustion. In 1980 Arizona forbade additional groundwater withdrawals around the cities of Phoenix and Tucson and required them to have a hundred-year supply of water for any further development. The intention was to stop overdrafting the groundwater. The solution was the free market. Says Bruce Babbitt, who as governor presided over the drafting of the law, "We deliberately drafted a provision that said groundwater will be freely transferable. We said, 'Let the market reign.' As a result, we triggered a rush to the Owens Valley. There were guys in black hats from Phoenix and Tucson roaming the backwoods of the state with open checkbooks." Phoenix, Tucson, Mesa, and Scottsdale acquired more than 50,000 acres of farms with the intention of retiring the fields, pumping the groundwater into the Central Arizona Project aqueduct, and delivering it to household taps nearly one hundred miles away. 17

Along Colorado's Front Range, municipalities have been buying up ditch companies and shares of stock in water supplies. The City of Aurora, a suburb of Denver, purchased a major irrigation ditch east of Pueblo, along the Arkansas River, and transferred the water to its municipal lines. Thornton, Colorado, secretly spent $52 million buying up 18 options and title to 20,000 irrigated acres in the Cache la Poudre River Basin.

In Utah, fast-growing Salt Lake City suburbs like Sandy City, South Jordan, West Valley City, and Bountiful came late to the trough, and are now seeking to buy older water rights. With their thirst in mind, the Salt Lake County Water Conservancy District is negotiating the purchase of rights to 40,000 acre-feet of water per year, water that now goes to farms. St. George, in southern Utah, is buying up irrigation rights to slake the thirst of a growing retirement community. Utah cities have acquired rights to more than 100,000 acre-feet of water by buying up shares of canal and ditch companies.

Around Reno, Nevada, municipalities and the local electric utility are buying up water rights to secure future development. Las Vegas has a standing offer to purchase water rights for $1,000 per acre-foot. The water agency for Castaic Lake, a rapidly urbanizing area north of Los Angeles, purchased land at Devils Den in the San Joaquin Valley and plans to transfer its water.

With the avid interest of municipalities, private brokers have gotten into the act. More than a decade ago a major insurance company bought shares of Colorado irrigation water and resold them for profit to Front Range cities. More recently, two Prudential-Bache investment funds bought more than $42 million worth of western water rights, intending to hold onto them for up to fifteen years before selling. William McLemore, a Colorado water broker, says he has put together more than $100 million in water deals in the past five years. Says Walraven Ketellapper of Stillwater Resources, a Denver investment firm, "I know one speculator who is holding 100,000 acre-feet, intending to sell it to the Denver area." Oil companies which purchased water rights in the 1970s for oil-shale developments are now holding onto those rights in the expectation they will increase in value. Ketellapper believes oil companies hold rights to as much as 400,000 acre-feet in Colorado alone. Says Steve Shupe, "There are a number of entrepreneurs knocking on the door of San Diego every week with a scheme to take water from Colorado and pipe it to San Diego." A Phoenix-area consortium, Agricom, purchased more than 30,000 acres in La Paz County, nearly a hundred miles away, expecting to use more than 100,000 acre-feet of groundwater a year in Phoenix land developments. Lincoln Savings, the Arizona savings and loan company run by Charles Keating, purchased 13,000 acres of farmland in La Paz County.

Increasingly, water is finding new owners. A study by researchers at the University of Arizona found that in the last twelve years there were about 6,000 transactions in Utah, 1,455 in New Mexico, 1,500 in Colorado. A 1985 study showed that half the available blocks of more than ten acre-feet of water around Reno were held by developers and half by owners waiting for higher bids. In Reno private brokers outbid the public agencies, repackage the rights, and sell them to developers for three times more than the agencies will pay. In Park City, Utah, rights were bought by a resort developer for $10,000 an acre-foot. With prices like that, farmers often inflate the value of their own water rights. Says Ketellapper, "In Colorado you might ask a farmer, 'What do you want for your water?' and he'll say twenty-five hundred dollars an acre-foot. You might then say, 'What do you want for your ranch?' and he might say half of that."

The trend makes farming communities nervous. They remember what happened to California's Owens Valley after the City of Los Angeles bought up its water rights. Los Angeles bought up to 75 percent of the agricultural lands and most of the urban real estate in the valley. In 1920 there were 140,000 irrigated acres on 521 farms in the valley. By 1950 only 30,000 acres remained, and the total declined to less than 5,000 during drought years. Today there are only a handful of commercial farms. Los Angeles and state and federal agencies own over 99 percent of Inyo County. "Because Los Angeles owns all the land in the valley, there's no opportunity for growth," says Greg James, Inyo County Water Department director. Inyo County and Los Angeles have sued each other back and forth, and they are still suing one another to determine who controls the valley's groundwater.

Residents of La Paz County, Arizona, fear they may be headed toward the same fate. Ninety-five percent of the land in La Paz County is federally owned, and therefore off the tax rolls. Half the remaining private lands have been bought by Arizona municipalities and speculators, "and that fifty percent of the land represents one hundred percent of the groundwater in the county," says Gene Fisher, chairman of the La Paz County Board of Supervisors. Under Arizona law, municipally held lands are exempt from taxes, so the county's tax base has shrunk and its ability to sell bonds has declined. Says Fisher, "The perception is out there that La Paz County is not a place where people would like to move and live and start a new industry."

Forty miles east of Pueblo, Colorado, 50,000 acres of irrigated Arkansas Valley farmland has already been retired by cities. Fifteen percent of the water formerly used in the valley has been moved to the cities. "That is wrecking the economy of the valley," says Frank Milenski, a seventy-six-year-old La Junta farmer. Milenski's family has grown alfalfa, onions, cantaloupes, corn, tomatoes, watermelons, peppers, and beans since the 1930s. Recently he has seen neighbors move off 4,000 acres of farmlands bought by the City of Aurora. Car dealerships and farm equipment stores have closed down. Neighboring

Crowley County begged for a new maximum-security prison to compensate for the loss of water. "They keep wanting to gather it up and gather it up," says Milenski, "and the rural areas don't have much representation anymore. It's an undesirable situation."

Chuck Howe, a University of Colorado economist, has studied the effects of water transfers on the Arkansas Valley. He concludes, "What we're getting is a phasing out of marginal lands, lands whose agriculture production has been low." He believes it is the lands that grow less-profitable crops, like alfalfa, pasture, and small grains, that go out of production. If a farm that grows cantaloupes or onions or tomatoes goes out, a neighbor is likely to see a market opportunity opening and switch from alfalfa to the more profitable crops. "Preliminary runs indicate the economic impact of the phaseout of 50,000 acres is really minimal," says Howe. He says the economy is changing anyway. "The area has been under strain for decades."

There were once at least fifteen sugar mills in the valley, but all had closed before the water transfers. "Young people have been moving out for thirty or forty years," says Howe. "You have an extraordinarily large percentage of welfare payments and retirement income in the valley." Howe guesses that as much as 20,000 acres more could go out of production, most of that now in pasture and small grains that support the valley's livestock industry. That would make the livestock operations "noncompetitive" and probably eliminate a quarter of the agricultural employment in the valley. Even then, says Howe, "the impacts on the state economy are pretty trivial."

"Sure the Arkansas Valley doesn't make a pimple on the whole damn business," says Milenski. "But it's a way of life and a damn good one. You look at Denver: Them bastards all get in a pile up there. You got smog all the time. I don't know why they couldn't bring the industry where the water is. It makes more sense to me."

Environmentalists and city water users have long pitted themselves against farmers in the West's water battles. They argue that while 85 percent or more of western water goes to agriculture, the farms contribute a disproportionately small benefit to the states' economies as a whole. For example, California advocates of water marketing point out that four crops—cotton, rice, alfalfa, and pasture—account for 45 percent of the state's total water use but provide only three-thousandths of one percent (.003 percent) of the state's economic return. City customers pay $250 or more per acre-foot of water, while farmers pay as little as $3.50 per acre-foot. By selling their water to cities instead of growing alfalfa or hay, the farmers might make even more money.

But at the same time it might put local seed, fertilizer, and equipment dealers out of business and thereby make life harder on those farmers who chose not to sell their water. Oregon recently passed a law requiring the State Department of Water Resources to take into account the needs of an area before water can be transferred out of it. Because of what Los Angeles did to the Owens Valley, California laws now require state and federal water projects to consider the local effects of withdrawals. But protections are not specified: Should there be a limit on how much water can be withdrawn? Ought the importing agencies to pay the counties of origin some in-lieu-of-tax payments and other compensations? Ought there to be severance taxes, such as are applied to timber, coal, or petroleum? When the Arizona Legislature wrestled with these questions last year, the rural counties felt the groundwater bill under consideration did not give them enough protections and defeated it.

One solution to the problem may be to allow farmers to sell their water without selling their water rights, in effect making them water ranchers. In Colorado there are thought to be more than a hundred active water ranches, farms whose owners intend to mine the water rather than the grass or soil. Franklyn Jeans, a Reno entrepreneur, bought a cattle ranch in Nevada for $2 million and found it much less valuable for its range than for its water, which he believes will be worth $120 million delivered to the Reno area. Since 1986 California's city-minded legislature has passed laws allowing transfers from the State Water Project, requiring irrigation districts to allow their canals to be used to move third-party water, and recognizing water transfers as beneficial uses so as to protect the long-term rights of districts and farmers who lease water to thirsty cities. The new laws were propelled in part by a deal inspired by the Environmental Defense Fund, under which the Metropolitan Water District (which provides water to 15 million Southern Californians) is paying to line Imperial Irrigation District canals and automate parts of its delivery system, and in turn each year will collect 100,000 acre-feet of water thus conserved. In 1989 the Yuba County Water Agency, finding its reservoir full when other parts of the state had experienced a drought, sold water to the urban Santa Clara Valley Water District. And the Metropolitan Water District has been negotiating with other water districts to purchase dry-year options, under which in dry years the MWD would pay farmers even more than their crops would bring to idle their lands and sell the cities their water. 31

But farmers are reluctant to sell even dry-year options. When the Berrenda Mesa Water District in California's Kern County sought to market water that some of its own farmers couldn't afford to buy, the county declared it illegal to transfer water out of the county. Kern, California's biggest farming county, has seen farmed acreage drop from 32

970,000 to 850,000 acres in the last fifteen years, but water officials there want to see any surplus water put into the ground to retire a 300,000-acre-foot overdraft.

State laws often make it illegal to transfer water very far. Texas outlaws interbasin transfers. New Mexico has tried to outlaw out-of-state transfers. When a Wyoming rancher decided he could make more money letting his water run into the Colorado River and selling his entitlements to a California city, the State of Wyoming—fearing the loss of the tax revenues that water might generate, and fearing a rush of other low-profit farms to the lucrative California water market—forbade him to sell his water out of the state.

Farmers fear that, under the western water doctrine of "use it or lose it," if they don't put the water on their fields they might lose their entitlement to it. Many farmers use more than they need because they are afraid that during a drought their entitlement may be cut to less than they need to make a profit. They may also fear increased costs for water as a free market for the commodity develops. Cheap water was the key condition for settlement of the West; because of that, western water law and custom have been shaped by the desire to protect farmers from market forces.

Farmers likewise are leery of the complexity of the market, which contrasts sharply with the simplicity of having a federal agency build and run a project and guarantee the supply of cheap water.

Nor do farmers want to sacrifice their rural independence to big city lawyers and bureaucrats. Says Milenski, "Water is supposed to be a property right. By God, I can't see why you've got to go into court three hundred sixty-five days a year to protect it."

As a result, "these guys don't like the idea of transfers," says California Assemblyman Phillip Isenberg, who has championed water markets in the legislature. "They're not used to it. It is a radically different way of thinking. It strikes at fundamental and deeply held attitudes."

But the growth of urban water demand and the decline of agriculture's economic weight are changing both attitude and law. With continuing population growth, says Colorado water broker William McLemore, "it's just a matter of where people want to live. When all the supplies have been adjudicated and there aren't any more available, that is going to necessitate transfers."

Concludes David Kennedy, director of the California Department of Water Resources, "Increasingly, water marketing is being recognized as one of the elements of water management that are going to solve our water needs."

It is unlikely that the developing water market will be entirely free. "To subject water to the forces of a free market is simplistic," says Jerri

Gilbert of Oakland's East Bay Municipal Utility District. "The other values—the instream uses, the social values—need to be sorted out in our political system." A host of questions need to be addressed as more and more water changes hands: Who—the farmer or the irrigation district—has the right to sell the water? Must ditch and canal companies make their facilities available to transport third-party water? What protections shall be required for local communities and the environment?

As answers to these questions are hammered out, there will be increasing participation in water issues. Isenberg expects that to protect rural interests, the California Legislature may require environmental impact reports on transfers. "Conceptually," he says, "that allows everybody to get in."

William McLemore cites a Colorado law that requires maintenance of vegetation on a piece of land when its water is sold and concludes, "Each time a transfer runs through the courts, they may carve off portions of those rights for the public and the environment."

Walraven Ketellapper observes that in Colorado, where anybody may object to a transfer before the Colorado Water Court, "there are more people involved. There are more laws. There are more water attorneys. In one, small, two-hundred-acre-foot transfer before the Colorado Water Court, there were twenty objectors. A few years ago there might have been only two."

The increased participation may have great benefits for the environment. Says California's David Kennedy, "Increasingly, the Department of Water Resources recognizes that environmental resources have not been dealt a fair hand. Both fisheries and waterfowl have not had adequate water reserved to them." Water marketing is beginning to restore some of that water. Isenberg recently persuaded the California Legislature to offer $60 million to the City of Los Angeles to buy water from San Joaquin Valley farmers to replace water it formerly took from Mono Lake. Private groups in California are trying to buy water rights to convey to the state and federal refuges that host nearly fifty percent of the Pacific Flyway's waterfowl in winter. Those refuges hold rights to almost none of the water they need to survive.

In Colorado the Chevron Corporation donated $7.2 million worth of water rights on the Black Canyon of the Gunnison to the Nature Conservancy, which passed the rights on to the State of Colorado. The state can legally reserve the water as instream flow for the benefit of the trout fishery and three endangered species—the Colorado squawfish, the razorback sucker, and the humpback chub.

In Idaho, when low water levels froze over a wintering area for trumpeter swans, the Nature Conservancy rented irrigation district water to run through the Upper Snake River, opening up essential water so the swans could feed. Congress recently appropriated more than $1 million to be spent acquiring water rights for the Stillwater Wildlife Management Area in Nevada.

Isenberg suggests that California conservationists write Secretary of the Interior Manuel Lujan "a letter offering to pay five dollars more an acre-foot than any farmer offers" for BuRec project water, and then use that water to benefit wildlife.

So, if profiting from water poses the risk that water will become a commodity, to be mined like timber and petroleum, it also offers a golden opportunity. Says water-policy consultant Steve Shupe: "Water is related to quality-of-life values that are not measured in economic terms. Having water nourishing wetlands or waterfalls sustains us in the West on some level other than dollars in the pocket. As water is treated more and more as a commodity, I think those values will grow."

But none of these changes is likely to come about quickly or without conflict. A measure of that growing conflict is the resignation, late in 1989, of Maurice Strong from AWDI's board of directors. Strong says control of the board had passed to a small group of investors who did not share his concerns for the welfare of the San Luis Valley and its environment, and that he had lost his influence over the company's decisions. He says he has faith in the company's management but "strong differences" with principal shareholders.

Strong himself owns a home on the edge of AWDI's Baca Ranch. Now that he no longer has a voice inside the company, he may find himself with the valley's farmers, arguing from the outside. "If I have to argue," he says, "I will argue."

And the turnabout makes him all the more eager to change western water policy. "With Colorado's use-it-or-lose-it law, there's no alternative but staking your claim before someone else stakes a claim," he says. "I think we've got to change our minds on water. You've just got to relate water to the larger issues of a conserver economy. Both rural and urban areas are guilty of using water extravagantly. If they use it wisely, they won't need these projects for years. If water could be saved for many, many years without development, I'd be happy."

Content Considerations

1. What are "water rights"? As you form a definition or understanding, consider how one obtains such rights, how one keeps them, and what it means to have them.
2. What have been some consequences of marketing water in the West?

3. In what ways is it true that "water and profit are the same word in the West"?

The Writer's Strategies

1. This article begins with a description of the desert area that is a focus of Steinhart's exploration of water use and ownership in the West. How is the introduction effective?
2. For what purposes might this article have been written?
3. What is the author's point of view regarding water issues? How is it revealed?

Writing Possibilities

1. Construct a dialogue between a farmer and a developer who is attempting to buy the farmer's water rights.
2. With classmates, discuss the reasons that people who do not live in the West should care about water policies there. Then construct an essay designed to present these reasons to an audience unaware of the issues.
3. Summarize the information in this article that was new to you, then respond by explaining what you think of it.

KIM HEACOX

A Poet, a Painter, and the Lonesome Triangle

Among the kinds of development that transforms desert regions from their natural state are ranching, mining, and a designation allowing off-road vehicle use. Not surprisingly, conservationists and naturalists often oppose such development because of their belief that it ruins the land.

Complicating the argument is the fact that some ranchers are currently working desert lands their families have worked for a hundred years or more. Mining companies and individual miners eagerly stake claims in areas they believe will yield precious metals. Off-roaders voice a right to enjoy their time and their country in the way they choose. Some argue that alternative land use—the establishment of national parks, monuments, and wilderness areas—is itself a form of development. In many areas residents and politicians are drawing up plans, arguing, and compromising in efforts to decide how best to use desert areas, how to resolve conflict fairly.

Kim Heacox, a writer and photographer who lives in Alaska, focuses on such a conflict in the following article, which appeared in Audubon *in May 1990. The conflict occurs in and about the Mojave Desert (parts of Nevada and Arizona and a large part of southern California). Before you read it, write about the view you would expect each of the following residents to have: a rancher, a miner, a painter, an employee of the Bureau of Land Management, a member of an off-road vehicle club. Then write your own position on what to do with the Mojave Desert, a position that may or may not be altered by the article.*

FIVE-THIRTY IN THE MORNING and Paul Harvey is on the radio extrolling the virtues of the American beef industry. Rob Blair, cowboy poet, turns the volume up and cracks another egg into the skillet. The coffee is hot and black, the toast, brown and buttered. Harvey's syncopated voice fills the kitchen and leaks out the window into the cool dawn of the East Mojave. Two dogs and five cats roam the courtyard, beyond which stands a second home, two trailers, a shed, a stable, a stack of hay, several water tanks, and a corral holding about one hundred head of cattle.

It is another working day at the Blair Brothers Ranch, a well-known outfit in these parts, home of Howard Blair, sixty-five, his son Rob, Rob's wife, Kate, and Rob and Kate's two young daughters. Seated for breakfast, they bow their heads as Rob gives thanks for family and friends and the desert they love, and, on this morning, for the company at their table, a fellow writing a story on the California Desert Protection Act for *Audubon* magazine. The Blairs recognize National Audubon Society as one of those conservation organizations with hundreds of thousands of members nationwide, most of whom live in cities, that advocate passage of two congressional bills, S. 11 and H.R. 780, the so-called "pro-wilderness, anti-jobs, great desert lock-up" sponsored by two California Democrats, Senator Alan Cranston and Representative Mel Levine. Nasty names out here.

But hospitality being second nature at the Blairs', Rob opens his door to a smiling stranger who comes knocking; not that he has to, but his father raised a good, decent son who believes in making a good, decent living. He passes the bacon around as sunrise spills over Wild Horse Mesa onto the Providence Mountains. The air begins to thaw, the shadows shorten. The promise of heat is everywhere.

So is the promise of change. If passed, S. 11 and H.R. 780 would rewrite the maps of southeastern California. Death Valley and Joshua Tree national monuments would be enlarged and converted into national parks. The 1.5-million-acre East Mojave National Scenic Area, which is located between Death Valley and Joshua Tree and is presently administered by the Bureau of Land Management (BLM), would become Mojave National Park, administered by the National Park Service (NPS), a sister agency of the BLM in the Department of the Interior. And an additional 4.4 million acres of BLM land would gain wilderness status. In all, the Cranston-Levine bills would create three national parks and eighty-one wilderness areas totaling 8.8 million acres, an area nearly twice the size of New Jersey. Off-road-vehicle use would be curtailed in some areas; mining claims would continue under valid existing rights arrangements unless the claims were condemned; and ranching would continue indefinitely in the BLM wilderness areas but be phased out in the new national park areas.

It so happens the Blair Brothers Ranch sits smack in the middle of the East Mojave National Scenic Area, the proposed Mojave National Park. For seventy-six years and five generations this family has lived here, raising their cattle, paying their taxes, sharing their meals, and, most recently, writing their poetry. Virtually everyone in the East Mojave will tell you, "Oh, the Blairs—good folks; a good outfit." Establish Mojave National Park, says S. 11 and H.R. 780, and they will have to leave in ten years or less, depending on the issue date of their last BLM grazing permit.

The East Mojave isn't Wisconsin; it takes a lot of desert to feed one cow. The Blairs range their 1,500 head on 250,000 acres of public land, fully one-sixth of the East Mojave National Scenic Area; land that belongs to you, me, and every other tax-paying U.S. citizen as much as it belongs to them. Drive the road to the Providence Mountains and you might see a black-tailed jackrabbit, or a roadrunner, or a red-tailed hawk. Or you might see a white-faced cow, perhaps a dozen of them. Cows eating brush and grass and yucca. Cows standing stone still, moon-eyed, and domestic. It's been this way for a hundred years. Rob reasons that if an area grazed for that long still qualifies as wilderness—obviously land "not all ripped up"—then the two must be compatible, cattle-ranching and wilderness, right?

On goes a dusty black hat. Out comes a can of chewing tobacco— just a pinch between the cheek and gum. His horse brushed, fed, and watered, Rob throws the saddle on and cinches it down. A shrike perches on a fencepost, a kestrel on the rooftop. Kate steps outside with their newborn daughter and waves goodbye. Rob waves back, smiling.

> *There's cattle to gather,*
> *   to corral and to brand.*
> *You got thirty miles to ride*
> *   over this lone desert land.*
> *The sands all shimmering,*
> *   a glistening white.*
> *The wildflower fields,*
> *   not a more beautiful sight.*

Up past Hole-in-the-Wall, several mesas and a dozen coyote calls to the north, Carl Faber sits on a three-legged stool with a painting on his lap and a straw hat on his head. The morning light shines through his whiskered face as he mixes paints to the tones of granite, juniper, and sage. "I'd like to see the day when people show as much respect for a sage as they do for an eagle," he says as he begins to brush the canvas lightly. A transcendentalist at heart, Carl subscribes to the Thoreauvian doctrine that a man is rich in proportion to the number of things he can afford to let alone. Walk with him through the desert and you take a serpentine route around everything, living and dead. Pausing at the roots of a deceased plant, he gets down on one knee and says, "Look at that: at the texture, the form, the color. This is what the desert is all about."

Few people tread so lightly in the Mojave. A former student of chemistry, shamanism, and street life who moved out here from Hollywood seventeen years ago, Carl has found in the desert his own Walden Pond. "I lived on rice and beans for years," he says. "No refrigerator, no bath, no company to speak of. I guess I was the ultimate desert rat. Others moved out here from the city but only lasted a year or two before moving back. It's one thing to visit this desert; it's another to live here."

A scrub jay nearly lands on his hat, and when a covey of Gambel's quail calls from afar Carl listens and takes heart in knowing the wild desert still has its own voice. "Sure beats the sound of off-road vehicles," he says. "I've had off-roaders tell me face to face how much they love this desert, and as they're telling me this they're standing right on top of a sage or a matchweed or some other plant. They're weekend warriors from places like Victorville, San Bernardino, and Riverside who might admire the desert on the horizon but have no concept or respect for the desert at their feet. I've spoken to hundreds of them, and they're all the same. They have off-road vehicles, so they're not happy until they're off the road."

In a land dominated by ranchers, off-roaders, and miners, Carl Faber is an ideological island. He has earned the reputation of being a likable guy, although some consider him wrongheaded on the desert conservation issue. But many enjoy dropping by to say howdy and to admire his paintings. The first portrait he did was of Rob Blair sitting on his horse with his hat pulled down over his eyes, coiling a rope in his hands. Carl gave it to him, and it hangs today in Rob's living room just around the corner from the kitchen and the smell of bacon.

As Carl's paintings improved, his reputation grew. Sales picked up, a dealer entered the picture, and prices climbed. Journalists arrived, looking for the "desert artist," and Carl ended up inside the *San Bernardino Sun*, the *Boston Globe*, and the *Los Angeles Times*. Sometimes they ran pictures of his paintings, sometimes they told his life story, sometimes they quoted him on the Cranston bill.

Then one night about three years ago while lying on his bed, Carl was shot. A single, small-caliber bullet entered the window, passed through his pillow, hit him just below the temple and lodged near his sinus cavity. He stumbled a couple of miles to a neighbor's home, made it to the hospital, and recovered. The sheriff's department decided it was attempted suicide, then dropped the case. Others said it was attempted robbery. A few suggested it was attempted murder because of Carl's ideas about desert conservation. He wasn't disabled, though he still suffers occasional headaches from the wound.

There's no lack of guns in the Mojave, or of people who use them.

And few debates in recent memory have fired as many tempers out here as the Cranston-Levine bills. One man was fatally stabbed in a Kern County campground while arguing with his assailant about what's best for the California desert. The stakes are high. Battle lines have been drawn, sides taken, and friendships frayed, for at risk are not only the homes, livelihoods, and recreation opportunities, but also the ecological integrity of the Mojave, the driest, loneliest, and perhaps loveliest corner of the most populous state in the nation.

The pot began to boil with the Federal Land Policy and Management Act of 1976, when Congress designated a whopping 25 million acres of California—one-fourth of the state—as the California Desert Conservation Area. It is an area rich in resources and economic opportunities, yet remarkably fragile—easy to scar, slow to heal—and located next to an exploding human population. The BLM had its orders: Prepare a comprehensive land-use plan in the desert "for the management, use, development, and protection of the public lands." The plan was to follow the BLM's principles of multiple use and sustained yield. Four years, $8 million, and 40,000 public comments later, the California Desert Conservation Area Plan was hatched and hailed as a prodigal child of the public planning process. Secretaries of the Interior Cecil Andrus and James Watt, philosophical antipodes in the Carter and Reagan administrations, approved the plan in 1980 and 1981, respectively.

Within this vast conservation area, the BLM administers nearly half the land, 12.1 million acres, while the Department of Defense administers 3.2 million, the National Park Service, 2.5 million, and the State of California, one million. The remaining six million are privately owned. Aerial maps of the area show a land of little rain compromised by powerlines, mining scars, slag heaps, railroads, tank tracks and motorcycle tracks, bomb craters, golf courses, highways, roadways, ranches, rambling towns, checkerboard properties. The local newspapers editorialize about there being room enough for local economies, weekend joyriders, *and* solace-seeking nature lovers, all in the Mojave, the same desert cursed by Kit Carson and the Mormon Brigade, scorned as a wasteland and feared as a death trap a century and a half ago, and today praised as a wonderland. Back then it was a place to avoid; now everyone wants a piece of the Mojave.

From Barstow, with a population of some 20,000, two freeways slice through creosote communities and head east: I-15 to Las Vegas, I-40 to Needles, diverging as they go, ushering traffic along at seventy and eighty miles an hour bound for the blackjack tables and the Wayne Newton Show. Together with the California–Nevada border these two freeways form the boundary of the "Lonesome Triangle," one of the least populated pieces of California, more than half of which became the East Mojave National Scenic Area with completion of the California Desert Conservation Area Plan in 1980. Administered by the BLM, it was the first such area established in the United States.

Only about five hundred people reside in and around the Lonesome Triangle, including Rob Blair and Carl Faber, and most didn't mind the creation of the new scenic area, for it posed no significant threat to their ways of life. Some even welcomed it. In 1987 the BLM developed a 208-page East Mojave NSA Draft Management Plan and Environmental Assessment to provide "the direction for preserving the scenic quality of this unique area" and "a comprehensive framework for managing and allocating resources in the management area for the next ten years." Ranching, mining, and off-roading would remain, and the BLM would enhance recreation opportunities, namely campgrounds and hiking trails, to "provide necessary services for up to 200,000 visitors by 1997."

Most folks in the Lonesome Triangle will tell you that everything was working fine until 1986, when Senator Cranston came along with his California Desert Protection Act, proposing vast changes in the desert, including the creation of a Mojave National Park. Then Representative Mel Levine jumped in, the national press groups got involved, and several major conservation organizations—including the National Audubon Society—intensified their efforts to save the Mojave Desert. Suddenly the Lonesome Triangle wasn't so lonesome anymore. It became a crucible of the controversy, and it still is. National park or national scenic area, wilderness or white-faced cows, sagebrush or slag heap, Gambel's quail or Kawasaki?

Proponents of the Cranston-Levine bill said the scenic quality of the desert was suffering at the hands of the BLM; that three special interests—grazing, speculatory mining, and off-road vehicles—received highest priority in too many places too often; that the BLM, serving the very interests they're supposed to regulate, might as well be called the Bureau of Livestock and Mining. They cited one example after another of habitat degradation, and presented compelling scientific evidence. Opponents of the Cranston–Levine bill, including the BLM, cried foul. Rich Fagan, the BLM area manager for the 4.5-million-acre Needles Resource Area, said, "We believe S. 11 is probably well intended, but it is unfair to the public in many aspects."

With each passing month the debate widened as California lawmakers divided along party lines, and with each succeeding Congress the honorable Mr. Cranston reintroduced his bill. Twenty-three of the state's forty-five congressmen gave their support, as did twelve California cities, including Los Angeles, San Francisco, and Sacramento.

Three-fourths of Californians surveyed in 1988 said they wanted more protection of the desert ecology; two-thirds supported creation of more national parks. But the desert districts themselves staunchly opposed the bill, and continue to oppose it.

In July 1989 the House Interior Subcommittee on Parks and Public Lands conducted a one-day hearing on the desert bill in Washington, D.C. More than forty groups and individuals testified. It was the season for fireworks, and in this case, for hyperbole as well. "Building a wall around the desert, as this legislation does," thundered Representative Jerry Lewis of Redlands, "will not serve the people. The California Desert Plan was built on public trust in a process that involved all the users of the desert. This [the Cranston–Levine bill] is a gross violation of that trust and an impractical circumvention of a long-standing congressionally mandated public policy process."

Dave Fisher, representing the California Cattlemen's Association, California Woolgrowers Association, and the California Farm Bureau, said it was incorrect to assume that if more of the desert becomes national parks "everyone will ride into the sunset and live happily ever after." He reminded the committee that taxpayers had invested $75 million in the California Desert Conservation Area Plan, that desert cattlemen had installed 285 miles of water lines for stock and wildlife, and that, given the way things were going, having to continually fight for what is rightfully theirs, the desert cattlemen were "starting to feel like the American Indian."

Living happily ever after feels less and less a reality to Kate Blair. While Rob rounds up cattle in the Providence Mountains, she feeds her daughters and writes the monthly High Desert Cattlemen's Association's newsletter. Raised in Camarillo and schooled in Santa Barbara, Kate never felt at home until she discovered Rob Blair and the Mojave Desert. "I couldn't go back to living around people," she says. "This is the only place I want to be. We're not trying to make a fast dollar ranching out here; it can't be done. We just want to make an honest living. I see nothing wrong with that. Ranchers don't always see eye to eye with miners and off-roaders, but in an issue this big we have to work together."

Kate slumps in her chair and tells me that bighorn sheep have declined everywhere cattle have been removed from the mountains because ranchers have then discontinued water maintenance. She says one match stream of water—a stream the diameter of a wooden match, running twenty-four hours a day from a pipe driven deep in the ground—will fill a five-hundred-gallon trough and sustain a lot of cattle and wildlife. Float valves installed by ranchers keep the troughs from overflowing. Ranchers even place wooden ties in the troughs for quail to perch upon as they drink. "We've been wrongfully accused of overgrazing the desert with our cattle," she says. "Cattle move from range to range and graze on a variety of plants, depending on the season and rainfall." She further tells me that three ounces of red lean beef have fewer calories than a chef's salad, that beef is a natural part of the human diet, that Americans are too preoccupied with cholesterol.

She gets up, disappears, and returns with a copy of an article written a few years ago by Edward Abbey, acidic crusader for desert wilderness and a great Belial among western cattlemen. The article originated as a speech at the University of Montana, then appeared in various forms in magazines and newspapers and in one of his books. The copy in Kate's hand is titled, "Free Speech: The Cowboy and His Cow." She throws it down. "Edward Abbey doesn't know what he's talking about," she says. "Everything in here is wrong." She slumps back in her chair with anger in her eyes.

"Western cattlemen are nothing more than welfare parasites," Abbey wrote. "They've been getting a free ride on the public lands for over a century, and I think it's time we phased it out. I'm in favor of putting the public lands livestock grazers out of business. First of all, we don't need the public lands beef industry . . . Only about two percent of our beef, our red meat, comes from the public lands of the eleven western states." He continued by saying most of the public lands in the West are overgrazed to the point of being "cowburnt," and concluded that "if all our 31,000 western public-land ranchers quit tomorrow, we'd never even notice . . . They've had their free ride. It's time they learned to support themselves."

Kate is making dinner as Rob comes through the door. He picks his daughter up and puts her on his lap. It was a long, hot day, as yesterday was and tomorrow will be. It's not easy rounding up cattle in a quarter-million acres of desert. He puts his daughter down, pulls off his dusty, well-worn boots, and leans back, looking tired.

I live in the days of yesteryear,
As time flies by I shed a tear.
I was born a hundred years too late,
Or a hundred years too soon.
What I mean, we're all doomed.

Visitors were calling on Carl Faber so often they were becoming a nuisance—writers, photographers, off-roaders, biologists, university students, curiosity seekers. "Can we see your bullet wound?" It was difficult to get any painting done. Then one day a writer named Adrienne Knute dropped by to meet Carl and to perhaps write about him.

She stayed awhile, they laughed and talked and, well, now they live together on Adrienne's thirty acres along Cedar Canyon Road.

Like Carl, Adrienne found a new beginning in the Mojave, and she likes her desert pristine. Each morning, she goes for a walk through the juniper and sage. If she sees a cow on her land she pulls out a shotgun and peppers it with bird shot. "The four-winged saltbush is coming back around here because we've chased the cattle out," she says. "And see all the matchweed and sage over there?—that's an area that's been heavily grazed; cattle hardly ever eat those plants. But they eat plenty else; I've seen them destroy flowering yuccas, and even yuccas that weren't flowering." An amateur botanist, Adrienne has collected, pressed, and keyed-out dozens of desert plants. She studies their distribution and abundance. "Heavens, yes," she says, "I think cattle grazing has had a definite effect on the plant life."

Steve Larson, a BLM supervisory range conservationist at the Needles area office, has studied grazing in the East Mojave for six years. "I think it's safe to say cattle have had an effect on the vegetation in the East Mojave," he says, "but to what extent is something we may never know. Cattle have been here since the eighteen seventies. They no doubt carried seeds on their hides and ate others that passed through their systems and germinated. But have cattle adversely impacted the desert plant ecology? Who can say? The BLM has established enclosures on land leased by the Blairs, and from casual observation it's difficult to see any difference. But we haven't had time to run transects."

In a world of give and take, Carl Faber and Adrienne Knute can live with cattle and cattle ranchers. It's the off-roaders and speculative miners they worry about.

"Scam miners," Carl calls them. "The only things they mine are their investors' pockets. They punch a road through the desert, rip into a mountain, find nothing, and move on to another scam. That's what happened at the Moon Star Mine. They said they were 'gonna tear that mountain down.' They started in nineteen eighty-four and still haven't produced a single ounce of gold. The BLM is trying to kick them out now, but I'd like to know why they let them in there in the first place. This is a scenic area, right? Well, that mountain will take centuries to be scenic again."

The road winds down past a hill and a sign, VISITORS ARE WELCOME. CLAIM JUMPERS ARE PROSECUTED, then crosses a wash and arrives at the Moon Star Mine. Caretaker Phil Catanzaro cradles a semiautomatic rifle in his arms. Two shotguns and two dogs lie next to him. Not much else is around, save the fruits of his labors: piles of junk, litter here and there, a mountainside scarred, a jackrabbit skinned for dog food. Four turkey vultures soar high overhead. Phil looks up.

"Condors," he says. "This is amazing country. I love it; been out here twenty-three months. Now the BLM says we gotta get out cause this is a wilderness study area. All this talk about wilderness pisses me off. If Alan Cranston sets foot around here, I've got orders to shoot him."

There are an estimated 18,000 mining claims in the California Desert Conservation Area, 8,000 of them in the East Mojave. Also in the East Mojave are twenty-three wilderness study areas, one of which includes the mountain blemished by the Moon Star Mine. When the BLM California State Office recently issued a new policy saying that all mines on wilderness study areas must be reclaimed, the Moon Star objected. Then one day BLM rangers and the Arson and Bomb Squad of the San Bernardino Sheriff's Department arrived, discharged five hundred pounds of Moon Star dynamite, issued another eviction notice, and left. Philip Catanzaro stood there fuming.

Shortly after the congressional hearing in Washington, D.C., in July 1989, the plot thickened with three new developments. First, the desert tortoise was placed on the endangered species list, its numbers having fallen more than fifty percent in four years in the western Mojave. A respiratory disease in domesticated tortoises released in the wild appeared to be the primary culprit. Blame also landed on cattle and off-road vehicles that crush tortoise burrows, if not tortoises themselves. And additional blame landed on ravens. As towns, dumps, landfills, sewage sites, roads, and road-killed rodents, snakes, and birds have become more prevalent across the Mojave, so have opportunistic, intelligent ravens. The *Los Angeles Times* reported that the raven population in the Mojave had increased 1,528 percent in the last twenty years. People said ravens preyed on young tortoises. Soon after the tortoise was placed on the endangered species list, forty ravens were found poisoned to death near Victor Valley.

Second, the General Accounting Office (GAO), the investigative arm of Congress, reported that the BLM had failed to protect wildlife in the Mojave, establishing off-road vehicle "free play" areas in desert tortoise habitat. It further reported the BLM had allowed excessive livestock grazing and mining in the Mojave. A shortage of money made things difficult for the BLM, the report acknowledged, but it concluded that "even if more funds were available . . . the BLM has not demonstrated the willingness to take actions necessary to protect wildlife interests." Predictable objections rained down from the BLM at every level as Senator Cranston brandished the GAO report and said the California desert is "a national heritage that should be preserved for present and future generations of Americans," rather than falling prey to "callous usage and the inroads of development."

Third, in repudiation of the Cranston–Levine package, Representative Jerry Lewis introduced his own version of the California Desert Protection Act as a "Voice of the People." The Lewis bill would establish no new national parks and terminate no grazing permits, and would establish 2.1 million acres of wilderness on BLM land, as opposed to Cranston's 4.4 million.

Rob Blair sits on the lawn outside the Barstow Civic Center, quietly listening to testimony from dozens of Californians seated inside before the House Interior Subcommittee on Parks and Public Lands. Their voices boom over loudspeakers raised on scaffolding in the parking lot. This is the day desert-dwellers have been waiting for; the day to set Congress straight on what is best for the Mojave, the day to defend or deride S. 11 and H.R. 780. Two thousand people have gathered—friends and foes and their indoctrinated children, waving placards, wearing armbands, yelling, watching, listening.

"NO-NO-NO-SEVEN-EIGHT-O," shout the miners and off-roaders.

"YES-YES-YES-WILDER-NESS," reply the environmentalists.

People complain that while a wall goes down in Berlin, Cranston and Levine attempt to put one up around the Mojave. Then Roy Rogers, America's second favorite cowboy next to Ronald Reagan, tells the crowd that no one loves the desert more than he does, but how can we enjoy it if it's turned into wilderness and the roads are closed?

Inside the civic center the testimony heats up. "For many people in the desert," says the mayor of Apple Valley, "S. 11 would be the possible endangerment or even extinction of our way of life. And the economic impacts on our area would be devastating. Grazing and cattle would cease to exist in these areas after current permits expire. The ranchers, ones who probably understand best the importance of conserving the desert, would be forced off their land. Mining exploration would be severely curtailed. Not only would the miners be hurt, but also the shopkeepers and other businesses that depend on mining to pump their dollars into the economy."

"Don't fence off our future," adds the mayor of Hesperia. The crowd cheers. Rob is expressionless.

Representative Mel Levine, the author of H.R. 780, reminds everyone that, of the more than 30,000 miles on maintained and non-maintained roads, routes, and ways in the desert conservation area, his bill would limit access to only 2,500 miles. And of the state's 806 operating mines, only six are within the boundaries of his legislation, and of those, only three *might* be closed. Representative Jerry Lewis counters that the Mojave holds a bonanza of rare-earth elements critical to our national defense and space-age technology, that the importance of mining in the Mojave lies "in those claims yet to be developed."

The testifiers file forward. "There are some desert canyons near my home where I can no longer go out and walk because of all the shooting and off-road vehicles."

"As you listen to the towheaded kids and adult off-roaders plead their case, I ask you to remember what we have learned. The intent of the driver is irrelevant to the Earth . . . even St. Francis of Assisi could not ride a motorcycle through this desert without damaging it."

"The proposed legislation, H.R. 780, is viewed by most as redundant, insulting, and an outrageous breach of faith to the citizens of California."

"I don't mind it when people use their vehicles to enjoy the desert, but it bothers me when they use the desert to enjoy their vehicles."

"Why create more national parks when the Park Service can't properly manage the ones they already have?"

And Peter Burk, president of the Citizens for Mojave National Park, says, "Let us imagine it is the year two thousand. Let me take you on a trip to Mojave National Park . . . There are seventy-five full-time employees of the park . . . the most accessible national park in the country, served by two interstate freeways. Last year over a million 'visitor use days' of people came to this park from all over the world, and sales tax from the park exceeded six million dollars . . . America has given the world two great original ideas, a written constitution and the national park . . . Mr. Chairman, we the people of the year two thousand and all future generations thank you for your visioning foresight."

The off-roaders slurp their sodas and howl when they hear this, and it occurs to me as I sit on the lawn, and Rob Blair sits nearby, that with every great idea comes a gaggle of mediocre minds who oppose it. It was so with the creation of virtually every national park in the United States, from Yellowstone to Yosemite, from Kobuk Valley to the Everglades. There have always been dull, unimaginative people who have clucked their provincial clucks and claimed the end is near; just as they do now in the Mojave. In some respects it *is* an end; but in other, brighter respects it is a new beginning. This is the 1990s, not the 1890s; the Moon Star Mine, moon-eyed cows, and joyriding off-roaders have had their way and their day and done their damage. It's the desert's turn now. A time to heal.

"NO-NO-NO-SEVEN-EIGHT-O."

"YES-YES-YES-WILDER-NESS."

Rob gets up. He's heard enough. He has supplies to buy and a ranch to run. He doesn't shout, chant, or complain; he just slips through the crowd and disappears.

Let your imagination run afar,
It's got to run to catch your star,
It'll never catch it standing still,
It's got to run to peak the hill.

Dusk settles over the Lonesome Triangle as Carl Faber sits outside his home, happy to have missed the hearing in Barstow. Sounds to him like it was a circus. He studies the final rays of sunlight leaving the juniper and sage. The air grows cool; the sky, magenta. Carl is in mid-sentence when he stops and cocks his head: "Shhh . . ." he whispers. "Did you hear that? In the distance? Gambel's quail." 56

Content Considerations

1. How does the "multiple use" policy for public lands create seemingly insoluble problems?
2. The Bureau of Land Management is under fire in this article from several sources. What are some complaints about this federal agency?
3. What conclusion can be drawn from the ending of this article? What particular point of view does the ending emphasize or strengthen?

The Writer's Strategies

1. In what manner does the author present the various sides in the debate over creating a Mojave National Park? How does he express his own views?
2. Describe the tone the author uses in this essay. How persuasive is it?

Writing Possibilities

1. Discuss ways to balance the economic rights of individuals with the leisure time rights of a majority.
2. Of the people described or interviewed in this article, select one with whom you sympathize and write about why.

Additional Writing and Research

Connections

1. Using the ideas presented in the essays by Haas, Steinhart, and Heacox, prepare a paper that explains how development has changed the desert.
2. Various peoples and their relationships to the desert are presented in articles by Nabhan, Haas, and Nichols. Integrate these presentations into a single essay about Hispanic, Native American, and Euro-American responses to desert regions.
3. Explore the idea that land might have rights of its own beyond its use for humans, that species and wilderness tracts may have the right to exist even if they have no economic value. You might want to use the essays by Alcock, Heacox, and Austin in a paper that explores this issue.
4. Compare or contrast the desert of Edward Abbey (Chapter 6) with the desert of Gary Paul Nabhan.
5. Using the information in any or all of the essays about deserts, write a paper that sets forth your original image of the desert and how it might have changed as a result of reading.
6. Both Nichols and Heacox write about federal land management policies. Combine the evidence presented in their two essays, then prepare an argument for or against continued management of public lands.

Researches

1. Deserts present special obstacles to human survival, yet Native Americans lived in undeveloped, natural desert regions for centuries. Find out how: Examine such factors as water, food, social and political responses, and community development.
2. Write an account of the earliest efforts to irrigate the American West.
3. Study other desert regions in the world and compare or contrast the development elsewhere with what has been done in the American West.
4. Find information about the kinds and amounts of natural resources used by one city in the desert in a year. How does this use affect desert ecosystems?
5. Learn where the water that flows from your kitchen or dormitory faucet comes from and how that source would be affected by diversion of its water supply. What is the likely future of water supplies in your area and in the region that supplies your water?
6. Compile a list of artists, writers, or photographers who have taken the desert as their subject. Read about one and study his or her work; try to form an opinion of what the work has added to human knowledge and understanding.

5

Oceans, Lakes, and Contamination

CONSIDER THE FOLLOWING PASSAGE from *Moby Dick,* Herman Melville's epic about whaling in the Pacific in the 1840s:

> Like noiseless nautilus shells, their light prows sped through the sea; but only slowly they neared the foe. As they neared him, the ocean grew still more smooth; seemed drawing a carpet over its waves; seemed a noon-meadow, so serenely it spread. At length the breathless hunter came so nigh his seemingly unsuspecting prey, that his entire dazzling hump was distinctly visible, sliding along the sea as if an isolated thing, and continually set in a revolving ring of finest, fleecy, greenish foam.

Moby Dick was more than a tale of whaling and adventure on the seas, of course, with its allegories of good and evil, of duality. But this passage illustrates one of the still-common views of the ocean: that it is a vast, horizonless, and nearly bottomless resource ready to be harvested. The creatures of such an ocean are merely one of many possible resources there for cultivation by humans. The ocean, in this scenario, is the hope of the world.

This vision of the ocean as infinite source of opportunity clashes with another, more apocalyptic view: that the oceans, the living brew of life, the source of life for all the planet, are in trouble, wounded deeply if not mortally by human ignorance, rapacity, and indifference. Nearly half the world's population live in coastal areas, and the stress on the oceans is severe.

Certainly, many of the images that come to us on the nightly news are disturbing: tankers aground, leaving vast clouds of crude oil billowing into the seas; beaches awash in garbage, including medical syringes and other potentially deadly debris; toxic and nuclear waste being dumped into the depths of the oceans, eventually seeping from sealed containers to surrounding waters; oil spreading its rainbow slickness across the Persian Gulf as an act of war—or ecoterrorism.

The poisoning of the oceans—including the inland seas—has been occurring for years. In the United States the oceans and the Great Lakes have been used for decades as dumping grounds for all manner of pollutants. Sometimes the sheer size of the body of water, combined with inadequate knowledge of what was being dumped, created the impression that no harm was being done, certainly nothing that nature couldn't handle. As William Ashworth points out in an essay on contaminants in the Great Lakes, the problem began benignly enough. Now PCBs and other chemicals have made parts of the Great Lakes highly toxic and bioaccumulative—that is, the toxins dumped there never leave the tissues of the organisms that ingest them, even as they move up the food chain. What to do with the sludge that forms on the bottom as toxins settle is an unsolved problem: There is no place to safely dispose of the poisons.

Pollution by dumping is not the only major problem. Rachel Carson writes about the great storehouses of minerals that are the oceans, as gold, silver, salt, bromine, magnesium, potassium, and other elements drive gradually to the sea. It would be prohibitively expensive to extract these trace elements with present technology. One of the greatest treasures, though, is the result of decomposition of organic matter—oil—and it can be profitably extracted. But oil exploration under the floors of the oceans often leads to spills and massive kills of present sea life: grim irony, indeed.

And despite the seeming abundance of life in the oceans, Paul Colinvaux writes that the sea environment will never be the savior of the earth; it is producing at its maximum now, and that is not enough for what some envision will be necessary in the future.

In fact, far from being the benevolent source of the future of humans, oceans are seen by some as potential foes, much as harpooners viewed Melville's whale. Because of a buildup of greenhouse gases—

products of human industrial activity—the temperature on earth is expected to rise between 2.5 and 5.5 degrees celsius in the next century. Such an increase would mean higher ocean levels, which would be catastrophic for some regions. Jodi Jacobson writes that a one-meter rise—well within most estimates—would result in vast changes and upheaval across the globe.

Oceans, whether friend, foe, or implacable nature, cover some two-thirds of the earth's surface and sculpt the outlines of the land upon which all humans and many other species live. But it is a fragile ecosystem, an aquatic environment whose volume does not protect it from devastating alteration.

In the following essays, learn more about this largely hidden world, what humans have been doing to it, and what it may do to us.

PAUL COLINVAUX

The Ocean System

In the preface to Why Big Fierce Animals Are Rare, *author Paul Colinvaux notes that ecologists, beginning with Darwin, have pondered and written about "how the world worked." Colinvaux does just that in the next essay: He explains the ways the ocean works, providing a good basis for understanding how problems can begin and how solutions can be structured.*

A professor of zoology, Colinvaux takes as a starting point the fact that "ecosystems have bewildering large numbers of moving parts." The numbers and kinds of moving parts, as well as the numbers and kinds of ecosystems, require an awareness of the complexitites that form a variety of relationships among living and nonliving things.

Such complexities certainly exist in the interrelationships that form the ocean. Before you read Colinvaux's article, write a brief summary of what you know about how the ocean works—especially how sunlight, rivers, and plants contribute to the ocean's workings.

THE WORLD OCEANS make up a vast desert, desperately short of nutrients and with living things spread most thinly through them. This is the shocking message of our inquiry into the blueness of the sea. I use the word "shocking" with care. Our generation has been treated to tales about the sea as the last frontier, as a place of wealth, of riches, of production. No journalist seems to be able to write on the subject of feeding the hungry without mentioning farming the oceans, as if they were some great untapped source of food for people. But they are not. The oceans are deserts with little more food in them than we are taking out already.

We can measure the efficiencies of plants by weighing and by totaling all the new tissues they make in a growing season or by monitoring the gases they take in or breathe out. We can measure in that way just how many food calories are made by the plants of a piece of land during the course of the year, and we find out how lamentably inefficient land plants are. Things are far worse in the ocean.

It is easier to measure the productivity of the seas than of the land because the plants live in water (an ideal medium for the chemist) and because their small size permits a nice population of them to be contained in a laboratory-sized bottle. The oceanographic ships of many nations, which now patrol the world oceans, routinely drag up bottles of sea water to look at what the plants are doing, and we now have good measurement from all the seas of the world. The results are grim.

All the seas of the world taken together produce about 92,000,000,000 tons of plant tissue a year, a figure that includes the fertile places with celebrated fisheries as well as the blue waters of the tropical oceans. This may seem a large figure but it needs to be compared with what the plants of the much smaller area of dry land can do. The gross production of all the plants of all dry land of the world is about 272,000,000,000 tons of plant tissue a year. And so we see that, although sea water covers nearly three-quarters of the surface of our planet, the plants in the sea account for only one-quarter of the calories fixed by living things.

The immediate reason for this appalling unproductiveness of the sea is, of course, the scarcity of chemical fertilizer. On land, fertilizing nutrients can sometimes be in short supply too, though wild vegetation usually manages to hoard and cycle the stuff to meet its needs. As a result, unless there is no water, or it is winter and too cold, the production of land plants is set by shortage of the raw material carbon dioxide. In the sea, plants get their carbon in solution as the bicarbonate, and usually they get more than they can use of it. And the reason the tiny sea plants cannot work up to *their* carbon limit is that they run out of chemical fertilizers such as iron, phosphate, or nitrate first.

A first reaction of technically minded man to the infertile sea is to fertilize the wet desert and make it bloom like a soggy rose. But this will not do. It is not that there is an actual shortage of chemical nutrient in the world oceans, because the actual reservoir of chemicals it contains is enormous. Indeed, we sometimes dream of "mining" the oceans for the minerals the waters hold. The problem is one of dilution. Plants must live in the top layer of the oceans where sunlight enters down to say a hundred meters (about three hundred feet) but often much less than this, and the only nutrients of much interest to them are those in this top, thin, lighted layer. All the tonnage in the vast deep beneath is out of reach of the plants. If we dumped more millions of tons of superphosphate and ammonium salts into the sea (assuming we had them), they would merely fall to those same inaccessible depths.

But we know that nature herself has made some patches of the sea fertile, for this is where we go fishing—the North Sea, the Newfoundland Banks, the waters off Peru–and it seems strange that some patches in this fluid thing, the sea, can be fertile while the rest remains a desert. But the explanation is simple enough. The waters that support the fisheries are not really all that different chemically from the desert waters, but they are continuously replaced. Where there are shallow banks or island arcs, the moving currents of the deep sea are forced to the surface, so that the water flows up from below. A similar thing can happen when two deep currents meet head-on so that water is forced upward, leading to some of the celebrated "upwellings." In all these places of banks and upwellings, life is provided with a vertical, slow-moving conveyor belt of water that brings endless supplies of fresh nutrients from the depths. Continuous replenishment of the thin nutrient broth at the bubbling head of a current is the secret of a productive fishery.

There are also fertile inshore waters—the narrow coastal strip or the turbid sea off the mouth of a great river. To these places good fortune comes in two ways, both because the slope of the land forces currents to rise, so that they convey nutrients endlessly as they break against the shore, and because the edge of the sea is fertilized by the debris the rivers bring. How important the fertile flow of a river is can be shown by stopping its flow, an experiment recently tried with the Nile. The Aswan dam has prevented the Nile from discharging its nutrient-laden silt into the sea, and the immediate consequence of this has been the collapse of the sardine fishery as the plants, which fed the plankton that fed the sardines, failed to grow. We have also sometimes been able to increase the amount of fertilizer in the sea near coasts and rivers to levels to which the local plants are not accustomed. The result is often called "pollution." 8

But, apart from narrow coastal strips, the only fertile patches of the oceans occur where currents from the cold depths bubble to the surface. Plants do not like the coldness of this water, but they put up with it for the sake of the nutrients it brings. There the little plants can live their short lives, bobbing in eddies so that a parent stock remains near the middle of the upwelling, passing through their fifty or sixty generations a month, stoking the food chains that support our fisheries. Only one tenth of one percent of the world oceans is a place of true upwelling, and about 10 percent is moderately productive coast. The immense remainder is a blue desert more useless to life than most of Arabia. 9

Thus, the fact is established: the oceans are cruel deserts, and the things they lack are the soluble plant nutrients. But this is a strange conclusion for one who broods about the history of the earth. The oceans have existed since almost the beginnings of early time, changing in shape, shoved from one part of the planet to another before the drifting continents, but always present in roughly the volume we know. And all the time soluble nutrients have been washed into them by the rivers coming from the land. The sea has been made salty by this process. And yet it lacks the nutrients needed for life. Strange. 10

Some details of the answer to this riddle still escape us, but the outlines are clear. The chemistry of the sea is controlled by its mud. Even as the endless-flowing rivers discharge their chemicals into the ocean, so the mud at the bottom soaks them up. The mud is selective. It has complicated minerals similar to the clays of temperate soils that hold metallic cations like calcium, potassium, and sodium. In its medium of salt, the surface of the mud allows slow crystals to grow, like the nodules of manganese that some mining corporations plan to dredge for. There are sites where calcium carbonate precipitates and collects into reefs of limestone, dragging with it other elements such as magnesium. The mud contains organic debris on which bacteria do their strange feeding, fixing some elements to their corpses and discarding others. In these ways the chemistry of sea water is warped away from the typical chemistry of the water of rivers, principally by chemical reactions on its bottom. The sea is not only more concentrated than fresh river water; it contains a quite different mixture of chemicals. This strange mixture is largely fashioned by sedimenting minerals and its mud.

Chemicals are removed from the ocean basins as fast as they arrive from the rivers. They flow back to the land with the writhing of the earth's crust. Every thrust of a mountain range and every emergence of a coastline from the sea brings the chemical-rich mud back to the land. All the sedimentary rocks, from limestone and sandstone to shale and schist, were once part of the ocean's mud. When they were lifted out of the sea, they took with them the chemical nutrients stored in them. At once these nutrients began to be washed out of the rocks again by rain, to be caught in the roots of land plants and held for a while in land ecosystems, then to escape in a slow leak to the rivers, and to be sent on another journey to the sea, another spreading through the fluid mass, and another sorting on the sea's bottom.

It is an enormous chemical machine that keeps the sea a desert. All the chemicals in the sea are cycled slowly through it. They come from the rivers, they spend a time in the oceans, diluted and in suspense, then they are taken in by the mud, held for a brief few million years, and then thrust back onto the land in a prison of rock. This is a system that holds the chemistry of the sea constant from eon to eon. It is not an ecosystem, for all that bacteria and other forms of life do some of the chemical things on the way, particularly influencing the deposition of carbonates. It is a passive physical and chemical system, driven by the sun because the sun is there, but not organized by life.

To live in the open sea, one must adjust to the lack of nutrients. A plant cannot grow large. The niches of small-planting in an unproductive ocean require that plants spend much of their time drifting in the lighted skylight of the sea. They can sink a bit into the dark, shutting down their factories for a while until some eddy brings them back up, and this helps them and their descendants get about. Their breeding strategy is forced on them by their size; they must swell and divide as fast as possible. And they are horribly vulnerable to the grazing animals that hunt them through that glowing open place.

An animal who would adopt the profession of hunting down these tiny plants faces a similar set of constraints. Either it takes the plants one by one, hunting them as a trout hunts a mayfly or a fox hunts a mouse, or it must filter the tiny things out of the water with some sort of sieve. The way the herbivore chooses its size is set by simple principles of mechanics. It must be very small and, as a direct consequence of this, constructed on tolerably simple lines. There can be no brain for cunning, no elaborate eyes to show proper pictures. It hunts through the skylit void according to simple mechanistic rules. Thus we have the herbivorous copepods and their kind, perhaps including animals up to the size of the krill that whales eat.

These "choices" of which I talk are, of course, made by natural selection. There is no conscious design or free will, but the options have been set by the size and habits of the plants and we can see what they must be. Once these options have been taken up, we can see what new possibilities open for natural selection. The first is the flesh-eating animal that will hunt down these humble hunters of the plants. He too must pick his way through the void, seeking his more scattered prey, needing better hunting techniques, being larger and more complex. And he must face the hard reality that natural selection will have provided fiercer animals still to hunt him through the lighted open places where there is no cover and nowhere to hide.

Fishes high on food chains are colored silver beneath and dark on top, no matter from which sea you take them or from what ancient stock they have evolved. Very clearly they are hunted (or hunt) animals that use eyes to locate their prey. Many of the larger animals spend their days deep in the sea where there is no food but where enemies cannot find them either, only coming to the surface at night to hunt the plankton and smaller fish herded there where the plants live.

Thus ecosystems build in the sea into patterns quite different from those of the land. These oceanic ecosystems can do much less to modify the harsh facts of physical existence. Not for them the regulation of the nutrient supply, of providing shade or physical structure and hiding places, as is done by the living things of a terrestrial forest. For nutrients and dwelling places, the life of the open sea must make do with what is provided by passive physical and chemical systems. What we have instead of regulation of the physical habitat is an ecosystem of hunt and be hunted, where plants and animals are ruthlessly adapted to living in a nearly desert but brightly lighted lens of water floating over a black immensity.

When people enter the sea in a quest of food, they harvest the produce of a desert, which means that they cannot get much, however large the desert may be. But matters are even worse than this implies because people cannot get at those tiny plants. Nor can they get at the small animals that eat those plants. For the most part, they cannot even get at the animals that hunt the animals that eat those plants, and must fish further up the food chains. With each link of the food chain, some 90 percent of the food calories originally present are burned away.

When we fish in the oceans we do not just harvest the meager produce of a desert. We get instead (when we are clever and lucky) some 10 percent of 10 percent of 10 percent of the harvest of a desert. The best estimates suggest that we are already fishing close to what these sums say the oceans can bear.

Content Considerations

1. What prompts the author to call the ocean a desert?
2. What makes some parts of an ocean more fertile than others?
3. In your own words, explain the "passive physical and chemical system" that regulates the chemistry of the sea.

The Writer's Strategies

1. What assertion does Colinvaux set out to prove? What is his evidence? Is it convincing?
2. What is the tone of this essay? How does it affect your reading?

Writing Possibilities

1. In an essay, explain how this piece has given you new information, corrected some beliefs, or reinforced and reminded you of what you already knew.
2. After reviewing Colinvaux's description of how the chemical machine of the ocean works, work with a classmate to discover how pollutants might affect that machine. Write a report of your suppositions.
3. Using the information in this article and whatever other sources you need, construct a physical model of the ocean food chain. Include a written explanation of your model.

RACHEL CARSON

Wealth from the Salt Seas

Rachel Carson (1907–1964) taught at the University of Maryland, worked as a marine biologist for the U.S. Bureau of Fisheries, and became chief editor for the U.S. Fish and Wildlife Service. Her books include Under the Sea-Wind *(1941),* The Sea Around Us *(1951),* The Edge of the Sea *(1955), and* Silent Spring *(1962). This last book made America aware for the first time of the damage done to land and people by the use of pesticides; that awareness led to a ban on DDT in America.*

Carson's writing illustrates her awareness of how parts connect, of how action in one part of an ecosystem will affect other parts. This awareness certainly holds in the following selection from The Sea Around Us. *In it Carson discusses what is to be found in oceans, what minerals and materials of value, and weaves through her discussion science, observation, and a strong belief in the spiritual ties that bind humans to nature.*

Before you read her essay, envision the old element chart from the chemistry classroom and let your imagination play. What elements might the ocean hold?

A sea change into something rich and strange.
—SHAKESPEARE

THE OCEAN is the earth's greatest storehouse of minerals. In a single cubic mile of sea water there are, on the average, 166 million tons of dissolved salts, and in all the ocean waters of the earth there are about 50 quadrillion tons. And it is in the nature of things for this quantity to be gradually increasing over the millennia, for although the earth is constantly shifting her component materials from place to place, the heaviest movements are forever seaward.

It has been assumed that the first seas were only faintly saline and that their saltiness has been growing over the eons of time. For the primary source of the ocean's salt is the rocky mantle of the continents. When those first rains came—the centuries-long rains that fell from the heavy clouds enveloping the young earth—they began the processes of wearing away the rocks and carrying their contained minerals to the sea. The annual flow of water seaward is believed to be about 6500 cubic miles, this inflow of river water adding to the ocean several billion tons of salts.

It is a curious fact that there is little similarity between the chemical composition of river water and that of sea water. The various elements are present in entirely different proportions. The rivers bring in four times as much calcium as chloride, for example, yet in the ocean the proportions are strongly reversed—46 times as much chloride as calcium. An important reason for the difference is that immense amounts of calcium salts are constantly being withdrawn from the sea water by marine animals and are used for building shells and skeletons—for the microscopic shells that house the foraminifera, for the massive structures of the coral reefs, and for the shells of oysters and clams and other mollusks. Another reason is the precipitation of calcium from sea water. There is a striking difference, too, in the silicon content of river and sea water—about 500 per cent greater in rivers than in the sea. The silica is required by diatoms to make their shells, and so the immense quantities brought in by rivers are largely utilized by these ubiquitous plants of the sea. Often there are exceptionally heavy growths of diatoms off the mouths of rivers. Because of the enormous total chemical requirements of all the fauna and flora of the sea, only a small part of the salts annually brought in by rivers goes to increasing the quantity of dissolved minerals in the water. The inequalities of chemical make-up are further reduced by reactions that are set in motion immediately the fresh water is discharged into the sea, and by the enormous disparities of volume between the incoming fresh water and the ocean.

There are other agencies by which minerals are added to the sea—from obscure sources buried deep within the earth. From every volcano chlorine and other gases escape into the atmosphere and are carried down in rain onto the surface of land and sea. Volcanic ash and rock bring up other materials. And all the submarine volcanoes, discharging through unseen craters directly into the sea, pour in boron, chlorine, sulphur, and iodine.

All this is a one-way flow of minerals to the sea. Only to a very limited extent is there any return of salts to the land. We attempt to recover some of them directly by chemical extraction and mining, and indirectly by harvesting the sea's plants and animals. There is another way, in the long, recurring cycles of the earth, by which the sea itself gives back to the land what it has received. This happens when the ocean waters rise over the land, deposit their sediments, and at last withdraw, leaving over the continent another layer of sedimentary rocks. These contain some of the water and salts of the sea. But it is only a temporary loan of minerals to the land and the return payment begins at once by way of the old, familiar channels—rain, erosion, run-off to the rivers, transport to the sea.

There are other curious little exchanges of materials between sea and land. While the process of evaporation, which raises water vapor into the air, leaves most of the salts behind, a surprising amount of salt does intrude itself into the atmosphere and rides long distances on the wind. The so-called 'cyclic salt' is picked up by the winds from the spray of a rough, cresting sea or breaking surf and is blown inland, then brought down in rain and returned by rivers to the ocean. These tiny, invisible particles of sea salt drifting in the atmosphere are, in fact, one of the many forms of atmospheric nuclei around which raindrops form. Areas nearest the sea, in general, receive the most salt. Published figures have listed 24 to 36 pounds per acre per year for England and more than 100 pounds for British Guiana. But the most astounding example of long-distance, large-scale transport of cyclic salts is furnished by Sambhar Salt Lake in northern India. It receives 3000 tons of salt a year, carried to it on the hot dry monsoons of summer from the sea, 400 miles away.

The plants and animals of the sea are very much better chemists than men, and so far our own efforts to extract the mineral wealth of the sea have been feeble compared with those of lower forms of life. They have been able to find and to utilize elements present in such minute traces that human chemists could not detect their presence until, very recently, highly refined methods of spectroscopic analysis were developed.

We did not know, for example, that vanadium occurred in the sea until it was discovered in the blood of certain sluggish and sedentary sea creatures, the holothurians (of which sea cucumbers are an example) and the ascidians. Relatively huge quantities of cobalt are extracted by lobsters and mussels, and nickel is utilized by various mollusks, yet it is only within recent years that we have been able to recover even traces of these elements. Copper is recoverable only as about a hundredth part in a million of sea water, yet it helps to constitute the life blood of lobsters, entering into their respiratory pigments as iron does into human blood.

In contrast to the accomplishments of invertebrate chemists, we have so far had only limited success in extracting sea salts in quantities we can use for commercial purposes, despite their prodigious quantity and considerable variety. We have recovered about fifty of the known elements by chemical analysis, and shall perhaps find that all the others are there, when we can develop proper methods to discover them. Five salts predominate and are present in fixed proportions. As we would expect, sodium chloride is by far the most abundant, making up 77.8

per cent of the total salts; magnesium chloride follows, with 10.9 per cent; then magnesium sulphate, 4.7 per cent; calcium sulphate, 3.6 per cent; and potassium sulphate, 2.5 per cent. All others combined make up the remaining .5 per cent.

Of all the elements present in the sea, probably none has stirred men's dreams more than gold. It is there—in all the waters covering the greater part of the earth's surface—enough in total quantity to make every person in the world a millionaire. But how can the sea be made to yield it? The most determined attempt to wrest a substantial quantity of gold from ocean waters—and also the most complete study of the gold in sea water—was made by the German chemist Fritz Haber after the First World War. Haber conceived the idea of extracting enough gold from the sea to pay the German war debt and his dream resulted in the German South Atlantic Expedition of the *Meteor*. The *Meteor* was equipped with a laboratory and filtration plant, and between the years 1924 and 1928 the vessel crossed and recrossed the Atlantic, sampling the water. But the quantity found was less than had been expected, and the cost of extraction far greater than the value of the gold recovered. The practical economics of the matter are about as follows: in a cubic mile of sea water there is about $93,000,000 in gold and $8,500,000 in silver. But to treat this volume of water in a year would require the twice-daily filling and emptying of 200 tanks of water, each 500 feet square and 5 feet deep. Probably this is no greater feat, relatively, than is accomplished regularly by corals, sponges, and oysters, but by human standards it is not economically feasible.

Most mysterious, perhaps, of all substances in the sea is iodine. In sea water it is one of the scarcest of the nonmetals, difficult to detect and resisting exact analysis. Yet it is found in almost every marine plant and animal. Sponges, corals, and certain seaweeds accumulate vast quantities of it. Apparently the iodine in the sea is in a constant state of chemical change, sometimes being oxidized, sometimes reduced, again entering into organic combinations. There seem to be constant interchanges between air and sea, the iodine in some form perhaps being carried into the air in spray, for the air at sea level contains detectable quantities, which decrease with altitude. From the time living things first made iodine a part of the chemistry of their tissues, they seem to have become increasingly dependent on it; now we ourselves could not exist without it as a regulator of the basal metabolism of our bodies, through the thyroid gland which accumulates it.

All commercial iodine was formerly obtained from seaweeds; then the deposits of crude nitrate of soda from the high deserts of North Chile were discovered. Probably the original source of this raw material—called 'caliche'—was some prehistoric sea filled with marine vegetation, but that is a subject of controversy. Iodine is obtained also from brine deposits and from the subterranean waters of oil-bearing rocks—all indirectly of marine origin.

A monopoly on the world's bromine is held by the ocean, where 99 per cent of it is now concentrated. The tiny fraction present in rocks was originally deposited there by the sea. First we obtained it from the brines left in subterranean pools by prehistoric oceans; now there are large plants on the seacoasts—especially in the United States—which use ocean water as their raw material and extract the bromine directly. Thanks to modern methods of commercial production of bromine we have high-test gasoline for our cars. There is a long list of other uses, including the manufacture of sedatives, fire extinguishers, photographic chemicals, dyestuffs, and chemical warfare materials.

One of the oldest bromine derivatives known to man was Tyrian purple, which the Phoenicians made in their dyehouses from the purple snail, Murex. This snail may be linked in a curious and wonderful way with the prodigious and seemingly unreasonable quantities of bromine found today in the Dead Sea, which contains, it is estimated, some 850 million tons of the chemical. The concentration of bromine in Dead Sea water is 100 times that in the ocean. Apparently the supply is constantly renewed by underground hot springs, which discharge into the bottom of the Sea of Galilee, which in turn sends its waters to the Dead Sea by way of the River Jordan. Some authorities believe that the source of bromine in the hot springs is a deposit of billions of ancient snails, laid down by the sea of a bygone age in a stratum long since buried.

Magnesium is another mineral we now obtain by collecting huge volumes of ocean water and treating it with chemicals, although originally it was derived only from brines or from the treatment of such magnesium-containing rocks as dolomite, of which whole mountain ranges are composed. In a cubic mile of sea water there are about 4 million tons of magnesium. Since the direct extraction method was developed about 1941, production has increased enormously. It was magnesium from the sea that made possible the wartime growth of the aviation industry, for every airplane made in the United States (and in most other countries as well) contains about half a ton of magnesium metal. And it has innumerable uses in other industries where a lightweight metal is desired, besides its long-standing utility as an insulating material, and its use in printing inks, medicines, and toothpastes, and in such war implements as incendiary bombs, star shells, and tracer ammunition.

Wherever climate has permitted it, men have evaporated salt from sea water for many centuries. Under the burning sun of the tropics the ancient Greeks, Romans, and Egyptians harvested the salt men and animals everywhere must have in order to live. Even today in parts of the world that are hot and dry and where drying winds blow, solar evaporation of salt is practiced—on the shores of the Persian Gulf, in China, India, and Japan, in the Philippines, and on the coast of California and the alkali flats of Utah.

Here and there are natural basins where the action of sun and wind and sea combine to carry on evaporation of salt on a scale far greater than human industry could accomplish. Such a natural basin is the Rann of Cutch on the west coast of India. The Rann is a flat plain, some 60 by 185 miles, separated from the sea by the island of Cutch. When the southwest monsoons blow, sea water is carried in by way of a channel to cover the plain. But in summer, in the season when the hot northeast monsoon blows from the desert, no more water enters, and that which is collected in pools over the plain evaporates into a salt crust, in some places several feet thick.

Where the sea has come in over the land, laid down its deposits, and then withdrawn, there have been created reservoirs of chemicals, upon which we can draw with comparatively little trouble. Hidden deep under the surface of our earth are pools of 'fossil salt water,' the brine of ancient seas; 'fossil deserts,' the salt of old seas that evaporated away under conditions of extreme heat and dryness; and layers of sedimentary rock in which are contained the organic sediments and the dissolved salts of the sea that deposited them.

During the Permian period, which was a time of great heat and dryness and widespread deserts, a vast inland sea formed over much of Europe, covering parts of the present Britain, France, Germany, and Poland. Rains came seldom and the rate of evaporation was high. The sea became exceedingly salty, and it began to deposit layers of salts. For a period covering thousands of years, only gypsum was deposited, perhaps representing a time when water fresh from the ocean occasionally entered the inland sea to mix with its strong brine. Alternating with the gypsum were thicker beds of salt. Later, as its area shrank and the sea grew still more concentrated, deposits of potassium and magnesium sulphates were formed (this stage representing perhaps 500 years); still later, and perhaps for another 500 years, there were laid down mixed potassium and magnesium chlorides or carnallite. After the sea had completely evaporated, desert conditions prevailed, and soon the salt deposits were buried under sand. The richest beds form the famous deposits of Stassfurt and Alsace; toward the outskirts of the original area of the old sea (as, for example, in England) there are only beds of salt. The Stassfurt beds are about 2500 feet thick; their springs of brine have been known since the thirteenth century, and the salts have been mined since the seventeenth century.

At an even earlier geological period—the Silurian—a great salt basin was deposited in the northern part of the United States, extending from central New York State across Michigan, including northern Pennsylvania and Ohio and part of southern Ontario. Because of the hot, dry climate of that time, the inland sea lying over this place grew so salty that beds of salt and gypsum were deposited over a great area covering about 100,000 square miles. There are seven distinct beds of salt at Ithaca, New York, the uppermost lying at a depth of about half a mile. In southern Michigan some of the individual salt beds are more than 500 feet thick, and the aggregate thickness of salt in the center of the Michigan Basin is approximately 2000 feet. In some places rock salt is mined; in others wells are dug, water is forced down, and the resulting brine is pumped to the surface and evaporated to recover the salt.

One of the greatest stock piles of minerals in the world came from the evaporation of a great inland sea in the western United States. This is Searles Lake in the Mohave Desert of California. An arm of the sea that overlay this region was cut off from the ocean by the thrusting up of a range of mountains; as the lake evaporated away, the water that remained became ever more salty through the inwash of minerals from all the surrounding land. Perhaps Searles Lake began its slow transformation from a landlocked sea to a 'frozen' lake—a lake of solid minerals—only a few thousand years ago; now its surface is a hard crust of salts over which a car may be driven. The crystals of salts form a layer 50 to 70 feet deep. Below that is mud. Engineers have recently discovered a second layer of salts and brine, probably at least as thick as the upper layer, underlying the mud. Searles Lake was first worked in the 1870's for borax; then teams of 20 mules each carried the borax across desert and mountains to the railroads. In the 1930's the recovery of other substances from the lake began—bromine, lithium, and salts of potassium and sodium. Now Searles Lake yields 40 per cent of the production of potassium chloride in the United States and a large share of all the borax and lithium salts produced in the world.

In some future era the Dead Sea will probably repeat the history of Searles Lake, as the centuries pass and evaporation continues. The Dead Sea as we know it is all that remains of a much larger inland sea that once filled the entire Jordan Valley and was about 190 miles long; now it has shrunk to about a fourth of this length and a fourth of its former volume. And with the shrinkage and the evaporation in the hot dry climate has come the concentration of salts that makes the Dead Sea a great reservoir of minerals. No animal life can exist in its brine; such luckless fish as are brought down by the River Jordan die and provide food for the sea birds. It is 1300 feet below the Mediterranean, lying farther below sea level than any other body of water in the world. It occupies the lowest part of the rift valley of the Jordan, which was created by a down-slipping of a block of the earth's crust. The water of the Dead Sea is warmer than the air, a condition favoring evaporation, and clouds of its vapor float, nebulous and half formed, above it, while its brine grows more bitter and the salts accumulate.

Of all legacies of the ancient seas the most valuable is petroleum. Exactly what geologic processes have created the precious pools of liquid deep within the earth no one knows with enough certainty to describe the whole sequence of events. But this much seems to be true: Petroleum is a result of fundamental earth processes that have been operating ever since an abundant and varied life was developed in the sea—at least since the beginning of Paleozoic time, probably longer. Exceptional and catastrophic occurrences may now and then aid its formation but they are not essential; the mechanism that regularly generates petroleum consists of the normal processes of earth and sea—the living and dying of creatures, the deposit of sediments, the advance and retreat of the seas over the continents, the upward and downward foldings of the earth's crust.

The old inorganic theory that linked petroleum formation with volcanic action has been abandoned by most geologists. The origin of petroleum is most likely to be found in the bodies of plants and animals buried under the fine-grained sediments of former seas and there subjected to slow decomposition.

Perhaps the essence of conditions favoring petroleum production is represented by the stagnant waters of the Black Sea or of certain Norwegian fiords. The surprisingly abundant life of the Black Sea is confined to the upper layers; the deeper and especially the bottom waters are devoid of oxygen and are often permeated with hydrogen sulphide. In these poisoned waters there can be no bottom scavengers to devour the bodies of marine animals that drift down from above, so they are entombed in the fine sediments. In many Norwegian fiords the deep layers are foul and oxygenless because the mouth of the fiord is cut off from the circulation of the open sea by a shallow sill. The bottom layers of such fiords are poisoned by the hydrogen sulphide from decomposing organic matter. Sometimes storms drive in unusual quantities of oceanic water and through turbulence of waves stir deeply the waters of these lethal pools; the mixing of the water layers that follows brings death to hordes of fishes and invertebrates living near the surface. Such a catastrophe leads to the deposit of a rich layer of organic material on the bottom.

Wherever great oil fields are found, they are related to past or present seas. This is true of the inland fields as well as of those near the present seacoast. The great quantities of oil that have been obtained from the Oklahoma fields, for example, were trapped in spaces within sedimentary rocks laid down under seas that invaded this part of North America in Paleozoic time.

The search for petroleum has also led geologists repeatedly to those 'unstable belts, covered much of the time by shallow seas, which lie around the margins of the main continental platforms, between them and the great oceanic deeps.'

An example of such a depressed segment of crust lying between continental masses is the one between Europe and the Near East, occupied in part by the Persian Gulf, the Red, Black, and Caspian seas, and the Mediterranean Sea. The Gulf of Mexico and the Caribbean Sea lie in another basin or shallow sea between the Americas. A shallow, island-studded sea lies between the continents of Asia and Australia. Lastly, there is the nearly landlocked sea of the Arctic. In past ages all of these areas have been alternately raised and depressed, belonging at one time to the land, at another to the encroaching sea. During their periods of submersion they have received thick deposits of sediments, and in their waters a rich marine fauna has lived, died, and drifted down into the soft sediment carpet.

There are vast oil deposits in all these areas. In the Near East are the great fields of Saudi Arabia, Iran, and Iraq. The shallow depression between Asia and Australia yields the oil of Java, Sumatra, Borneo, and New Guinea. The American mediterranean is the center of oil production in the Western Hemisphere—half the proved resources of the United States come from the northern shore of the Gulf of Mexico, and Colombia, Venezuela, and Mexico have rich oil fields along the western and southern margins of the Gulf. The Arctic is one of the unproved frontiers of the petroleum industry, but oil seepages in northern Alaska, on islands north of the Canadian mainland, and along the Arctic coast of Siberia hint that this land recently raised from the sea may be one of the great oil fields of the future.

In recent years, the speculations of petroleum geologists have been focused in a new direction—under sea. By no means all of the land resources of petroleum have been discovered, but probably the richest and most easily worked fields are being tapped, and their possible production is known. The ancient seas gave us the oil that is now being drawn out of the earth. Can the ocean today be induced to give up some of the oil that must be trapped in sedimentary rocks under its floor, covered by water scores or hundreds of fathoms deep?

Oil is already being produced from offshore wells, on the continental shelf. Off California, Texas, and Louisiana, oil companies have drilled into the sediments of the shelf and are obtaining oil. In the United States the most active exploration has been centered in the Gulf of Mexico. Judging from its geologic history, this area has rich promise. For eons of time it was either dry land or a very shallow sea basin, receiving the sediments that washed into it from high lands to the north. Finally, about the middle of the Cretaceous period, the floor of the Gulf began to sink under the load of sediments and in time it acquired its present deep central basin.

By geophysical exploration, we see that the layers of sedimentary rock underlying the coastal plain tilt steeply downward and pass under the broad continental shelf of the Gulf. Down in the layers deposited in the Jurassic period is a thick salt bed of enormous extent, probably formed when this part of the earth was hot and dry, a place of shrinking seas and encroaching deserts. In Louisiana and Texas, and also, it now appears, out in the Gulf itself, extraordinary features known as salt domes are associated with this deposit. These are fingerlike plugs of salt, usually less than a mile across, pushing up from the deep layer toward the earth's surface. They have been described by geologists as 'driven up through 5000 to 15,000 feet of sediments by earth pressures, like nails through a board.' In the states bordering the Gulf such structures have often been associated with oil. It seems probable that on the continental shelf, also, the salt domes may mark large oil deposits.

In exploring the Gulf for oil, therefore, geologists search for the salt domes where the larger oil fields are likely to lie. They use an instrument known as a magnetometer, which measures the variations in magnetic intensity brought about by the salt domes. Gravity meters also help locate the domes by measuring the variation in gravity near them, the specific gravity of salt being less than that of the surrounding sediments. The actual location and outline of the dome are discovered by seismographic exploration, which traces the inclination of the rock strata by recording the reflection of sound waves produced by dynamite explosions. These methods of exploration have been used on land for some years, but only since about 1945 have they been adapted to use in offshore Gulf waters. The magnetometer has been so improved that it will map continuously while being towed behind a boat or carried in or suspended from a plane. A gravity meter can now be lowered rapidly to the bottom and readings made by remote control. (Once an operator had to descend with it in a diving bell.) Seismic crews may shoot off their dynamite charges and make continuous recordings while their boats are under way.

Despite all these improvements which allow exploration to proceed rapidly, it is no simple matter to obtain oil from undersea fields. Prospecting must be followed by the leasing of potential oil-producing areas, and then by drilling to see whether oil is actually there. Offshore drilling platforms rest on piles that must be driven as far as 250 feet into the floor of the Gulf to withstand the force of waves, especially during the season for hurricanes. Winds, storm waves, fogs, the corrosive gnawing of sea water upon metal structures—all these are hazards that must be faced and overcome. Yet the technical difficulties of far more extensive offshore operations than any now attempted do not discourage specialists in petroleum engineering.

So our search for mineral wealth often leads us back to the seas of ancient times—to the oil pressed from the bodies of fishes, seaweeds, and other forms of plant and animal life and then stored away in ancient rocks; to the rich brines hidden in subterranean pools where the fossil water of old seas still remains; to the layers of salts that are the mineral substance of those old seas laid down as a covering mantle over the continents. Perhaps in time, as we learn the chemical secrets of the corals and sponges and diatoms, we shall depend less on the stored wealth of prehistoric seas and shall go more and more directly to the ocean and the rocks now forming under its shallow waters.

Content Considerations

1. In what ways are materials exchanged between land and sea?
2. How did the minerals found in the sea get there?
3. What do oceans have to do with the creation of petroleum?

The Writer's Strategies

1. How does the title of this piece relate to its content?
2. How is this essay organized? It may help to map it—that is, construct a short phrase or sentence for each paragraph, then examine them in order to determine the organization.
3. What seems to be Carson's attitude toward offshore drilling? How do you know?

Writing Possibilities

1. What conclusions can you draw about the wealth of the ocean from this article? Write them.
2. With a classmate, explore how the information Carson presents affects your understanding of natural geological processes.

3. This essay was published in a revised edition of *The Sea Around Us* in 1961. Select parts of the essay to revise as a class by locating facts or information within it you want to confirm. Use various sources for updated information; write your update in a report to be shared with other class members.

Peter Sears

Oil Spill

Oil spills and other environmental incidents often rouse fierce emotions: the source of those emotions may be both the incident and the human activities and beliefs that caused it. In the following, teacher and poet Peter Sears expresses some of that emotion. As you read his words, compare your own feelings about environmental harm to the ones he expresses.

The ocean is leather in slow motion.
Waves don't break, they squat and slide.
Stick a finger in—thick as fudge
up and down the beach;

and the fixed, brown bubbles
mean death by suffocation.
May the men who did this
boil and roll in a sea coat of oil;

and may their superiors
join us and our neighbors
on our knees
to scrub this beach to the bone.

If they are unwilling,
may they stand naked facing a mirror
and hear behind them, forever,
the fawning and sniveling of underlings.

—Peter Sears

ART DAVIDSON

In the Wake of the Exxon Valdez: Marine Birds

On March 24, 1989, the tanker Exxon Valdez ran aground off Alaska, spilling almost 11 million gallons of crude oil into the coastal waters. That event set a myriad of others into motion; government officials, company employees, and teams of rescue and cleanup workers began their tasks almost immediately. But they didn't all share the same view of those tasks.

Art Davidson reports on the spill and its aftermath in his book In the Wake of the Exxon Valdez *(1990). The following excerpt details the initial planning of those who were to locate, gather, and clean the oiled birds that were still living. Before you begin reading, write for a few minutes about what kinds of action are required in cleaning up an oil spill.*

"THERE ARE WHOLE POPULATIONS of birds endangered out there, and I can't save them all," Jay Holcomb said. Alyeska Pipeline Service Co. had called Holcomb, of the International Bird Rescue Research Center, in the first hours following the March 24, 1989, *Exxon Valdez* oil spill. His organization would spearhead the effort to rescue oiled birds.

Holcomb first became involved with oiled birds in 1971, when two oil tankers collided under San Francisco's Golden Gate Bridge. Then 21, he watched pictures of oiled birds on television and rushed down to the bay to help. "Thirteen thousand live birds were collected," Holcomb remembered. "They were stacked in cages, and most of them died."

While struggling through the chaos of that spill, Holcomb met Alice Berkner, who was then beginning to see a need for a worldwide organization to care for birds caught in oil spills. Her dream was eventually realized in the International Bird Rescue Research Center, based in Berkeley, Calif. Although largely funded by oil companies, the bird rescue center maintains a volunteer spirit and its own sense of purpose. "When we go to a spill, we don't stand there and talk about how horrible the oil company is. We take care of the birds," Holcomb said. "Through the years, we've been on a lot of spills and have developed some pretty good expertise in treating birds."

Realizing that the bird rescuers would face great difficulties in

Alaska, Holcomb brought along Jessica Porter, a veterinarian with a lifelong passion for helping wild creatures.

"We flew over the spill our first morning in Valdez and were stunned by how much oil was out there," Porter said. "This spill was unlike any we had ever seen."

Not only were Porter, Holcomb and the Alaskans they hired going to have to treat lots of oiled birds, but as soon as they landed in Valdez, Holcomb and Porter faced the problems of authority and responsibility that plagued so many aspects of the spill response. On other spills, the U.S. Fish and Wildlife Service had always brought oiled birds to the cleaning centers. The bird rescue center people had always taken birds at the door, cleaned them, and decided when they were ready for release. This time Holcomb and Porter were told they would have to find the birds and bring them in.

"It was bizarre. This lady, Pam Bergman, with the Department of the Interior, was calling the shots, telling us we had to go look for the birds," Porter said. "This was crazy! We should concentrate on what we do well—cleaning, feeding and medical treatment of birds. We can't be out there trying to find birds in places we've never seen. But Bergman kept saying, 'The Fish and Wildlife Service can't do anything. The party that spilled the oil is responsible. Exxon is responsible, period.'"

With the USFWS assuming a strictly monitoring role, no one had clear authority over Exxon in the area of bird rescue. As Porter said, "We couldn't tell Exxon what to do. Exxon wouldn't listen to us—they'd get ticked off. The Fish and Wildlife Service should have been there telling Exxon to do this and do that. The bird rescue center is not a lead agency. We are not the trustees of America's wildlife. Interior put us in a really bad situation. Bergman dumped all the responsibility on us, but gave us none of the authority."

As the first oiled murres and loons washed ashore near Bligh Reef, the newly created rescue effort stalled, caught between the priorities of these two women from two different worlds. Porter was used to working long hours caring for wild creatures. Bergman spent her life in interagency meetings, coordinating environmental policy for the USFWS and other Interior Department agencies in Alaska.

"Up here it's a completely different ball game than the bird rescue people have ever dealt with before. I think they felt a little confused," Bergman said. "It's not the Fish and Wildlife Service's mission to mass an army to rescue birds. And it really rankles me—why are they called experts if they can't do that sort of thing? The point is, Porter's bird center is contracted to clean birds. Maybe they got a little spoiled, got used to people bringing in birds for them on other spills. Now they have to go out and get them."

Porter said that being forced to capture oiled birds was "totally against" the national contingency plan established under the federal Clean Water Act. "We ended up in the middle, writing contracts, bartering for days. It was wasting our time and raising our frustration level. We needed help from the Fish and Wildlife Service more than ever before."

"We *were* helpful," Bergman countered. "We gave them all the help in the world. We gave them the names of people to call. The bird center's fallback is to blame the Fish and Wildlife Service. They should have been more aware of how things work up here."

The tension and confusion were fueled by the federal government's interagency regional response team's new wildlife guidelines, which had been designed to shift much of Fish and Wildlife's trustee responsibility for wildlife to the oil industry in the event of a spill. The guidelines state that, "The Fish and Wildlife Service will assume lead responsibility for capturing oiled birds and transporting them to a cleaning and rehabilitation center(s). Organizations that clean and rehabilitate oiled birds generally will not be responsible for those tasks."

This corresponded exactly to the type of assistance Porter had experienced on other oil spills and had expected in Alaska. However, the guidelines go on to stipulate that the Fish and Wildlife Service would assume this active role *only if* the spiller did not accept responsibility for the spill and respond adequately. This qualifying clause, adopted the previous December, had never been pointed out to anyone at the bird rescue center. Now, in the midst of the spill, Bergman explained that this obscure clause was why Fish and Wildlife wouldn't help rescue oiled birds.

The pivotal question became, "What constitutes an adequate response?" With no guidelines for determining the adequacy of a spiller's response, it fell to Bergman and her supervisor, Paul Gates, to decide whether the response was inadequate and federal assistance was needed. Exxon itself didn't have the waterfowl expertise to do the rescue work that the USFWS had always done. As Alyeska's, and later Exxon's, designated bird cleaners, Porter and Holcomb felt overwhelmed by the difficulty and danger of capturing oiled birds in 10,000 square miles of coastal terrain they had never seen. However, Bergman and Gates determined that Exxon was responding adequately and chose to monitor rather than mitigate Porter and Holcomb's struggles. Said Gates, "I don't think the Fish and Wildlife Service is on the hook on this thing."

Dr. Calvin Lensink, who had worked for the USFWS for 33 years, expressed dismay at their lack of involvement in what he called the "biological aspects" of the spill—the wildlife rescue efforts. Commenting on what he observed to be a gradual deterioration of the government's ability to protect wildlife, he said, "You'd think we would have been on top of the biological aspects. But we aren't. I was thoroughly disgusted, as were a lot of refuge people. If a problem isn't urgent, the Fish and Wildlife Service tends not to address it. So when a real crisis like this comes along, they aren't ready for it. Fish and Wildlife people were ready to go out to the sound, but they weren't being given the authority to go, so they just ended up sitting around."

Lensink was far and away the agency's most experienced biologist on the spill. However, he had retired from the service the previous November and, after the spill, went to Valdez as a volunteer. "I couldn't possibly have sat in Anchorage and done nothing," Lensink said. "I told Fish and Wildlife I'd just as soon not be on salary so I could do what I thought had to be done the most. And I didn't take the nice safe jobs, either. I took some of the nasty jobs. Working with dead animals is a nasty job."

While Lensink set up a makeshift lab to study dead birds, Porter and Holcomb tried to find a way to rescue those still living. "Interior was forcing us into areas we knew nothing about," Porter said. "All of the sudden, we had to find boats and planes—organize an armada, get equipment, train people to catch birds. We knew nothing about Prince William Sound. Fish and Wildlife knew all the bays and coves, all the estuaries, currents, rivers, rookeries and marshes. But they wouldn't help us.

"Some Fish and Wildlife people wanted to help, but their hands were tied by Interior," Porter continued. "We started calling this the 'oil spill from hell'—not because of the dying birds, the long hours of cleaning, the rugged geography, or even the amount of oil, but because of the attitudes of Fish and Wildlife and Exxon."

At this point, the frustration was so great, the bird center considered calling off their rescue efforts altogether. Alice Berkner, who set up the Valdez rescue center, said it was too dangerous to send people out. Holcomb replied, "Alice, this spill is killing thousands of birds. We can't just sit back and let this happen. Somebody has to go out there and try to get them. I know how to catch birds. We'll just have to find somebody who really knows the sound."

Enter Kelly Weaverling, a long-haired, bearded Cordova bookseller anxious to put his unusual skills and experience to work. Weaverling was probably the only person who had walked every oil-threatened beach of Prince William Sound. He knew where the winter birds were, and where and when the migrants would arrive. When the spill hit, Weaverling called friends in Valdez to say, "I'm here. I know Prince William Sound. And I want to help. Call me."

Weaverling came to Alaska in 1976. He spent his first summers guiding sea kayakers through Prince William Sound, and after he married, he and his wife, Susan, decided simply to live out there during the summer months. "What we really like to do is camp out in the sound, just be there and kayak from place to place," Weaverling said. "We'll be out there from the first of May until the end of August and always explore places we haven't seen before. I've kept notes and records of every beach I was on, every stretch of coastline I've paddled."

On March 29, five days after the spill, somebody told Jay Holcomb about Kelly Weaverling, and Holcomb called the Cordova bookstore. "Weaverling, can you help us find and collect oiled birds?" he asked.

Weaverling immediately put out the call for bird-catching equipment. "People started bringing things in lickety-split," he said. "They gave us their plastic dog kennels for cages. If they didn't have plastic kennels, they brought in cardboard boxes. It all came together just like that."

By nightfall, Weaverling and his companions were headed to Valdez with three boats. They arrived at 6:30 a.m. and roused the bird rescue people.

"Weaverling was real defensive toward us at first," Porter recalled. "He viewed us with suspicion because we were under contract to Exxon. At that time, anyone working with Exxon was the bad guy. And he came from Cordova, where tempers were hot. But he could see that we wanted to save birds as much as he did.

"When he stood there looking at us," Porter mused, "I'm sure he was thinking, 'These people don't know a damn thing about catching birds in Prince William Sound.' And he was right. He became our main contact with the sound."

Porter and Holcomb went straight to the harbor with Weaverling to check out his boats. Since the Cordova boats didn't have enough heated space for bringing back birds, Weaverling began looking for a boat with more enclosed area.

"I glanced around the harbor, and sitting right there in front of us was just the kind of boat we wanted," Weaverling said.

"We talked to the boat's skipper and found out he was already on contract. I looked at his ship's log, and for the last four days all it said was 'standing by for Exxon bird rescue.' The bird rescue people didn't even know that boat was available to them. Anyway, we didn't have to negotiate with the skipper. He was already working for us."

They took off for Knight Island, which had been heavily oiled. Holcomb went along to teach Weaverling and his team how to catch oiled birds. On the ride out, Weaverling described what he and his wife loved about coming to the sound in spring. "We'd always try to get out there before the bird migration begins in early May. Twenty million birds go by in a period of about two or three weeks. Sometimes we'd see 150,000 go by in an afternoon—golden plovers, sandpipers, snow geese, swans, all kinds of ducks and small shorebirds."

That's how Weaverling remembered the sound. But on this cold April evening when they tied up in Snug Harbor, oil was 18 inches deep on the surface of the water. "There was oil all over the place," Weaverling said. "Dying animals were floating around. Just the worst. You can't tell what it's like from television, personal accounts, flying over, or from a boat. You have to reach down into that oily water and pull out a bird.

"We cried a lot," Weaverling said. "All of us did, at least once, maybe twice a day. You'd just have to stop and sit down. It's just beyond imagination. Oil everywhere. Snow falling. Dead otters. Dead deer. Dead birds."

As oil spread beyond the sound, the surface of the sea swirled with an iridescent sheen. The oil missed some rookeries and hit others. In many places, frothy, ankle-deep "mousse" washed ashore, coating and recoating the intertidal zone with each change of the tide.

Reports of dead seabirds and eagles began coming in from beyond the sound like news of fatalities from some distant war.

"It's absolutely wrenching in some of those heavily oiled areas," said a biologist documenting the oil's impact on birds. "Once, I set a tape cassette on a black beach rock, and the rock began to crawl away. It was a seabird caked with oil."

Sheets of oil reached Montague Island, and the leading edge of the spill approached the murre and puffin rookeries on the seawalls of the Chiswell Islands, at the outer reaches of Kenai Fjords National Park. To cope with the spread, Weaverling took out his map and sectioned off seven bird-capture areas. Each was larger than the state of Delaware.

In addition to finding vessels and crews, Weaverling had to broker their contracts with Exxon, a process that often tested his patience. "I feel like I've been jacked around," he said. "We've had eight different people in charge of Exxon operations in Cordova alone. They come up here in cowboy boots with names like Skeeter and Bubba, and work here for a week or two, and then they're gone. We never see them again. I don't know whether there's a plan to avoid accountability or not. But it certainly works that way."

Porter: "We didn't feel Exxon really wanted us here. Oh, some of their people were helpful, but the thing that was missing from most of them was remorse. It's understandable that they didn't want to be so far away from their families. But they just didn't seem to care about Alaska or the birds or other people trying to help.

"Personally, I've never been treated so rudely by an oil company in my life," she added. "At one public meeting, we were ridiculed when an Exxon manager said, 'We have to deal with birds, so we've got these tree huggers taking care of them. But, we'll keep them in line . . . I'll kick their butts.'"

"But the local people and fishermen were absolutely fantastic to us," Porter added. "They came in and gave, even if it was just encouragement. And kids in schools wrote us letters: 'Don't worry. We know you are sad. Save as many birds as you can.'" 41

Content Considerations

1. How were government agencies, Exxon, and the rescue center supposed to work to rescue the birds after the oil spill? What happened?
2. Why was responsibility for capturing the birds such a stumbling block?
3. Describe the process of rescuing birds in the aftermath of an oil spill.

The Writer's Strategies

1. Most of this article comprises statements by people involved in the cleanup. What is the effect of this method on you as you read?
2. For what reasons might the author have chosen to focus on marine birds?

Writing Possibilities

1. Write an essay explaining where your own sympathies lie among the people mentioned in this article.
2. Using this article and any other sources you need, explain the ideal plan for rescuing and rehabilitating birds affected by an oil spill.
3. Respond to what this article suggests about the willingness of volunteers to help in a crisis.
4. In your own view who bears responsibility for a major cleanup operation? How should such operations proceed?

PETER NULTY

What We Should Do to Stop Spills

Millions of gallons of oil are shipped across the ocean every year. Transported by huge tankers, the oil usually reaches its destination. But when it doesn't—when the tanker runs aground as did the Exxon Valdez *in 1989 or catches fire as did the* Mega Borg *in 1990—several million gallons of oil pour into the ocean.*

Then the oil must be cleaned up—no easy task. Marine animals must be rescued and saved; the carcasses of those who are not saved must be removed. The shore must be scoured and the estuaries cleaned so that life in those fragile environments may continue in some fashion. The damage done by a large spill is great and long-lasting; no one seems to know exactly how deeply into the future the damage cuts.

In the next article, reprinted from a July 1990 issue of Fortune, *writer Peter Nulty explains that the means to stop spills are available—only the will is lacking. Before you read his perspective, write a paragraph about how this could be so.*

EXXON VALDEZ. Kill Van Kull. *Mega Borg.* The names of oil spills are beginning to carry the kind of emotional charge usually stored in the names of infamous military defeats: Dunkirk, Pearl Harbor, Tet. One mention and most listeners feel a rush of anger, humiliation, and disappointment. The oil industry in the past 15 months is awash with such defeats. [1]

The time has come to dam the spillage of oil. On this much we all agree. After that, the facts about oil spills and what we can do about them are pretty well misunderstood. Two points are paramount: First, the tanker fleet is not spilling more oil than it used to; and second, there's no mystery about what's needed to improve the present record—it's a matter of finding the will. [2]

Despite the impression created by wide coverage of such mishaps as the fire aboard the Norwegian tanker *Mega Borg* off Galveston, Texas, last month or last year's *Exxon Valdez* grounding in Alaska, U.S. Coast Guard records show no clear trend of increase or decrease in the amount of oil being spilled since the early 1970s. The median annual spill rate is just over four million gallons a year. [3]

Like airline disasters, though, monster oil spills seem to come in bunches, and when they do, they skew the record wildly upward. The past 15 months is such a time. The *Exxon Valdez*, which spilled 10.7 million gallons of crude, accounted for about 80% of the spillage in 1989. A series of spills in New York harbor and the *Mega Borg* incident, which may have put four million gallons into the Gulf of Mexico, have made the first half of 1990 thoroughly messy. Other bad years include 1975 and 1985, which were even worse than 1989.

If the recent plague of accidents represented a new norm, it might justifiably be called a crisis. But nothing in the evidence indicates that is so. The industry's critics are right when they point out that the tanker fleet is aging. Orders for new vessels dropped off in response to reduced demand for oil in the early 1980s. And tanker crews are smaller, replaced in part by electronics. If these changes have resulted in deteriorating seamanship or unseaworthy ships, it doesn't show in the records yet.

On the contrary, the number of pollution-causing tanker accidents recorded by the Coast Guard has fallen by over 50% in the 1980s, suggesting that safety is improving. In spite of the bad start, 1990 could yet turn out to be a cleaner year than 1989, and 1991 might be better still. That would fit the pattern of sudden spikes and relative lulls.

Another spate of horrendous spills a few years down the road would also fit the pattern, however, so don't lower the sludge alert banner yet. In fact, run up another one just for emphasis. While chances are that accidents like the *Exxon Valdez* and the *Mega Borg* are not becoming annual events, expect the long-term average rate of spillage to begin climbing—unless even more is done to improve tanker operations.

Oil imports have increased almost 30% since 1986, and they are likely to continue growing. Those imports will arrive in an ever-thickening flow of tanker traffic. Says Sean Connaughton, a marine transportation expert at the American Petroleum Institute, a lobbying organization: "In general, the greater the traffic, the greater the risk."

As imports rise, so will the use of supertankers too large to enter many existing ports. That will result in more "lightering" operations, the transfer of oil to smaller ships for the last part of the journey. The Coast Guard says lightering, which entails some risk of fire or explosion during the transfer, has already doubled since 1988. The *Mega Borg* fire, whose cause remains undetermined, began during lightering.

While the full impact of the Alaskan spill is not yet known, the U.S. has probably not yet suffered a worldclass, worst-case spill disaster. Pray it doesn't. Ranked by size, the *Exxon Valdez* was only the world's 21st-largest spill, according to the *Oil Spill Intelligence Report,* a newsletter in Arlington, Massachusetts. The world's biggest was probably the blowout in 1979 of an offshore Mexican well called Ixtoc 1, which spewed at least 140 million gallons of crude into the Gulf of Mexico. Ixtoc 1 was 15 times larger than the *Valdez* spill and 36 times larger than *Mega Borg*. The environmental impact of Ixtoc was never fully evaluated.

Scientists don't know a lot about the damage caused by spills, but one thing they feel sure of: Size is not the most important factor. The type of oil, and weather and sea conditions, count for more. Light oils are more toxic than heavy crudes, although the light varieties evaporate more rapidly. Wind and currents may disperse a spill, which is helpful. Or they may mix it with the sea, thereby poisoning marine life. They may also drive the oil into wetlands and estuaries teeming with wildlife, a worst-case event that for a while seemed possible in the *Mega Borg* accident.

As it happened, the fire was extinguished, saving most of the 38-million-gallon cargo. And most of the oil that spilled burned off or evaporated. Perhaps more significant for the future were the results of the first "bioremediation" experiment conducted on a spill in open waters. Alpha Environmental Inc. of Austin, Texas, spread naturally occurring oil-eating microbes it had gathered from around the world on the *Mega Borg's* slick. The microbes devoured the oil, in effect eating themselves out of a job and a life. When the oil was gone, they died.

A better approach, of course, is to keep the oil in the ships in the first place. There is surprising unanimity on what steps are needed to do so. What is lacking is the congressional willpower to sweep aside the endless niggling of interest groups intent on gaining some last-minute advantage. Here are some commonsense things to do:

- First, upgrade the traffic-control system run by the Coast Guard. Says Captain Antonio Valdes, vice president of Conoco Shipping Co.: "Europe has tremendous traffic control. The port of Rotterdam is run like an airport. We need more electronic guidance, more buoys, clearer lanes of traffic."

- Second, train crews better. Roughly 80% of the accidents are caused by human error. An international convention called the Standards for Training and Certification of Watchkeepers sets educational and licensing rules for crew members operating ships in most countries. The Senate has not ratified the treaty, apparently because of pressure from U.S. ship operators who would be required to spend a lot more to train their crews.

- Third, license managers of tanker companies, the owner-operators. Says Arthur McKenzie, founder and director of Tanker

Advisory Center, Inc., which compiles the safety records of all tankers: "The most important thing about a tanker is who owns it. Do something about the guys in the office." McKenzie suggests that licenses be renewed every few years based in part on the holder's accident and pollution record.

- Fourth, upgrade the fleet with safer tankers. Both houses of Congress have passed bills, calling for double-hulled tankers, that will probably become law by the end of the year. The industry generally accepts that double hulls will spill less in an accident than single hulls. But an important principle may be lost. The law should leave open the path to future technologies that may prove superior to double hulls. Congressman Don Young of Alaska says he wants the law to call for "double hulls or better." There just might be some better ideas. Shell International Marine and Mitsubishi have proposed designs that promise significant improvements over double hulls.

- Fifth, establish a clear central authority to direct emergency salvage and cleanup operations. The oil industry is organizing an entity called PIRO (petroleum industry response organization) that plans to set up five spill stations around the country, each with enough equipment—booms, skimmers, and the like—to handle nine-million-gallon spills. The cost: $500 million in the first five years.

    PIRO is a good idea, but in an emergency authority over it should lie with the Coast Guard alone. The cleanup of the *Exxon Valdez* spill was slowed by a disagreement between the state of Alaska and the Coast Guard over whether to stabilize the stricken ship first or go after the spreading slick. Moreover, PIRO should be granted immunity from damage suits except in cases of gross negligence. Some members of Congress prefer unlimited liability. But PIRO President Jack Costello fairly asks: "How good would the fire department be if it got sued every time the men broke a door or a window while putting out a fire?"

- Finally, a large fund should be established to pay for damages and cleanup, not to let companies off the hook but to be sure the job gets done if an incident involves a company with shallower pockets than Exxon's. There are numerous schemes for how this might be handled. One is to join a set of international protocols. The protocols were vigorously promoted by the U.S. in the early 1980s, then rejected because of opposition in Congress to the concept of limited liability that they embody: $78 million for the owner of a vessel causing a spill. In addition, the protocols provide up to $260 million from a fund created by cargo owners.

These amounts are woefully inadequate. The limit on liability should be high enough to give tanker operators incentive to avoid accidents. Some in Congress favor joining the protocols and then working to enlarge the amounts. Others would create a backup fund of about $1 billion in the U.S., which would be available after the international fund was exhausted.

Whatever solution the lawmakers arrive at should have the following characteristics: The money in reserve should be large, several billion dollars at least. (Exxon has spent more than $2 billion to clean up Prince William Sound and the meter is still running.) This doesn't have to cost consumers very much. A $2 billion fund could be financed by a 5-cents-a-barrel fee on all oil consumed in the U.S. Assuming the cargo owner passes on the cost, roughly one-tenth of a cent would be added to each gallon of fuel. That's both bearable and worth the price.

The issue of spills is coming to a head not because the industry is getting sloppier but because public tolerance is shrinking fast. That is as it should be. Earth's ability to recover from industrial insult is also running down, and only we can recharge it.

Content Considerations

1. What beliefs about oil spills does the article attempt to correct?
2. Explain the six strategies the article offers as ways to prevent and clean up oil spills.
3. Why haven't the strategies been implemented, especially since the author claims universal support for them?

The Writer's Strategies

1. What kinds of evidence does Nulty provide for his assertions? Are they convincing?
2. What is the author's perspective? In what ways does he make it known?

Writing Possibilities

1. With a classmate, think of ways an oil spill affects you. Write about them.
2. Examine again the strategies Nulty outlines; search your own thoughts and experience for additional material, then draft a letter to an official in which you urge adoption of the strategies you support.
3. Find a passage in the article with which you disagree and construct an argument against it.

JODI L. JACOBSON

Holding Back the Sea

The "greenhouse effect" refers to the ability of the Earth's atmosphere to hold heat. It is a natural phenomenon that makes Earth warm enough to sustain life. But when the Earth emits more methane, carbon dioxide, chlorofluorocarbons, and other gases than its systems can absorb, the gases are trapped in the atmosphere, holding the heat the Earth's surface radiates. This increases the greenhouse effect and thus increases surface air temperatures. (Some of these gases also affect the layer of ozone in the stratosphere that normally blocks the sun's ultraviolet rays; increased ultraviolet radiation poses threats to human health and agriculture.)

Among the many possible effects of global warming is that the level of the seas will rise. Researcher Jodi L. Jacobson discusses the causes and effects of such a rise in the following article.

Before you read her discussion, write a few sentences about sea level—what the term means and how such a thing as sea level may affect a society.

QUICK STUDY OF A WORLD MAP illustrates an obvious but rarely considered fact: Much of human society is defined by the planet's oceans. And the boundary between land and water determines a great deal that is often taken for granted, including the amount of land available for human settlement and agriculture, the economic and ecological productivity of deltas and estuaries, the shape of bays and harbors used for commerce, and the abundance or scarcity of fresh water in coastal communities.

For most of recorded history, sea level has changed slowly enough to allow the development of a social order based on its relative constancy. Global warming will radically alter this. Increasing concentrations of greenhouse gases in the atmosphere are expected to raise the earth's average temperature between 2.5° C and 5.5° C over the next 100 years. In response, the rate of change in sea level is likely to accelerate due to thermal expansion of the earth's surface waters and a rapid melting of alpine and polar glaciers and of ice caps. Although the issue of how quickly oceans will rise is still a matter of debate, the economic and environmental losses of coastal nations under most scenarios are enormous. One thing is clear: No coastal nation, whether rich or poor, will be totally immune.

Accelerated sea-level rise, like global warming, represents an environmental threat of unprecedented proportion. Yet most discussions of the impending increase in global rates obscure a critical issue: In some regions of the world, the local sea level is already rising quickly. Egypt, Thailand, and the United States are just a few of the countries where extensive coastal land degradation, combined with even the recent small incremental changes in global sea level, is contributing to large-scale land loss. These trends will be exacerbated in a greenhouse world.

Low- to middle-range estimates by the U.S. Environmental Protection Agency (EPA) indicate a warming-induced rise by 2100 of anywhere from a half-meter to just over two meters. A one-meter rise by 2075, well within the projections, could result in widespread economic, environmental, and social disruption. G.P. Hekstra of the Ministry of Housing, Physical Planning, and Environment in the Netherlands asserts that such a rise could affect all land up to five meters in elevation. Although only a small percentage of the world total—about 3%—this area encompasses one-third of global cropland and is home to a billion people.

As sea level rises, coastal communities face two fundamental choices: retreat from the shore or fend off the sea. Decisions about which strategy to adopt must be made relatively soon because of the long lead time involved in building dikes and other structures and because of the continuing development of coasts. Yet allocating scarce resources on the basis of unknown future conditions—how fast the sea will rise and by what date—entails a fair amount of risk.

Protecting beaches, homes, and resorts can cost a country with a long coastline billions of dollars, money that is only well spent if current assumptions about future sea level are borne out. Assessing the real environmental costs is difficult because traditional economic models do not reflect the fact that structural barriers built to hold back the sea often hasten the decline of ecosystems important to fish and birds. Moreover, protecting private property on one part of the coast often contributes to higher rates of erosion elsewhere, making one person's seawall another's woe.

GLOBAL CHANGES, LOCAL OUTCOMES

A higher global average temperature can alter sea level in several ways: The density can decrease through the warming and subsequent expansion of seawater, which increases volume. The volume also can be raised by the melting of alpine glaciers, by a net increase in water as the fringes of polar glaciers melt, or by more ice being discharged from ice caps into the oceans.

The slight variations in global climate of the last 5,000 years are responsible for correspondingly small fluctuations in sea level, on the order of 1–10 centimeters every century. Over the past 100 years, however, global sea level rose 10–15 centimeters (4–6 inches), a somewhat faster pace. Scientists continue to debate the cause of this rise, many arguing that there is no evidence that it is due to human-induced warming, while others are not so sure.

Uncertainties abound on the pace of all the possible changes expected from global warming. The most immediate effect will probably be an increase in ocean volume through thermal expansion. The rate of thermal expansion depends on how quickly ocean volume responds to rising atmospheric temperatures, how fast surface layers warm, and how rapidly the warming reaches deeper water masses. The pace of glacial melt and the exact responses of large masses such as the Antarctic shelf are equally unclear. Over the long term, however, glaciers and ice caps will make the largest contribution to increased volume if a full-scale global warming occurs.

Over the past five years, a number of scientists have estimated the possible range of greenhouse-induced sea-level rise by 2100. Gordon de Q. Robin projects an increase of anywhere from 20 to 165 centimeters. Computations by other scientists yield projections as high as two to four meters over the next 110 years. Widely cited EPA estimates of global mean sea-level rise by 2100 range from 50 to 200 centimeters depending on various assumptions about the rate of climate change. Most models do agree that initial rates of increase will be small relative to the much more rapid acceleration expected from 2050 on. In any case, even the low range of estimates portends a marked increase over the current global pace.

What is important about the sea-level rise expected from global warming is the pace of change. The rate expected at the global level in the foreseeable future—one meter by 2075 is certainly plausible—is unprecedented on a human time scale. Unfortunately, with today's level of population and investment in coastal areas, the world has much more to lose from sea-level rise than ever before.

LANDS AND PEOPLES AT RISK

Intense population pressures and economic demands are already taking their toll on deltas, shores, and barrier islands. Rapid rates of subsidence and coastal erosion ensure that many areas of the world will experience a one-meter increase in sea level well before a global change of the same magnitude.

The ebb and flow of higher tides will cause dramatic declines in a wide variety of coastal ecosystems. Wetlands and coastal forests, which account for most of the world's land area less than a meter above mean tide, are universally at risk.

According to EPA estimates, erosion, inundation, and saltwater intrusion could reduce the area of present-day U.S. coastal wetlands up to 80% if current projections of future global sea level are realized. The Mississippi Delta, the Chesapeake Bay, and other vital estuaries and wetland regions would be irreparably damaged.

No one has yet calculated the immense economic and ecological costs of such a loss for the United States, much less extrapolated them to the global level. Yet as global sea level rises, these problems will surely become more severe and widespread in ecosystems around the world.

Land subsidence is a key issue in the case of river deltas, such as the Nile and Bengal, where human activities are interfering with the normal geophysical processes that could balance out the effects of rising water levels. These low-lying regions, important from both ecological and social standpoints, will be among the first lost to inundation under global warming.

Under natural conditions, deltas are in dynamic equilibrium, forming and breaking down in a continuous pattern of accretion and subsidence. Subsidence in deltas occurs naturally on a local and regional scale through the compaction of recently deposited river-borne sediments. As long as enough sediment reaches a delta to offset subsidence, the area either grows or maintains its size. If sediments are stopped along the way, continuing compaction and erosion cause loss of land relative to the sea, even if the absolute level of the sea remains unchanged.

Large-scale human interference in natural processes has had dramatic effects on both relative rates of sea-level rise and on coastal ecosystems in several major deltas. Channeling, diverting, or damming of rivers can greatly reduce the amount of sediment that reaches a delta, as has happened in the Indus, the Mississippi, and many other major river systems, resulting in heavier shoreline erosion and an increase in local relative water levels.

Worldwide, erosion of coastlines, beaches, and barrier islands has accelerated over the past 10 years as a result of rising sea level. A survey by a commission of the International Geophysical Union demonstrated that erosion had become prevalent on the world's sandy coastlines, at least 70% of which have retreated during the past few decades.

Increased erosion would decrease natural storm barriers. Coastal floods associated with storm surges surpass even earthquakes in loss of life and property damage worldwide. Apart from greater erosion of the barrier islands that safeguard mainland coasts, higher seas will increase flooding and storm damage in coastal areas because raised water levels would provide storm surges with a higher base to build upon. And the higher seas would decrease natural and artificial drainage.

A one-meter sea-level rise could turn a moderate storm into a catastrophic one. A storm of a severity that now occurs about every 15 years, for example, could flood areas that are today only affected by truly massive storms once a century.

In Bangladesh, storm surges now reach as far as 160 kilometers upriver. In 1970, this century's worst storm surge tore through the countryside, initially taking some 300,000 lives, drowning millions of livestock, and destroying most of Bangladesh's fishing fleet. The toll climbed higher in its aftermath. As the region's population mounts, so does the potential for another disaster.

Studies indicate a dramatic increase in the area vulnerable to flooding in the United States as well. A one-meter rise would boost the portion of Charleston, South Carolina, now lying within the 10-year floodplain from 20% to 45%. A 1.5-meter rise would bring that figure to more than 60%, the current area of the 100-year floodplain. As a result, "once-a-century" floods would then occur on the order of every 10 years. Sea-level rise will also permanently affect freshwater supplies. Miami is a case in point. The city's first settlements were built on what little high ground could be found, but today most of greater Miami lies at or just above sea level on swampland reclaimed from the Everglades. Water for its 3 million residents is drawn from the Biscayne aquifer, which flows right below the city streets. That the city exists and prospers is due to what engineers call a "hydrologic masterwork" of natural and artificial systems that hold back swamp and sea.

Against a one-meter rise in ocean levels, Miami's only defense would be a costly system of seawalls and dikes. But that might not be enough to spare it from insidious assault. Fresh water floats atop salt water, so as sea levels rise the water table would be pushed nearly a meter closer to the surface. The elaborate pumping and drainage system that currently maintains the integrity of the highly porous aquifer could be overwhelmed. The higher water table would cause roads to buckle, bridge abutments to sink, and land to revert back to swamp. Miami's experience would not be unique. Large cities around the world—Bangkok, New Orleans, Taipei, and Venice, to name a few—face similar prospects.

Most Vulnerable, Least Responsible

The social and environmental costs of sea-level rise will be highest in countries where deltas are extensive, densely populated, and extremely food-productive. In these countries, most of which are in the Third World, heavy reliance on groundwater and the completed or proposed damming and diversion of large rivers—for increased hydropower and agricultural use, for flood control, and for transportation—have already begun to compound problems with sea-level rise. Almost without exception, the prognosis for these vulnerable, low-lying countries in a greenhouse world is grim.

The stakes are particularly high throughout Asia, where damming and diversion of such systems as the Indus, Ganges-Brahmaputra, and Yellow rivers have greatly decreased the amount of sediment getting to deltas. As elsewhere, the deltas reliant on these sediments support sizable human and wildlife populations while creating protective barriers between inland areas and the sea. Large cities, including Bangkok, Calcutta, Dhaka, Hanoi, Karachi, and Shanghai, have grown up on the low-lying river banks. These heavily populated areas are almost certain to be flooded as sea-level rise accelerates.

The U.N. Environment Programme's 1989 global survey represents the first attempt to analyze systematically the regions most vulnerable to sea-level rise. Ten countries—Bangladesh, Egypt, The Gambia, Indonesia, the Maldives, Mozambique, Pakistan, Senegal, Surinam, and Thailand—were identified as "most vulnerable." These 10 share many characteristics, including the fact that they are, by and large, poor and populous.

UNEP estimates based on current population size and density show that 15% of Bangladesh's land area, inhabited by 15 million people, is threatened by total inundation from a primary rise of up to 1.5 meters. Secondary increases of up to three meters would wipe out over 28,500 square kilometers (20% of the total land area), displacing an additional 8 million people.

Pressures to develop agriculture have quickened the pace of damming and channeling on the three giant rivers—the Brahmaputra, the Ganges, and the Meghna—that feed the delta. As a result, subsidence is increasing. This situation is being made worse by the increasing withdrawal of groundwater.

In the Nile Delta of Egypt, extending from just east of the port city of Alexandria to west of Port Said at the northern entrance of the Suez Canal, local sea-level rise already far exceeds the global average due to high rates of subsidence.

At least 40% of Indonesia's land surface is classified as vulnerable to sea-level rise. In terms of both size and diversity, the country is home to one of the world's richest and most extensive series of wetlands. Here, too, population pressures are already threatening these fragile ecosystems.

A one- to two-meter rise in sea level could be disastrous for the Chinese economy as well. The Yangtze Delta is one of China's most heavily farmed areas. Damming and subsidence have contributed to a continuing loss of this valuable land on the order of nearly 70 square kilometers per year since 1947. A sea-level rise of even one meter could sweep away large areas of the delta, causing a devastating loss in agricultural productivity for China.

Paying by the Meter

Assuming a long-run increase in rates of global sea-level rise, societies will have to choose some adaptive strategies. Broadly speaking, they face two choices: fight or flight.

Along with the intensified settlement of coastal areas worldwide over the past century has come a belief that human ingenuity could tame any natural force. As a result, people have been inclined to build closer and closer to the ocean, investing billions of dollars in homes and seaside resorts and responding to danger by confrontation.

Nowhere in the world is the battle against the sea more actively engaged than in the Netherlands. Hundreds of kilometers of carefully maintained dikes and natural dunes keep the part of the country that is now well below sea level—more than half the total—from being flooded.

The Dutch continue to spend heavily to keep their extensive system of dikes and pumps in shape and are now protected against storm surges up to those with a probability of occurring once in 10,000 years. But the prospect of accelerated sea-level rise implies that maintaining this level of safety may require additional investments of up to $10 billion by 2040.

Large though these expenditures are, they are trivial compared with what the United States would have to spend to protect its more than 30,000 kilometers (19,000 miles) of coastline. Preliminary estimates by EPA of the total bill for holding the sea back from U.S. shores—including costs to build bulkheads and levees, raise barrier islands, and pump sand—range from $32 billion to $309 billion for a one-half-meter to two-meter rise in sea levels.

Many countries have made vast investments reclaiming land from the sea; witness the efforts in Singapore, Hong Kong, and Tokyo. Political pressures to maintain these lands through dikes, dams, and the like will be high, but political support for subsidizing coastal areas may be undercut by competing fiscal demands over the long run. With increasing competition for scarce tax dollars, property owners in the year 2050 may find the general public reluctant to foot the bill for seawalls.

Moreover, what may seem like permanent protection often turns out to be only a temporary measure. While concrete structures may divert the ocean's energies from one beach, they usually displace it onto another. And by changing the dynamics of coastal currents and sediment flow, these hard structures interrupt the natural processes that allow wetlands and beaches to reestablish upland, causing them to deteriorate and in many cases disappear.

Beach nourishment is a relatively benign defensive strategy that can work in some cases. And comparing the costs and benefits illustrates that it is not usually as prohibitively expensive as other approaches. Sand or beach nourishment, for example, can cost $620,000 per kilometer, but these costs are often justified by economic and recreational use of the areas. A recent study of Ocean City, New Jersey, found it would cost about 25¢ per visitor to rebuild beaches to cope with a 30-centimeter rise, less than 1% of the average cost of a trip to the beach. And the fact remains that most beach replenishment is temporary at best, indicating the need for continuous investment.

The legal definitions of private property and of who is responsible for compensation in the event of natural disasters are already coming into question. As sea-level rise accelerates, pushing up the costs of adaptation, these issues will likely become part of an increasingly acrimonious debate over property rights and individual interests versus those of society at large.

Site-specific studies of several towns in the United States suggest that incorporating projections of sea-level rise into land-use planning can save money in the long run. Projections of costs in Charleston, South Carolina, show that a strategy that fails to anticipate and plan for the greenhouse world can be expensive. Depending on the zoning and development policies followed, including the amount of land lost and the costs of protective structures built, the costs of a one-meter sea-level rise may exceed $1.9 billion by 2075—an amount equal to 26% of total current economic activity in this area. If land-use policies and building codes are modified to anticipate rising sea levels, this figure could be reduced by more than 60%. Similar studies of Galveston, Texas, show that economic impacts could be lowered from $965 million to $550 million through advanced planning.

Whatever the strategy, industrial countries are in a far better financial position to react than are developing nations. Debates over land loss may be a moot point in poorer countries like Bangladesh, where evacuation and abandonment of coastal land may be the only option when submergence and erosion take their toll and when soil and water salinity increase. As millions of people displaced by rising seas move inland, competition with those already living there for scarce food, water,

Planning Ahead

An active public debate on coastal-development policies is needed, extending from the obvious issues of the here and now—beach erosion, river damming and diversion, subsidence, wetland loss—to the uncertainties of how changes in sea level in a greenhouse world will make matters far worse. Raising public awareness on the forthcoming changes, developing assessments that account for all future and present costs, and devising sustainable strategies based on those costs are all essential.

Taking action now to safeguard coastal areas will have immediate benefits while preventing losses from soaring higher in the event of an accelerated sea-level rise. Limiting coastal development is a first step, although strategies to accomplish this will differ in every country. Governments may begin by ensuring that private-property owners bear more of the costs of settling in coastal areas.

A new concept of property rights will have to be developed. Unbridled development of rivers and settlement of vulnerable coasts and low-lying deltas mean that more and more people and property will be exposed to land loss and potential disasters arising from storm surges and the like. Governments that plan over the long term to limit development of endangered coasts and deltas can save not only money, but resources as well. Wherever wetlands and beaches are not bordered by permanent structures, they will be able to migrate and reestablish further upland, allowing society to reap the intangible ecological benefits of biodiversity.

Of course, protection strategies will inevitably be carried out where the value of capital investments outweighs other considerations. But again the key is to plan ahead. As the Dutch discovered, more money can be saved over the long term if dikes and drainage systems are planned for before rather than after sea levels have risen considerably.

A cap on problematic dam-building and river-diversion projects in large deltas would lessen the ongoing destruction of wetland areas and prevent further reductions in sedimentation, thereby minimizing subsidence as well.

Additional money is needed to do more research on sea level globally and regionally. Funds are needed to support studies of beach and wetland dynamics, to take more-frequent and widespread measurements of global and regional sea levels, and to design cost-effective, environmentally benign methods of coping with coastal inundation.

The majority of developing nations most vulnerable to sea-level rise can do little about global warming independently. But they have a clear stake in reducing pressures on coastal areas by taking immediate actions. Among the most important of these is slowing population growth and, where necessary, changing inequitable patterns of land tenure in interior regions that promote coastal settlement of endangered areas. Furthermore, the governments of Bangladesh, China, Egypt, India, and Indonesia, to name just a few, are currently promoting river development projects that will harm delta ecosystems in the short term and hasten the date they are lost permanently to rising seas.

The issue of how to share the costs of adaptation equitably may well be among the hardest to resolve. Industrial countries are responsible for by far the largest share of the greenhouse gases emitted into the atmosphere. And no matter what strategies poorer nations adopt to deal with sea-level rise, they will need financial assistance to carry them out. The way industrial countries come to terms with their own liability in the face of accelerated sea-level rise will play a significant role in the evolution of international cooperation.

Content Considerations

1. How does global warming make sea levels rise?
2. Summarize the ecological and social consequences of a substantial (1-meter, or 39.37-inch) rise in sea level.
3. What might countries do to reduce the harm predicted to occur if sea levels rise?

The Writer's Strategies

1. Rate the objectivity of this piece. How is the author's opinion expressed?
2. How does Jacobson speak with a voice of authority in this article?

Writing Possibilities

1. Challenge or support Jacobson's assertion that many people believe "human ingenuity could tame any natural force" (paragraph 34), using what you have observed or read to illustrate how widespread that belief is.
2. With a classmate, explore the likely consequences of the two choices that may face countries if sea levels do rise: "retreat from the shore or fend off the sea."
3. Using this article and the one by Anne W. Simon, explain how coastal practices may exacerbate the problems created by a rising sea.

WILLIAM ASHWORTH

Sludge

The Great Lakes—Superior, Michigan, Huron, Erie, and Ontario—are large enough to hold 18 percent of the earth's liquid fresh water; they supply drinking water to 24 million people. In recent years much of the pollution that has threatened the lakes has slowed—sewage disposal, for example, has been greatly reduced. But reducing the levels of toxic chemicals in the lakes is still a major concern; in 1986 bordering states and provinces signed the Toxic Substances Control Agreement, an agreement of planned reductions in toxic disposals in the lakes.

The Great Lakes have long served as a dumping ground, however, some more than others; the wastes have killed fish, have led to a ban on the sale of fish because of toxicity, and are suspected to have caused numerous health problems.

William Ashworth examines these and other conditions in The Late, Great Lakes: An Environmental History *(1986), from which the following selection is taken. Before you begin, recall what you know about disposing of toxic wastes and write about it.*

MOVING WATER carries sand, silt, and small particles of soil; standing water drops them. From this simple geologic truth comes the wearing down of mountains and the slow infilling of lakes and oceans. It is a natural process that has always gone on and will do so until the end of time. Rivers move; they carry loads of silt with them, eroded from the uplands. Entering still water, they slow down and stop, letting go of their loads, forming great fan-shaped deltas off their mouths. Currents in the deep pick silt and sand from the deltas and transport them further, spreading them in layers throughout the lake or ocean. Some currents, aided by gravity, carry their loads downslope to the profundal depths of the bed; here, because they can go no further, they stop. Other currents move parallel to the shore. The sands and silts move with them. Back eddies swirl in coves and bays and behind peninsulas; the shore currents slow, the sands and silts drop out once more. Bars form. The process is continuous, subtle, and much more rapid than most people realize. More than eleven million metric tons of silt and sediment enter the Great Lakes from their tributaries each year; another forty-nine million tons are moved about by shore drift. That is sixty million tons of sediment deposition and rearrangement annually—nearly 1.2 million tons per week.

Because light objects tend to remain suspended in water longer than heavy ones, the sediments are sorted as they are deposited: big particles on the bottom and close to shore, small particles on top and further out. It is the small particles that are the primary problem. At a certain size—two microns in diameter, roughly the size of fine-grained clay—they become remarkably efficient at picking up and holding chemicals on their surfaces, a process known technically as "adsorption." Falling slowly through the water, these particles act like thousands of tiny vacuum cleaners, sweeping up pollutants as they go. Dioxins, phenols, phosphorus, mercury, halozines, pesticides, and nearly everything else that happens to be dissolved or suspended in the water are all adsorbed molecule by molecule and carried to the bottom. This is good for the water but bad for the bottom. Like an animal accumulating pollutants in its body fat, the lake accumulates them in its sediments—a process that is every bit as dangerous for the lake as it is for the animal.

Animals exposed to high concentrations of pollutants often develop localized malignancies—tumors. Lakes, too, have localized malignancies from high concentrations of pollutants. These are the limnetic equivalent of tumors—the trouble spots, the places where the symptoms of poisoning first begin to show up. And along the Fifth Coast, the worst of them all is at Waukegan, Illinois.

The story actually begins more than 100 years ago, in the summer of 1882. Chester A. Arthur was President of the United States; John Peter Altgeld had just been elected Governor of Illinois. The Chicago Cubs were winning the National League pennant for the third year in a row (there was as of yet no American League). And thirty miles to the north of the pennant celebrations, up by the Wisconsin border, the citizens of the little coastal community of Waukegan—having sought and secured the necessary appropriations from Congress—were beginning the excavation of a commercial harbor in the sand flats bordering Lake Michigan in front of their city.

It was not a simple job. There were no natural openings in the coastline in the Waukegan vicinity—no coves, no bays, no river mouths, hardly even a curved beach. The little stream the city had been built on was inadequate to float anything larger than a small canoe, and an earlier attempt to enclose a harbor by building a breakwater had met with failure when the half-completed structure washed away in one of the big Lake's oceanic storms. Waukegan was a waterfront city with no waterfront; it was a federally designated United States Port of Entry, but it had no port to enter.

It did, however, have the sand flats. They lay at the base of the clay bluff the city was built on and extended as much as a half mile out to the edge of the boundless water, where the ships came up over the horizon from Mackinac and steamed right past to the honest-to-goodness Port of Entry at Chicago. A harbor could be dug in that sand; it could be connected to the Lake by a narrow channel, and the mouth of the channel could be protected by a smaller and more secure breakwater than the one the storm had run away with earlier. Then perhaps some of the ships would slip in past the breakwater and up the channel instead of steaming past, and the docks would fill with commerce, and Waukegan, which had begun to show signs of stagnation, would grow and prosper once again.

That, at any rate, was the scenario the city had managed to sell to Congress in the spring of 1882. Twenty years later the work was done; and during the next half-century Waukegan, which had stabilized at around 7,000 people before the harbor project was completed, grew rapidly—just as had been predicted—to a population of over 50,000. Manufacturers came, drawn by the new port facility; for the most part they built their factories at the base of the bluff, on the flats near the water.

Among these manufacturers was a boatbuilder named Johnson, who diversified into outboard motors in the early years of the twentieth century. The company expanded rapidly. In 1929, Johnson Motors merged with Evinrude and several others to form the Outboard Motors Corporation; and in 1936 a further merger resulted in the formation of a company called Outboard Marine, which by 1970 had grown to dominate the American pleasure-boat business, as well as holding large shares of the market for chain saws, snowmobiles, and similar equipment powered by small gasoline engines. Through all these changes, though, the company continued to build its increasingly popular Johnson outboard motors at the old plant location on the sand flats beside Waukegan's artificial harbor. The Johnson plant was progressively run; it was a major employer of the Waukegan work force; and it was generally looked upon as a boon to the community.

An assessment, as it turns out, that—due in part to a curiously coincident set of occurrences halfway around the world and in a completely unrelated field—was eventually going to be subject to a great deal of change.

In 1881—the year before Waukegan began construction of the harbor that led to all this activity—a group of organic chemical compounds called araclors was synthesized in a laboratory in Germany for the first time. Araclors were a byproduct of the investigation of coal tar, which was transforming the science of chemistry during the latter half of the nineteenth century. Long thought of as merely an ugly nuisance, coal tar—a sticky black residue left over from the manufacture of illuminating gas from coal—had turned out to be a virtual cornucopia of exotic hydrocarbon-based compounds with unusual and often mystifying chemical properties. Most of these compounds were built around the benzine ring—six hydrogen atoms and six carbon atoms, arranged in the shape of a closed hexagon.

Many benzine derivatives were useful as artificial dyes—a characteristic that had been the original impetus for their study. The araclors were not among these. They were thick, colorless, odorless fluids, chemically inert and almost impossible to break down by heat or pressure. There were more than 100 varieties of them, and they all appeared quite useless. The German chemists noted their properties, set the compounds aside, and went on in their search for dyes and other more interesting materials.

In 1929—the year of Johnson's merger with Evinrude—a use was finally discovered for the "useless" araclors. Their high stability and resistance to heat and pressure, it turned out, made them excellent lubricants and hydraulic fluids for industrial applications such as aluminum die-casting, which requires presses capable of exerting pressures of more than 2,000 tons per square inch. Once one application was found, others soon followed: as insulators for electrical transformers and capacitors, as constituents of ink and of "carbonless" carbon papers, as ingredients of paint and plastics, as road-dust suppressors, as hardeners in certain metallurgical processes. The Monsanto Corporation, a chemical firm founded in St. Louis in 1915 by a pharmaceuticals clerk who was tired of seeing everything in the store brought in from Europe at tariff-inflated prices, sought and received permission to manufacture araclors in America. The company sold them under that name, but by that time chemists were calling them something else. The new name was based on the compounds' chemical composition. Since the basic araclor molecules were composed of two modified benzine rings, they were properly called biphenyls; since each benzine ring had two or more chlorine atoms replacing its normal hydrogen atoms, they were called polychlorinated. And since "polychlorinated biphenyls" was too much of a mouthful even for an organic chemist, the araclors (or Araclors, since that was now properly a trade name) became known among scientists by a three-letter abbreviation: PCBs.

In 1954, the Outboard Marine Corporation—which had recently converted to die-cast aluminum construction in its outboard motors—

purchased the first small shipment of what would eventually amount to approximately ten million pounds of this "useless" fluid for use in their plant on the flats beside Waukegan's artificial harbor. It was cheap, and the chemists had assured everyone that it was almost completely inert, so it seemed pointless to be particularly careful with it. By the company's own estimates, approximately 20 percent of the material—some 100,000 tons of it—leaked from the machinery, spilled, or was otherwise lost in the plant, where it was washed down drains and discharged into Waukegan's proud artificial harbor. PCBs are slightly heavier than water. Most of those 100,000 tons sank to the bottom of the harbor and stayed there.

Meanwhile, elsewhere in the world, the doubts were beginning. 14

At first they were just hunches. DDT, the widely used "miracle" 15 pesticide of World War II, had turned out to be perilously dangerous when loose in the environment—building up alarmingly in the body fat of predators, interfering with the reproductive abilities of birds and mammals, killing shellfish and other invertebrates, disrupting food chains, and generally wreaking havoc. DDT and PCBs were closely related to each other; both were chlorinated hydrocarbons originally derived from coal tar. Would PCBs, too, turn out to be dangerous? Laboratory work in Sweden in 1966 suggested that they would; and two years later that suggestion was tragically confirmed by an incident in Japan. In Kitakyushu City, on the northern tip of the island of Kyushu, more than 1,000 people came down with moderate to severe symptoms of a mysterious poisoning: acne-like skin lesions, jaundice, nausea and vomiting, abdominal cramps, swollen limbs and joints, and severe lassitude. Mothers gave birth to gray-skinned, weakened, undersized infants with running discharges from their eyes. Analysis of the victims' body fat pointed to the villain: PCB concentrations averaging in the neighborhood of seventy-five parts per million (ppm). The PCBs, it eventually turned out, had come from a leaking transformer in a local rice-milling plant. A batch of rice oil had become contaminated by the leaked compound; the two fluids are strikingly similar in texture and color, and the PCBs had no taste or odor to give them away, and the contamination had gone undetected until it was too late. Five people died.

Since PCBs, like DDT, are bioaccumulative—meaning that they 16 tend to remain in the body, building up to higher and higher levels as each successive dose adds to the amount already in storage—long-term consumption of food containing even very tiny amounts of the material could be predicted to have the same eventual effects on the consumer as the unfortunate citizens of Kitakyushu City had suffered from short-term consumption of high amounts. Clearly, PCBs were not the benign material they had been made out to be. Monsanto voluntarily limited production of the compound in 1970, selling it after that only to manufacturers of high-load electrical equipment; and two years later the federal government banned the stuff altogether and initiated a survey of possible sites of serious PCB contamination around the country.

Four years later, in the fall of 1976, they came to Waukegan Harbor.

What they found there was classified information for a number of years, subject to a protective court order issued at the request of the Outboard Marine Corporation, but the protective order was lifted in 1981, and a year later the Water Quality Board of the International Joint Commission on Boundary Waters, in its 1982 *Report on Great Lakes Water Quality,* was able to spell out for the first time the singularly complete extent of the contamination of Waukegan Harbor. The harbor, the board noted with careful understatement, was "grossly contaminated with PCB." The bottom sludge was a nightmarish concoction of chemicals with a PCB content of as much as 500,000 milligrams per kilogram (mg/kg); the water itself was carrying as much as three micrograms of the supposedly insoluble stuff per liter. Fish from the harbor showed whole-body accumulations of as much as seventy-seven ppm—more than the victims of the Kitakyushu City incident had shown in their fatty tissues alone (PCBs, like many other chlorinated hydrocarbons, are strongly lipophilic—meaning that they tend to concentrate in their victims' body fat—so whole-body concentrations of the compound are normally considerably lower than fat concentrations). Fishing was banned from the harbor; dredging also had to stop, because dredging activities would stir up the sludge and resuspend it in the water. The Environmental Protection Agency and the State of Illinois sued Outboard Marine and were in turn sued by the company. Attempts began to find ways to remove the sediments and dispose of them safely; but no really satisfactory method turned up.

Today, four years later, it still hasn't.

"Nobody knows what to do about it," says Jane Elder, midwest representative for the Sierra Club, with a territory covering Wisconsin, Michigan, Minnesota, Indiana, Illinois, and Iowa. She works out of a windowless basement office near the University of Wisconsin in downtown Madison. "There are some fancy high-tech things they may be trying," she explains, "where they sort of set up a vacuum system so that when you're dredging, the water doesn't leave the exact dredge area. But even if you dredge it successfully, the thing you've got is a bunch of toxic sludge that nobody wants to landfill. And the question of what you do with toxic sludge once you get it is one that's never been satisfactorily answered."

That is the crux of the matter. What do you do with the stuff? Do you place it in a diked-disposal area on the edge of the harbor? Do you transfer it to railroad cars or tank trucks—with all the hazards the transfer entails—and take it off somewhere and bury it? Do you take it out to the middle of the Lake, close your eyes, and let go? Not one of these "solutions" seems satisfactory. The argument goes on; the dredges sit idle. The silt, which up until the PCBs were discovered was dredged out regularly, builds up. The water is at pleasure-boat depth now, and soon it will not be even that. The artificial harbor Waukegan developed so proudly out of its sand flat a century ago is well on its way back to being a sand flat again.

There are many ways for pollutants to enter the bottom sediments of a lake. There are industrial waste pipes and sewage outfalls; there are shipping discharges, ranging from massive accidents to bilge cleaning and even deck hosing. There is "nonpoint runoff" from streets and parking lots, and upland sediment derived from sprayed and fertilized fields and forests and carried into the coastal waters by rivers like the Maumee and the Fox and the Saginaw. There are mine tailings. All these things end up on the Lakebeds, where they are covered by "normal" sediments and mixed into the bottom mud by the turbulence from storm currents and passing ships. The result is "in-place pollutants"—toxic sludge. Waukegan Harbor has the worst known case of it, but it is not alone: There are plenty of others just a notch or two down the scale.

There is, for example, the lower end of Green Bay, Wisconsin, near the outfall (this sewer-engineering term seems particularly appropriate here) of the Fox River. The bottom sediments in this area, says the IJC's Great Lakes Water Quality Board, are "grossly polluted, with high concentrations of volatile solids, chemical oxygen demand, total Kjeldahl nitrogen, oil and grease, mercury, phosphorus, lead, zinc, and ammonia . . . [and] with PCB in excess of 10 mg/kg [milligrams per kilogram]." The main sources of these contaminants are the sixteen paper plants and seven municipal sewage outfalls along the lower Fox River. Several of these sources—Appleton Paper, Consolidated Paper, Fort Howard Paper, Nicolet Paper, the Green Bay Metro sewage plant, and three or four others—were still discharging ammonia and/or phosphorus in excess of permit standards at least as late as 1982; three (Fort Howard, Wisconsin Tissue, and the Bergstrom Paper Company) were recycling printed papers containing, in the printers' ink, significant quantities of PCBs (Wisconsin law specifically exempts paper recyclers from most restrictions on working with PCBs). The PCBs were supposed to be contained within the plants, but they often were not. Measurements of the effluent from the Fort Howard mill in 1981 found it running a PCB concentration of 4 micrograms per liter, well above clean-water standards.

In Milwaukee Harbor the situation is much the same. There are phosphates, oil and grease, Kjeldahl nitrogen, PCBs. There is not as much ammonia as in Green Bay, but there are significant quantities of lead, zinc, cadmium, and copper. There is chlordane and DDT. The principal current source of contamination seems to be the Milwaukee sewer system, an antiquated combined design that overflows regularly in wet weather. Milwaukee is working on its sewers, but if the current plan is followed, it will be 1996 before it comes into full compliance with federal and state effluent standards. Meanwhile the toxic sludge continues to build.

At Collingwood, Ontario, on the south shore of Georgian Bay—a major winter-sports center and one of Canada's loveliest small cities—harbor dredging is under severe restrictions owing to sediment contamination by heavy metals and organochlorides, including PCBs. At Jackfish Bay and at Nipigon, on the wild north shore of Superior, harbor sediments are so thickly coated by wastes from pulp and paper manufacturing that dredge spoil cannot be disposed of in open water but must be confined, for reasons of health, behind dikes and levees. And at Port Hope, on the north shore of Lake Ontario, "contingency spills" from the Eldorado Nuclear Plant on the edge of the Port Hope Turning Basin have added to the severe sediment contamination already in place from industrial and shipping discharges prior to 1945 to create a deadly mix of heavy metals, uranium, radium, and PCBs that cannot, with current technology, be safely dredged at all.

Canada's worst in-place pollutants problem—and one of the worst anywhere—is in Hamilton Harbour, at the west end of Lake Ontario. Tucked in behind a three-mile-long sand and gravel bar (deposited by the glaciers but shaped by the Lake), the five-square-mile harbor serves its nation's greatest concentration of heavy industry, a place known to its neighbors, somewhat derisively, as Steeltown. In that confined space the results have been devastating. Hamilton Harbour's sediments, the Great Lakes Water Quality Board reports,

> . . . exceed the provincial guidelines for open-water disposal with respect to iron, lead, arsenic, zinc, copper, nickel, mercury, chromium, total phosphorus, total Kjeldahl nitrogen, ammonia, ether extractables, and oil and grease. . . . PCB levels in sediment exceed provincial guidelines for open-water disposal along the south shore and in the deep water areas. . . . Organochlorine pesticides and their metabolites have been detected in sediments at average levels close to 10 µg/kg [micrograms per kilogram].

In 1982, the principal continuing sources of pollutants were two steel firms, Stelco and Dofasco, which exceeded Ontario Ministry of the

Environment effluent standards for a variety of microcontaminants, including cyanide (Stelco); suspended solids, ammonia, and ammonium thiocyanate (Dofasco); and phenols (both plants); before the inauguration of a new sewage treatment plant at the beginning of that year, the Hamilton sewers had been a third major contributor. All effluents are expected to be in compliance with government standards by 1987, but the problem will remain. Trapped behind that confining bar, the sludge is going nowhere. "The sediment contamination problem," notes the Water Quality Board with typical understatement, "will persist over the long term." For Hamiltonians, cyanide, ammonia, and phenols—like diamonds—apparently are forever.

Next to Waukegan Harbor, the worst problem spot on the Lakes is probably the Indiana Harbor and Ship Canal at Gary, sixty-odd miles to the south. Here the sediments are, in the International Joint Commission's words, "heavily polluted for all conventional pollutants and heavy metals." Bottom samplings by the Indiana Stream Pollution Control Board and the U.S. Environmental Protection Agency have turned up oil and grease levels as high as 175,000 mg/kg; phosphorus and lead levels of 15,000 mg/kg each; iron concentrations of 325,000 mg/kg; and volatile solids concentrations of a whopping 609,000 mg/kg, or more than 60 percent of that particular sample. The source of these contaminants is the Hammond/Gary industrial ganglion of steel plants, chemical factories, plastic plants, coal-fired generators, and others—probably the world's largest single concentration of heavy industry. Most of these plants currently meet all state and federal effluent standards, but this is misleading, for two reasons. Since the Grand Calumet River here has almost no natural flow—the whole river is currently composed of effluent—the standards are significantly lower than they are elsewhere. And many of the industrial plants meet these reduced standards by simply not releasing any effluent of their own at all, at least directly. Instead, they have connected themselves to the overloaded municipal sewage facilities at East Chicago, Hammond, and Gary. None of these plants is currently in compliance with either state or federal standards at the best of times—and all of them are combined systems with little storm runoff capacity, which release massive amounts of heavily contaminated effluent in the worst of times. The "floatables" from the effluent drift up on nearby beaches, closing them (the beach at Hammond's Lake Front Park has had to be posted so often for "temporary" hazardous conditions by the health authorities that they have finally closed it permanently); the rest of the effluent—cyanides, phenols, ammonia, halogens, old Uncle Tom Cobbley and all—becomes part of the sludge.

In-place pollutants are a sticky problem—in more ways than one.

Bottom muck is, after all, part of the ecosystem—a natural phenomenon that water-borne life has adapted to over the billions of years of geologic time. It is the home of mayfly nymphs, oligochaetes, and other invertebrates. Close to shore, plants root in it. Fish eat the invertebrates and the plants, and are in turn eaten—by larger fish, by birds, by animals, by humans. Pollutants in the muck are thus transferred up the food chain right to the top, bioaccumulating as they go. In this way do PCBs, picked up by sediments and carried "harmlessly" out of the water, still end up on our plates.

They may also end up once again in our water. It is known that sediments readily adsorb many organic chemicals and some heavy metals; it is not known how readily they give them up again. Sludge may be a sink for pollutants, but how full can the sink get before it overflows and the pollutants return to the water? This is a problem that is just beginning to receive serious study. The bottom sediments of Lake Erie are apparently contributing to the Lake's continuing phosphorus problem: There are still elevated levels of PCBs in the water of Waukegan Harbor, where no releases have taken place for over ten years. Some recycling of pollutants is clearly going on. How much is it? How dangerous? The answers are not yet available, but the hints we have seen have not been very comforting.

The most difficult thing about in-place pollutants, however, is not what they do *down there;* it is what they do when they are brought back *up here.* This is not a problem that is likely to be solved soon. Toxic dredge spoil is like any other toxic waste: a thing virtually impossible to get rid of with current technology. About all that can be done with it is to stockpile it, and that doesn't solve the problem, it merely transfers it. What does one do, eventually, with the stockpiles?

"In some cases the best thing to do may be to just leave it down there," states Jane Elder somewhat wearily. "Landfills are certainly an inadequate solution. The idea that just by burying it you take care of the whole problem is so primitive—and yet, that's what we do with most of our toxic materials." And in Windsor, Ontario, Pat Bonner of the IJC agrees. "What do you do with the sludge?" she asks rhetorically. "Do you stir it up and remove it? Do you cover it up? If you're going to remove this stuff, where is it safe? What makes it safe? What are the conditions? We just don't know. So is it safer to leave it where it is? I still don't know that, either."

And there is this little problem with much toxic sludge: It *has* to be moved. It often cannot be merely secured in place, although this is probably—as Elder suggests—the preferable method. Harbors are usually built at the mouths of rivers, so that eleven-million-metric-ton wallop of silt that enters the Lakes each year down their tributaries does so through the harbors. They must be dredged to keep them usable. If they are too dirty to dredge—as the citizens of Waukegan are finding to their sorrow—they soon fill in. Ships cannot enter; docks stand idle. And in a waterfront town, idle docks spell economic disaster.

So in most places along the Fifth Coast, the dredging goes on. The toxic spoil comes up; the barges fill and are towed away. Most are currently towed outward. In the center of the Lake, in deep water, the tows stop and the contents of the barges are released. This is the time-honored way of doing things. It makes more sense, certainly, with toxic sludge—a thing of lake bottoms, anyway—than with other forms of toxic wastes for which the same sort of fate has been suggested. But what is it doing down there?

No one knows.

And as if to emphasize our ignorance—nature has a way of highlighting these things—certain disturbing reports have lately come from the Delaware Basin. There, in the salt-water estuary where the Delaware River meets the Atlantic, chlorinated hydrocarbons for which there are no apparent landward sources are beginning to show up in the bottom sediments. According to Dr. Gerald Hansler, the former EPA engineer who is currently chairman of the Delaware River Basin Commission, the most likely source of these chlorinated hydrocarbons is the sediments themselves. Chlorinated effluent from city sewage systems—a result of the almost universal practice of adding a little chlorine to city water as a biocide—has combined with chemicals in the sediments to actually create the toxins right there in the bottom muck. The estuary has become a vast chemical factory, and it is slowly poisoning itself.

Now, consider this: Twenty-four million people drink the Great Lakes. Virtually every drop of that drinking water is chlorinated first—and then returned to the Lakes. What is happening on the Delaware is surely happening here as well. What are we doing? What sort of Frankenstein's Monster are we building down there, deep in the once-beneficial sediments beneath the beds of the greatest reservoirs of fresh water on the face of the earth?

Content Considerations

1. Explain the natural process that distributes silt, sand, and soil. What happens to pollutants in this process?
2. How did PCBs end up in Waukegan Harbor?
3. How is toxic sludge formed?

The Writer's Strategies

1. How does the author build his essay around the central problem of toxic sludge?
2. What is the author's position on the sludge problem?

Writing Possibilities

1. Respond to Ashworth's essay—what kinds of concern does it provoke in you?
2. With a classmate, discover and write about other societal practices whose possible effects are not known.
3. Discuss the economic causes and effects of the building of Waukegan Harbor, then discuss the causes and effects of its current condition. What warning might society heed from the harbor's predicament?

Joseph S. Levine

The Tainted Cup

What people do affects what the oceans and lakes are. And, according to author Joseph S. Levine, the waters are polluted and becoming more so because of what people continue to do. Like others, he makes the point that human activities in the last century or two have altered marine ecosystems more dramatically than in any other period of history. Compounding the problem is that so little about the ocean is known.

Humans don't really know the full effect of their actions, but there are signs: species endangered or driven to extinction as their habitats are ruined; toxic substances found in drinking water; high levels of elements that impair health and reproduction saturating animals that are a part of the human diet.

The next essay is the last chapter of Levine's Undersea Life. *Before you begin it, write a couple of paragraphs about the human activities you suspect may harm marine ecosystems.*

SCENE 1:
The warm, clear waters of the Caribbean
A young leatherback turtle paddles its 90-kilogram body lazily along the surface, munching on jellyfish and other gelatinous zooplankton. Leatherbacks usually live for many years and weigh as much as 450 kilograms. This one will not. It mistook several floating plastic sandwich bags for jellyfish and swallowed them. In a few days its intestines will be blocked by a twisted mass of indigestible plastic, and it will die.

SCENE 2:
The cold, turbid North Atlantic
Beneath the surface, cod and pollock spawn. Each female releases between 200,000 and 1 million buoyant eggs that float near the surface. These eggs usually hatch into larvae that join the zooplankton to feed and grow before metamorphosing into their adult forms and joining their parents near the bottom. But this batch has encountered an oil slick from a broken tanker. Many eggs have become fouled with a mixture of petroleum hydrocarbons. Twenty percent of the cod embryos and nearly half of the pollock embryos will die before hatching—from five

to ten times the mortality rate of uncontaminated eggs observed in the laboratory.

SCENE 3:
The placid waters off the Yucatán Peninsula
An offshore oil well has blown out of control and spews crude oil into the Gulf of Campeche. Where winds and tides carry the oil shoreward, it drifts into estuaries where it coats exposed mangrove roots and adheres to sediments in tidal creeks. For the next three days, oysters, other root-dwelling filter feeders, and many animals that burrow in the upper layers of sediment will die en masse. They will decay rapidly, and their soft tissues will disappear without a trace. A few weeks later, fouling mangrove roots will begin to blacken, and the trees most severely affected will begin to die. Beneath the surface, the oil will consolidate in the sediments, forming a dark, poisonous ooze that will remain, hidden from casual observers, for years after the spill.

One dead turtle or a few thousand damaged fish eggs do not endangered species make. Rarely, if ever, does a single oil spill anywhere in the world damage a healthy ecosystem beyond its ability to recover.

But several dozen dead turtles can make a big difference to a species whose numbers have been reduced from thousands to hundreds by excessive hunting. Repeated damage to eggs and planktonic larvae can adversely affect the reproductive ability of fishes already pushed to their limits by heavy exploitation. Repeated oil spills can have significant, long-term effects on the relatively few salt marshes and mangrove swamps left undisturbed by coastal development in both the temperate zones and the tropics. And whenever estuaries are adversely affected, populations of migratory animals that depend on them for breeding or nursery grounds begin to suffer.

Though knowledge of the sea and its ecosystems is growing rapidly, the ability of human beings to alter those ecosystems permanently is growing even faster. As human population grows, ecologically critical coastal areas disappear under bulldozers and coastal pollution increases exponentially. Within a split second of geologic time, the human species has changed the face of the continents and has begun to affect marine ecosystems.

The species of plants and animals in those ecosystems have evolved their shapes, colors, defenses, behaviors, and physiological characteristics through thousands of years of coexistence with one another and with their environment. Each is extremely well adapted to life in the world to which it is accustomed. But if that accustomed world changes too suddenly or too severely, the adaptations may become useless or even counterproductive to its continued survival. Ecological webs are only as strong as their weakest strands, and sudden changes in only a few plant or animal populations may have major repercussions elsewhere in the biosphere.

Until the current century, man-induced changes have had little overall effect on the open sea. Ecosystems are, after all, reasonably robust. A slight perturbation in the natural order may have existed for a while, but some plants absorbed the extra nitrogen and phosphorus, while bacteria and other microorganisms were able to break many of the poisons down into harmless components.

But today the perturbations keep coming harder and faster. While under temperature stress from the unnaturally warm water discharged from a power plant, an estuary is inundated with oil. Chemicals in coastal waters discourage the growth of more and more species. Organisms heavily exploited by people go first, but others follow. Significantly, the shallow continental shelves, which produce over 95 percent of our seafood at present and hold the only real promise for supporting marine aquaculture in the future, bear the brunt of civilization's impact on the sea.

People have always clustered near the shore, both for business and for pleasure, and they show no signs of changing this preference. By the year 2000, more than fifty cities around the world will contain more than 5 million people each. Half of those cities are located on estuaries, including seven of the ten largest—New York, Tokyo, London, Shanghai, Buenos Aires, Osaka, and Los Angeles. Outside the megalopolises, coastal resort development chews up salt marshes and mangrove swamps in the United States, while the "reclamation" of land for farming and the cutting of trees for lumber and firewood imperil mangrove swamps in developing nations throughout the tropics.

As many as 27 percent of American estuarine areas classified by the United States Fish and Wildlife Service as commercial shellfishing sites are closed because of pollution. The Caribbean coast of Mexico supports important fisheries and offers vast potential for producing medically and industrially useful compounds, but it is also the backbone of one of the busiest petroleum-related industrial areas in the world and is exposed to an ever-increasing number of oil spills and exploration-related disturbances. As industry runs out of terrestrial sites to use for disposing of chemical wastes, both corporations and governments look to coastal waters for convenient disposal sites. In Japan, for example, although shrimp and *nori* producers can never supply market needs, the pollution of shallow areas suitable for aquaculture hampers aquacultural expansion and in fact threatens the industry with slow decline.

Present knowledge of marine systems is still so fragmentary that it is difficult even for marine scientists, much less members of the general public, to determine in advance just what dangers a particular human activity poses to marine life. People may have heard brief newspaper reports about chemical contaminants such as mercury in seafood, and they may have seen the remnants of an oil spill on their favorite beach, but their understanding of the full story of chemicals and oil in the ocean is far from accurate or complete.

A case in point is the story of human-introduced mercury in the sea, a story that illustrates the difficulty of both prediction and action in such matters. Mercury, an element that in high concentrations is severely toxic to many organisms, is naturally present in the ocean in reasonably large amounts; marine chemists estimate the world ocean's total mercury content at about 70 million tons. Mercury, like ciguatoxin, is absorbed from sea water by many planktonic organisms, and biological magnification steadily concentrates it in the higher trophic levels. Top-level carnivores such as tuna and swordfish have therefore exhibited mercury levels of nearly 0.4 parts per million (ppm) since the late nineteenth century.

Given the natural mercury content of the oceans, industrial additions of mercury to the sea are small overall and probably of minor global importance. Locally high mercury concentrations are a different matter. Mercury, which is required as a catalyst in a variety of chemical reactions, is released into rivers and estuaries by chemical plants around the world. In 1938, a plant in Minamata Bay, Japan, began discharging mercury into the sea. Because the mercury was released in a form that was water-insoluble, plant officials and local authorities probably gave it little thought, imagining that the compound would simply sink to the bottom of the bay and stay there.

The mercury did sink to the sediment surface as expected, but then it was unexpectedly altered chemically into methyl mercury by bacteria in the sediments. The methyl mercury dissolved easily in the water and at once began rising through the food chain, accumulating at levels of between 10 and 55 ppm in the fish consumed in large quantities by local fishermen. By the mid-1950s, people nearby began to lose their sight, hearing, and coordination. Some were unable to speak. Others regressed into infantile helplessness. Over 100 people were affected, and as many as 50 of them died from what became known as Minamata disease. After a similar incident in Niigata, the Japanese government recognized that mercury poisoning was the cause of the problem and closed the plants.

The stage was set. In 1969, mercury levels between 1.6 and 5.0 ppm were found in Great Lakes pike. Because local residents in both the United States and Canada ate far less fish than the Japanese of Minamata, no human poisonings were reported. Nonetheless, the findings triggered a panic. All fishing in the area was stopped, and the United States Food and Drug Administration announced that existing levels of mercury pollution represented "an intolerable threat to the health and safety of Americans." Safety standards for mercury content were hurriedly set, and quick testing of canned fish samples resulted in removal of more than a million cans of tuna from the market. The tuna industry reeled under the impact.

Within a year, closer examination of available data led to establishment of different, less stringent safety guidelines. Most of the previously banned tuna was put back on the market, although swordfish—which was also included in the original ban—remained prohibited.

The FDA clearly acted on the best available evidence in issuing the original ban; it was forced to decide speedily, on the basis of fairly limited data, what move was best to protect the public interest. In this case, the agency seems to have erred initially on the side of safety, and the decision ultimately damaged both the tuna industry and the FDA's credibility.

But it is still not certain how scientists, physicians, and policymakers can best help such agencies make accurate judgments in the future. The procedure for establishing health guidelines is hotly debated in both political and scientific circles. The topic of permissible contamination levels is an important one because biological magnification significantly concentrates numerous industrial and agricultural chemicals, from DDT to PCBs, even when initial pollutant concentrations are very low. Once toxic chemicals accumulate in higher trophic levels, they persist for long periods even when the original source has been eliminated. Unfortunately, many equally important (if not more important) decisions on food and water contamination must be made on the basis of even less data than were available in the case of mercury.

Petroleum operations and oil spills present different problems. Questions remain about just how harmful normal oil operations and routine, small oil spills really are to marine life. Oil company scientists are quick to point to increases in fish populations around many offshore oil rigs: the submerged metal structures provide a hard substrate for algae and filter feeders, helping to attract fish normally associated with rocky areas that would otherwise not be found far from shore. But other impacts of oil exploration and transportation are not so benign. During the drilling process, finely powdered material called drilling mud, which is used to lubricate the drill, is discharged into the water around the rig. The effect of this extra sediment on both bottom-dwelling and planktonic organisms is not yet known. Onshore terminal development in mangrove areas, near-shore dredging operations, and offshore introduction of drilling mud all change water circulation patterns and increase the water's load of suspended material. Some organisms can tolerate this extra rain of silt and peat, but others, including many corals, are accustomed to clear water and have trouble cleansing themselves.

Then there is the question of oil spills themselves. Open-water spills do not directly affect bottom communities very much, but can cause abnormalities or death among plankton. Because plankton populations in places such as George's Bank are both home and food for commercially valuable fish and shellfish, repeated spills and repeated discharges of drilling mud might have unforeseen results on heavily fished stocks. A great deal more research is necessary before scientists sufficiently understand the effects of oil spills in such situations.

When oil washes into a salt marsh or mangrove swamp, on the other hand, the effects are immediate and obvious. Massive kills of up to 95 percent of the bottom-dwelling organisms usually occur within days. These kills strike public sensibilities the hardest: helpless seabirds dripping with oil, crabs staggering around and collapsing, mussels agape and rotting at low tide. But evidence of this mass mortality lasts only for a few days or weeks. Animals die, decompose, and are carried out by the tide. Their smells and their carcasses disappear. Soon, a television crew could enter the marsh and take a few shots of pristine-looking tidal flats at sunset, and a viewer could hardly believe anything was wrong.

But other, longer-lasting consequences of oil intrusion are not very obvious. Microorganisms in the sediments, vital links in the marsh's nutrient cycles, become coated with hydrocarbons. Some bacteria alter their metabolism, some plants die, and some—the natural hydrocarbon decomposers—proliferate. Sea-grass populations smother at their base, out of sight below the lowest tides, and the sediments they previously held in place begin to shift.

Animals such as lobsters and catfish, which depend on their sensitive chemical senses to inform them of what to eat and when and with whom to mate, suffer from gustatory hallucinations; confused, they prefer oil-soaked debris to their natural foods, and mating success suffers. Eggs and the planktonic larvae of organisms whose adults readily survive the presence of a little oil become fouled, develop abnormally, die, or refuse to settle and metamorphose.

Oil mixes with loose sediment and slowly oozes down tidal channels, into the estuary bottom and beyond. Spreading slowly for months after a spill, oil and oil-soaked sands may remain for as long as several years, preventing the growth of any but the most pollution-tolerant organisms.

Most insidious of all is the low-level, long-term uptake of hydrocarbons by animals and plants. Hydrocarbons accumulate in living tissues, where they may cause slight (or not so slight) changes in metabolic processes. Mussels not obviously harmed by a spill may become sterile. In young animals of all sorts, hydrocarbons synergize with other sources of stress to cause significantly more damage than either stress would cause alone. Some particularly carcinogenic, colorless, odorless, tasteless hydrocarbons persist in animal tissue for months.

Ultimately, fouled areas probably do recover from the effects of a spill, and isolated spills may have little permanent ecological effect. But whether on George's Bank, in the Gulf of Mexico, in the Red Sea, or off the coast of Australia, repeated incidents are bound to cause trouble. Oil spills in coastal waters, it should be noted, already top 100 million barrels each year.

Unfortunately, as ecologist William Murdoch wrote, "large corporations not only have the power to pollute, they have the economic and political power to prevent, delay, and water down regulatory legislation. They also have the power and connections to ensure that the regulatory agencies don't regulate as they ought to." To that impressive list of powers one might add that big corporations have sufficient resources and talented public relations departments to confuse and mislead an underinformed public.

Yet the problems of pollution and habitat destruction are hardly restricted to capitalistic societies. The bureaucracy that controls industrial development in the Soviet Union seems just as deaf to the pleas of biologists as major corporations in the United States are. Lake Baikal, the deepest lake in the world, is in serious trouble from pollution, and the Caspian Sea is, in places, little better than an open sewer for industrial wastes. The approach that China and Third World nations will take toward the natural environment as they hurtle into industrialization has yet to be seen. China's long experience with the ecological results of dense human settlement offers hope that it will proceed with caution: In many Third World countries, however, severe economic crises couple with skyrocketing human populations to impel governments toward drastic, short-term solutions that seriously compromise water pollution controls and undermine the integrity of vital coastal ecosystems.

In Western democratic society, at least, responsibility for pollution rests not only with international corporations and the government, but with every citizen. It is an easy, self-satisfying exercise to point an angry finger at the sneering corporate scoundrels of satirists' cartoons who twiddle their handlebar moustaches as they bulldoze coastal wildlife sanctuaries and pour toxic chemicals into estuaries. But it is the voting public that empowers elected and appointed officials to whom "environment" and "conservation" are dirty words and to whom pollution control represents little more than an obstacle to economic growth. Consumers in free-market societies purchase goods and services provided by pollution-creating industries, and, until recently, consumers have been unwilling to shoulder the cost of pollution control associated with those industries.

There is, unfortunately, no simple fix. Pollution control has its costs, and under the free-market system, it is the public that must pay those costs. No sensible environmentalist today would argue that cleaning up polluting industries and equipping new ones with environmental safeguards is cheap. Even preserving wilderness areas seems expensive; offshore wildlife sanctuaries generate no tax revenues and—aside from preserving industries such as fisheries that most people away from the coast tend to take for granted—generate far fewer jobs than oil and gas exploration and development.

But long-term costs of marine pollution far outweigh the short-term economic benefits to be gained by ignoring proper controls. No one would suggest that halting progress, ceasing oil exploration, or shutting down every polluting industrial operation is a viable option. But research into long-term environmental effects is desperately needed. Caution is essential in directing development to avoid expensive, inconvenient, and occasionally health-threatening environmental mistakes. For every penny saved now by ignoring the need for pollution control, people will pay back many dollars in the future. Critical spawning and nursery areas, from coastal estuaries to offshore banks, were not selected by conservationists. They were adopted by marine organisms in the course of their long and wandering evolutionary relationships with one another and with their environment.

Human beings cannot choose which areas are ecologically important; we can only do our best to identify those we must protect from befoulment. Like the evolutionary interactions among all other organisms in the biosphere, the relationship of *Homo sapiens* with the sea and its inhabitants has been a long and wandering one. We have made mistakes, we have hit several dead ends, and we have made several surprisingly felicitous discoveries. But our power as intelligent and accomplished beings makes the future course of our wanderings critically important. Mindful of the weight of our immense numbers, we must tread lightly as we wander, and we must examine our options at each juncture before rushing ahead. The sea, although far from bottomless, can provide food and drugs in abundance. We can, however, maintain and expand our harvest of the oceans' bounty only if we understand and preserve the beautiful, resilient, yet vulnerable habitats of the diverse marine organisms that constitute the oceans' living legacy to mankind.

Content Considerations

1. How do ecosystems react to change?
2. How does the extended example about mercury illustrate the level of knowledge about marine ecosystems?
3. Summarize how oil production and spillage affect the ocean.

The Writer's Strategies

1. Levine begins this essay with three scenes, or vignettes. How do they prepare you for the content and tone of the essay?
2. In what ways is this essay persuasive? What does Levine want the reader to believe and do?

Writing Possibilities

1. Use the information from this piece to construct a series of vignettes similar to those with which it begins.
2. Write an essay delineating the sources of ocean contamination; conclude it by recommending whatever action you believe worthwhile.
3. Respond to and expand upon Levine's statement in the last paragraph that "Human beings cannot choose which areas are ecologically important; we can only do our best to identify those we must protect from befoulment."

Additional Writing and Research

Connections

1. Using any four essays, discuss how the social order is linked to natural processes.
2. What do the essays by Colinvaux, Carson, Jacobson, and Ashworth tell you about how the global water system works?
3. Using information from the pieces by Levine, Ashworth, Colinvaux, and Jacobson, explore how pollution affects large bodies of water.
4. Compare the comments on oil spills by Sears, Levine, Davidson, and Nulty. Focus on the ways the problem affects individuals, companies, government, and nature's ecosystems.
5. What do Carson, Colinvaux, and Levine say about what can be harvested from the ocean? Center your essay on their divergences of opinion.
6. Isolate the problems or possibilities in this chapter that concern you most. Explain your concern and how you might act on it.

Researches

1. Focus on global warming—its causes, predicted consequences, and the world's response to the problem. In June of 1990, for example, ninety-three nations agreed to ban fluorocarbons and other chemicals that contribute to global warming. Find out more about this and other moves toward solutions.
2. What dangers to marine wildlife do various kinds of pollution pose? How serious is the danger? What are its consequences? You may want to narrow your investigation to one species of marine life, perhaps one on an endangered list.
3. Discover how different kinds of toxic waste are disposed of and how their disposal affects oceans or other bodies of water. Besides reading books, newspapers, and magazines, try to interview professors, EPA and other government officials, and spokespersons for local companies that produce or dispose of toxic waste.
4. Find out how the resources of one ocean or coastal region are used. What does the region produce? How is it used? How does it contribute to the economy?
5. Become familiar with offshore drilling. How is oil located? How is it removed? What precautions and standards have been set by industry, government, or international law?
6. Investigate what is not known about ocean or lake ecosystems. (A good place to start might be by asking experts what mysteries fascinate them.) How does a lack of knowledge interfere with solving problems? How does it create problems?

6

Wilderness and Intrusion

THERE IS IN ALL OF US, perhaps, a secret place away from worry, away from work, away from . . . everything. A remote mountain peak, an isolated beach, a canyon in the American desert, a coastal island, a trout stream in the woods. Our own spot where we can observe the dances of nature without fear of being summoned by the time clock.

Wilderness—whether in our minds or in some spot on Earth—is necessary for humans, it seems. According to novelist and teacher Wallace Stegner, the idea of wilderness, of some place not controlled or modified or affected by humans, is only mystical or magical yet is essential to our well-being. In 1960 Professor Stegner wrote a letter to then-Secretary of the Interior Stewart Udall in which he championed the idea of wilderness apart from any value as recreation:

> Something will have gone out of us as a people if we ever let the remaining wilderness be destroyed; if we permit the last virgin forests to be turned into comic books and plastic cigarette cases; if we drive the few remaining members of the wild species into zoos or to extinction; if we pollute the last clear air and dirty the last

clean streams and push our paved roads through the last of the silence. . . . And so that never again can we have the chance to see ourselves single, separate, vertical and individual in the world, part of the environment of trees and rocks and soil, brother to the other animals, part of the natural world and competent to belong in it.

The thirty years since that letter was written have seen a marked reduction in the areas of the world that could be called wilderness, as an oil pipeline is stretched across Alaska, as the rain forest in Brazil is bulldozed and burned, as the forests of Malaysia fall to chainsaws, as rivers are dammed in the American West and fill pristine canyons.

Those three decades have also witnessed a burgeoning new industry: eco-tourism, adventure travel, remote-area expeditions. The thirst of humans for wilderness, for contact with endangered species and threatened areas, has caused a trickle of visitors to widen into a flood, so that in some areas, as in Yellowstone, it has become necessary to limit the numbers of visitors.

We are loving our wilderness to death.

The question many are now asking is how to balance the delicate ecosystems of wilderness with the desire of humans to visit. Wilderness as an idea might be necessary to humans, as Professor Stegner asserted. It most assuredly is necessary to animals and plants who can live only in wild areas. To whom does the wilderness belong?

An argument has erupted among segments of the human population that goes something like this: Only the young and rich have the stamina and the money to take a true wilderness outing. Therefore, since present wilderness is only a playground for the elite, pristine areas should be opened with roads and campgrounds and conveniences so that more people can enjoy them. And by the way, how about opening them to logging and mining interests too?

Those who try to counter the call for opening the wilderness sound something like this: Such a move would destroy the wildness of wilderness. Virgin areas of the Earth should not be the playgrounds of anyone; rather, they should be preserved in a natural state for the idea of wilderness, for the unique habitat and environment that it is. Human contact and impact should be severely limited.

There are other arguments, of course. Multiple-use advocates get in the faces of ethicists, hard-line nature lovers blast proposals of energy and resource advocates. The writer Barry Lopez, in an essay in this chapter, makes an arresting point when he talks about how federal law may force some wilderness settlers to leave. Settlers coerced Native

Americans to move because they stood between development and the land; now, at a time when many seek rapprochement with the Earth, we are about to do the same thing to different people, Lopez writes.

Included in this chapter are essays that consider wilderness in environs we often overlook. Edward Abbey reminds us that deserts are wilderness and should be trod lightly, or better yet, not trod at all, to protect the delicate ecosystem that functions so well there. Brooks Atkinson explores Great Swamp, a leftover from the last glacial age that survives intact within sight of New York City. Sue Hubbell tells of encounters with snakes in the Missouri Ozarks. And Anna Quindlen talks of her sense of trespassing when she spots a bear in her backyard.

All of these writers explore what wilderness is and what it may become.

ANNA QUINDLEN

Our Animal Rites

Raccoons and squirrels live in the trees over city streets and in the attics of houses. Occasionally a possum lumbers across a backyard or hangs on a fence. Black snakes coil under the steps; lizards sun on a windowsill. Most of the creatures with whom people share living quarters are pesky rather than dangerous, and most of them are very small.

But what about a bear out back?

In the next essay syndicated columnist Anna Quindlen describes a close-enough encounter with a bear that lives someplace in the hills behind her country home. Quindlen, who writes on a wide variety of topics, muses about the encounter, and her musings lead to questions of intrusion and trespass, of who belongs and who wins.

Before you read what she says, recall what you know of creatures who live in populated areas that once were wilderness. Write about them for a few minutes, then add a few sentences about how people respond to the presence of such creatures.

THE BEAR HAD THE ADENOIDAL BREATHING of an elderly man with a passion for cigars and a tendency toward emphysema. My first thought, when I saw him contemplating me through tiny eyes from a rise just beyond the back porch, was that he looked remarkably bearlike, like a close-up shot from a public television nature program.

I screamed. With heavy tread—pad, pad, pad, harrumph, harrumph—the bear went off into the night, perhaps to search for garbage cans inexpertly closed and apiaries badly lighted. I sat on the porch, shaking. Everyone asks, "Was he big?" My answer is, "Compared to what?"

What I leave out when I tell the story is my conviction that the bear is still watching. At night I imagine he is staring down from the hillside into the lighted porch, as though he had a mezzanine seat for a performance on which the curtain had already gone up. "A nice female, but not very furry," I imagine him thinking. "I see the cubs have gone to the den for the night."

Sometimes I suspect I think this because the peace and quiet of the country have made me go mad, and if only I could hear a car alarm, an ambulance siren, the sound of a boom box playing "The Power" and its owner arguing with his girlfriend over whether or not he was flirting with Denise at the party, all that would drive the bear clear out of my head.

Sometimes I think it is because instead of feeling that the bear is trespassing on my property, in my heart I believe that I am trespassing on his.

That feeling is not apparent to city people, although there is something about the sight of a man cleaning up after a sheepdog with a sheet of newspaper that suggests a kind of horrible atonement. The city is a place built by the people, for the people. There we say people are acting like animals when they do things with guns and bats and knives that your ordinary bear would never dream of doing. There we condescend to our animals, with grooming parlors and cat carriers, using them to salve our loneliness and prepare us for parenthood.

All you who lost interest in the dog after the baby was born, you know who you are.

But out where the darkness has depth, where there are no street lights and the stars leap out of the sky, condescension, a feeling of supremacy, what the animal-rights types call speciesism, is impossible. Oh, hunters try it, and it is pathetic to consider the firepower they require to bring down one fair-sized deer. They get three bear days in the autumn, and afterward there is at least one picture in the paper of a couple of smiling guys in hats surrounding the carcass of an animal that looks, though dead, more dignified than they do.

Each spring, after the denning and the long, cold drowse, we wait to see if the bear that lives on the hill above our house beat the bullets. We discover his triumph through signs: a pile of bear dung on the lawn, impossible to assign to any other animal unless mastodons still roam the earth. A garbage box overturned into the swamp, the cole slaw container licked clean. Symmetrical scratch marks five feet up on a tree.

They own this land. Once, long ago, someone put a house on it. That was when we were tentative interlopers, when we put a farmhouse here and a barn there. And then we went nuts, built garden condos with pools and office complexes with parking garages and developments with names that always included the words Park, Acres, or Hills. You can't stop progress, especially if it's traveling 65 miles an hour. You notice that more this time of year, when the possums stiffen by the side of the road.

Sometimes the animals fight back. I was tickled by the people who bought a house with a pond and paid a good bit of money for a little dock from which to swim. It did not take long for them to discover that the snapping turtles were opposed to the addition to their ecosystem of humans wearing sunscreen. An exterminator was sent for. The pond was dredged. A guest got bit. The turtles won.

I've read that deer use the same trails all their lives. Someone comes along and puts a neo-Colonial house in the middle of their deer paths, and the deer will use the paths anyway, with a few detours. If you watch, you can see that it is the deer that belong and the house which does not. The bats, the groundhogs, the weasels, the toads—a hundred years from now, while our family will likely be scattered, their descendants might be in this same spot. Somewhere out there the bear is watching, picking his nits and his teeth, breathing his raggedy bear breath, and if he could talk, maybe he'd say, "I wonder when they're going back where they belong."

Content Considerations

1. What does Quindlen suspect has caused her to become almost obsessed with the bear?
2. Why is "condescension" or "speciesism" impossible once one is out of the city?
3. Summarize the relations between humans and animals in the city and in rural areas.

The Writer's Strategies

1. What irony does the essay contain? How does the author express it?
2. How does Quindlen illustrate the clash between civilization and wilderness?

Writing Possibilities

1. With classmates, discuss how your own wilderness experiences parallel Quindlen's or differ from them. Then write about them.
2. Explore the consequences of the idea that "The city is a place built by the people, for the people." Is the city also "of the people"?
3. Using a region that you know well, explain the battles that have erupted between nature and the human urge to civilize. Include ideas for establishing a truce—or for weighting the balance in one direction or the other.

BARRY LOPEZ

Yukon-Charley: The Shape of Wilderness

Like most countries, America lost much of its wilderness as its population increased. Farms, towns, and then entire cities forced wilderness to recede, although, as many of the readings in this chapter suggest, it may not go calmly or completely. Remnants exist even in the most developed of places—animals, plants, and natural forces that at times seem to make a joke of civilization's efforts.

But less physical remnants also exist, it seems. Many writers see the landscape of land and its wilderness as a shaper of history, of daily life, of ideas and emotions. Barry Lopez is one such writer. Author of several books as well as essays and stories, and winner of the National Book Award in 1986, Lopez often travels to remote places, bringing them back home in words and imagination.

In this essay Lopez writes of a canoe trip on the Charley and Yukon rivers in Alaska. The region is populated by moose and bears, salmon and beavers, cranes and caribou, and very few humans. His immersion in wilderness leads to questions about it—how we use it, how we value it, how it may be essential. Before you begin reading, ask yourself these questions about wilderness and write your answers.

I KNOW THAT COCOON FEELING, wearing wool socks, long underwear, jeans, hip boots, several shirts, a down vest and windbreaker. Sitting motionless in the bow of the nineteen-foot Grumman as it cuts cold water transparent as glass, I imagine I can pull my skin back from the innermost wall of fabric, pretend that I have found, by some inexplicable and private adventure, the tunnel to a strange window: I look out on what the map calls the Eastern Intermontane Plateau physiographic province, deep in Alaska's gut. Some of the peaks in the distance have no names; the water of the Charley River rides like glycerin up my fingers and over my palm, feels frigid against my wrist.

Behind us, eight miles to the east, the Charley enters the upper Yukon. Sixty miles to the north, the Yukon will pass the town of Circle, from where we have come. Another sixty miles farther on, at Fort

Yukon, the Porcupine will come in. Then the Yukon will turn sharply southwest and line out a thousand riverine miles to Norton Sound and the Bering Sea. From this cartographer's sense of isolation, as though I had vibrissae or other antennae extended, I surface, aware we will hit the gravel bar ahead, haul out, and build a fire to cook, to dry socks soaked in leaky boots.

I look up, as if drawn by puppet strings, to find a bald eagle dipping its hunting arc over us. Working the river. It breaks away and heads north and west into low mountains. The bow of the canoe rides up hard, rattling over the stones.

It is hours between such noises.

On shore we find again what we have known with such pleasure for days—signs of animals. Fresh moose tracks, much older bear tracks, and the bones of a grayling from some animal's meal. With 10 x 40 Leitz glasses, elbows pressed to my knees, I can see back down the river: warbonnet heads of red-breasted mergansers, long-necked pintails, the green bandit masks of male widgeon, and white, quarter-moon slashes on the faces of male blue-winged teal. Some of the birds are so far away I have to guess.

Our trip had begun that morning from a base camp on a gravel bar in the middle of the Yukon at the mouth of the Charley. The Charley's luminous black surface, crinkling in the wind, foundered in the silt-laden, warhorse current of the larger river, as far across as a Dakota wheat field, a bold child slipping into a twilight clearing.

The first morning in camp on the Yukon we found fresh grizzly tracks only twenty feet from the tent, an errand that, thankfully, hadn't included us. In the benign light of an arctic summer—the two of us stood about, waking up with mugs of coffee, our shoulders to a cool wind—a beaver arrived. Huge, he slacked his stroke, circling back in the leaden sweep of the river, slapped his tail twice and moved on, his head riding the current like the bow bumper of a tugboat. I watched a marsh hawk alone far to the west, a harrier, unload itself repeatedly against the wind in somersaults and chandelles and remembered the erroneous summary of the field guide: ". . . the flight low, languid, and gliding." He flew as if his name were unrecorded.

That morning, too, as I pulled on my boots, a wolverine walked into camp, looked us over, stabbing the air with his nose to confirm what he saw: dead end. Bob and I stood up, as though someone important had walked in, his arrival as unexpected as the smell of cinnamon.

The Charley and the Yukon, the beaver, the widgeon, the grizzly, lie within a federal preserve, created by emergency presidential order under provisions of the Antiquities Act and designated Yukon-Charley

Rivers National Preserve on December 2, 1980. Among the reasons for this laying by: an exemplary weaving of interior Alaska flora and fauna. Untouched by Ice Age glaciation, the land has a high potential as a reservoir for undisturbed early aboriginal sites. The area is rich in fur-trapping and gold-mining history. And it offers protection for an entire watershed, that of the Charley, and for peregrine falcons, who nest on the high bluffs that rise along this part of the upper Yukon.

The landscape itself, however, the pattern of birch and spruce and creeks spreading over the hills and up into the steep mountains where sheep dwell, shows no sign of the designation. There are no green park buildings, no managers, campfire circles or roads. Instead there is the trace evidence of thirty or so people living here, placer miners and subsistence trappers whose predecessors have been in the vicinity for more than eighty years. Their cabins are spotted every six or seven miles along the Yukon; the colored floats on gill nets bob close by on the khaki-brown surface of the river—salmon nets, winter food for their dogs. At Coal Creek and Woodchopper Creek, the two working mines, crude landing strips have been bulldozed near a few buildings; the mills and the rest of the improbably heavy machinery were brought into the country by barge. Supplies come to this part of the river now by skiff, canoe, dogsled, or snow machine, depending on the season and one's circumstances, down from the roadhead at Eagle, 105 miles above the mouth of the Charley, or up from Circle.

Until someone forces them to change their way of life, to give all this up and quit the country, the Yukon-Charley Rivers National Preserve will, in residents' minds, remain little more than a colored panel on someone else's map. The formal setting aside of this land, in fact, represents to them an incomplete understanding of the country.

When we shoved off from camp that morning on the Yukon we took most of our gear in the canoe with us, a nuisance, but a necessary precaution against bears, who might shred it or drag it off into the river.

The sight of the broad back of the Yukon in mid-June triggers a memory of the Nile or the Amazon; but breakup ice is piled in shattered rafts the size of freight cars along the shores. The banks have been deeply gouged, the bark scraped from trees to reveal gleaming yellow-white flesh beneath. Silt boils against the canoe, a white noise that is with us until we cross half a mile of open water and hit the Charley. Its transparent flow, turned against the downstream bank of the bigger river, narrows to nothing after a thousand yards, absorbed.

The mouth of the Charley—its name oddly prosaic, a miner's notion, fitting the human history of the region—is a good place to fish for pike or burbot. (We bait a trotline for burbot, a freshwater cod that looks like a catfish. We will check it on the way out.) The flats of the river's floodplain, several square miles, are dense with an even growth of willow, six to eight feet high. The undersides of the long, narrow leaves are a lighter shade of green than that above; their constant movement, a synaptic fury in the wind, makes them seem all the more luminous. Moose are bedded down among them, beyond the reach of our senses. Their tracks say so.

The river's banks, flooded with an aureate storm light underneath banks of nimbus cloud, are bright enough to astonish us—or me at least. My companion's attention is divided—the direction of the canoe, the stream of clues that engage a wildlife biologist: the height at which these willows have been browsed, the number of raven nests in that cliff, a torn primary feather which reaches us like a dry leaf on the surface of the water. Canada goose.

What is stunning about the river's banks on this particular stormy afternoon is not the vegetation (the willow, alder, birch, black cottonwood, and spruce are common enough) but its *presentation*. The wind, like some energetic dealer in rare fabrics, folds back branches and ruffles the underside of leaves to show the pattern—the shorter willows forward; the birch, taller, set farther back on the hills. The soft green furze of budding alder heightens the contrast between gray-green willow stems and white birch bark. All of it is rhythmic in the wind, each species bending as its diameter, its surface area, the strength of its fibers dictate. Behind this, a backdrop of hills: open country recovering from an old fire, dark islands of spruce in an ocean of labrador tea, lowbush cranberry, fireweed, and wild primrose, each species of leaf the invention of a different green: lime, moss, forest, jade. This is not to mention the steel gray of the clouds, the balmy arctic temperature, our clear suspension in the canoe over the stony floor of the river, the ground-in dirt of my hands, the flutelike notes of a Swainson's thrush, or anything else that informs the scene.

A local trapper advised us against the Charley. Too common, too bleak. Try the Kandick, he said, farther up the Yukon. I did not see a way in the conversation—it was too short, too direct—to convey my pleasure before mere color, the artifices of the wind. My companion and I exchanged a discreet shrug as we left the man's cabin. Differing views of what will excite the traveler.

At Bonanza Creek, while our socks dried by the fire, we fished for arctic grayling. Our plan had been to go twenty-five or thirty miles farther up the Charley, to where mountains rise precipitously on both sides and we might see Dall sheep with their lambs, or even spot a new species of butterfly (a lepidopterist in Fairbanks, learning of our destination, had urged a collecting kit on each of us). We abandoned the plan. Mosquitoes got to us as the river narrowed, and it was a banner year for them. Step a few feet into the bush and hundreds were on you. Insect repellent only kept them from biting—and they quickly found any unsprayed spot where cloth hugged skin close enough to let them drive home, so often the inside of a thigh. Nothing to be done about their whining madly in the ears, clogging the throat, clouding one's vision.

The memory of our windswept campsite on the Yukon, far from shore, a gravel bar without vegetation and so mosquito-free, passed wordlessly between us. We put on dry socks, cleaned two grayling, folded our 1:250,000 physiographic maps, and swung the Grumman downriver.

In the summer of 1967 grizzly bears killed two people in Glacier National Park. The Park Service killed the bears and ignited a controversy about the meaning and importance of wilderness in America. The timber industry, for its part, said wilderness was the private playground of a young, upper middle-class elite. This castigation obscured their own considerable peevishness: according to the prevailing federal mandate—to seek to manage public lands for recreation, logging, grazing, mining, hunting, fishing, and watershed protection—wilderness failed to serve only logging.

An argument for wilderness that reaches beyond the valid concerns of multiple-use—recreation, flood control, providing a source of pure water—is that wild lands preserve complex biological relationships that we are only dimly, or sometimes not at all, aware of. Wilderness represents a gene pool, vital for the resiliency of plants and animals. An argument for wilderness that goes deeper still is that we have an ethical obligation to provide animals with a place where they are free from the impingements of civilization. And, further, an historical responsibility to preserve the kind of landscapes from which modern man emerged.

The Reagan administration regards such arguments as these from science and ethics as frivolous. It wishes to reduce "the wilderness controversy" to economic terms, which is like trying to approach the collapse of a national literature as primarily an economic problem. The administration's attitude reveals an impoverished understanding of the place and history of the physical landscape in human affairs—of its effect, for example, on the evolution and structure of language, or on the development of particular regional literatures, even on the ontogeny of human personalities. Such observations have been offered by writers and artists recently to make a single point: as vital as any single rationale for the preservation of undisturbed landscapes is regard for the profound effect they can have on the direction of human life.

The insistence of government and industry, that wilderness values be rendered solely in economic terms, has led to an insidious presumption, that the recreational potential of wild land, not its biological integrity, should be the principal criterion of its worth. This, in turn, has shifted public attention away from concerns about wilderness that are harder to define (or price out) and created problems of its own, by making personal risk and physical exertion more the hallmarks of a wilderness experience than finding humility and serenity. The spiritual, aesthetic, and historical dimensions of wilderness experience, at least in congressional hearings, have become subordinate.

Wilderness travel can be extremely taxing and dangerous. You can fall into a crevasse, flip your kayak, lose your way, become hypothermic, run out of food, or be killed by a bear. Far less violent events, however, are the common experience of most people who travel in wild landscapes. A sublime encounter with perhaps the most essential attribute of wilderness—falling into resonance with a system of unmanaged, non-human-centered relationships—can be as fulfilling as running a huge and difficult rapid. Sometimes they prove, indeed, to be the same thing.

America, at least in its written law, is uncertain what it values in wilderness, beyond recreational utility. Some who might succinctly address the larger issues—child psychiatrists, geneticists, theologians—don't view themselves as spokesmen. That wilderness can revitalize someone who has spent too long in the highly manipulative, perversely efficient atmosphere of modern life is a widely shared notion; but whether wilderness experience has a clear therapeutic value remains scientifically uncertain.

These more subtle arguments for the preservation of wilderness point directly to our ethical and psychological well-being as a country. Though they lag light-years behind in having any legal standing they are as critical as economic arguments and much of this suggests that the real wisdom of the Wilderness Act of 1964, setting wilderness aside, has yet to fully emerge.

Against the backdrop of such lofty thoughts the Yukon-Charley Rivers National Preserve seems somewhat anomalous because it lacks spectacle. Set for comparison against the Valhalla of Denali National Park or the Cambrian silence of Grand Canyon it seems commonplace. But for the presence of relatively numerous peregrines, its river headlands differ little from those along the Hudson, pleasing but not distinguished. Its hills remind one of the Blue Ridge, a reclining countryside. The Yukon itself at this point is without rapids or picturesque waterfalls and thick with silt. In the spring of 1981 there was also a general scarcity of large animals. There is nothing at all, in fact, very remarkable about the place—except that it is largely uninhabited and undisturbed. That people live here is unusual. Theoretically their presence is to be regarded as a drawback to wilderness, though they actually give the land a pleasing dimension.

The loss of animals is puzzling. There is a curiously persistent popular belief that if wild animals are left alone they will flourish in wilderness areas and reach an "optimum population" in balance with other species in the ecosystem; but this is not so. Like any other landscape, the Yukon-Charley country is susceptible to forest fires, hard winters, epizootic disease, and predator pressure. It has had good as well as lean years; no one knows enough wildlife biology to determine precisely why recent years have been ones in which the populations of moose, wolverine, caribou, marten, bear, and wolf have dwindled. The hunting pressure from outside the region has been fierce, and it is partly to blame; but the situation is as complicated a subject as schizophrenia or genetic drift. A visitor is not unwarranted, therefore, in speculating that the recent demise of animals may in some way be tied to the intense and proprietary scrutiny that the land has been subject to—I sense the imposition of analytic thought, the imputation of logic on this guileless land, in my own notebook.

One morning on the Yukon we had to grit our teeth against a driving rainstorm. We put into shore when it showed no sign of letting up, only to be driven out by mosquitoes. We shoved off, wrapped in raingear, feeling the edge of that depression that comes with relentless bad weather. Two hours later we arrived at a place identified as a cabin on the map. Shelter, we thought. It would likely be a trapper's unoccupied winter residence, bearing a familiar notice—use it as your own, leave it as you found it.

The cabin reveals itself as a trapper's cabin but there is another kind of sign here, in store-bought, iridescent pink on black: PRIVATE PROPERTY, NO TRESPASSING. And on brown cardboard in hand-lettered black: OUR HOME IS PROTECTED WITH SET GUNS.

So does wild country change. There probably are no set guns, but we are not eager to test the idea. What the sign says is: *The country is filling up with people. I came here to get away from you. I'm not a backwoods nut shooting at peregrines. I don't own a bulldozer or contemplate building roads or condominiums and I don't pose a threat to anyone's peace of mind. This part of the nation's heritage will do just fine without any wilderness boundaries, without any rangers, or any tourists. Get lost.*

I understand, I think, the position. It's a gulf that divides old-guard Alaskans from those with urban concerns and perceptions. But I bear the man ill will in the pouring rain and swarming mosquitoes. There's not been this kind of rancor, this level of belligerence in the country before.

An hour later the rain stops. We haul up on a gravel bar, beneath a sun that grows brighter and hotter. For the next few hours we sleep sprawled on the warm stones while our equipment dries, then we eat, bail the canoe, and set up the spotting scope to watch a nest of young ravens on the opposite shore. Gulls swivel above, acrobatic in the blustery winds. Which are they? Without a bird guide we depend on each other to know, but are not distressed at forgetting the distinction between herring and mew gulls. We will look it up later. For now it is enough to be resting here in the streaming light.

I am drawn later to the water's edge, a primal attraction. Bent over like a heron I start upriver, searching for stones, lured by the sparkling quartzes and smooth bits of glistening debris: maroon and blue, wheat colors, speckled birds' eggs colors, purple, coal—I can settle twenty on the back of my hand, each one a different shade. I could poke here until I dropped of old age. My pockets slowly fill with stones, each tied vaguely to pleasure. It's ten-thirty at night. The sun, low on the northwest horizon, throws light across to a full moon in the southeast sky.

We fall asleep to a clatter of sandhill cranes, the running screech of a belted kingfisher. A breeze brings the spermish odor of balsam poplars over the water to lie punk in our noses.

On the river the following day, the rain squalls behind us, I wonder about violence in a place like this. A caribou cow swatted to its knees by a bear. Lightning-caused fires rage out of control twenty miles north of us in an unnamed valley. Frozen lenses of underground ice leer like a dark Norwegian secret from beneath the brows of the river's banks. A peregrine snatches a teal from the sky like a paper bag. A cow moose, driven mad by insects in her face, thrashes in willows. A wolf carcass lies rotting on the shore. The images whisper to me of the fullness of the land, of the tentativeness of my visit.

In search of water free of silt we turn up Sam Creek. Flocks of goldeneye with their high foreheads explode vertically off its tannin-colored waters. Mallards sweep past and arctic loons labor by, dip-necked and humpbacked like seals. It's as though we had barged in. Thousands of multicolored feathers float amid stalks of horsetail fern at the creek's edge. Snagged on one of the scabrous sheaths is the hooked, iridescent tail feather of a male mallard. My hands are slick with river silt, stiffened by wind and sun. I can hardly grasp it.

We often come to wilderness to find animals; we are less sure about the presence of people. In the Wilderness Act humans are construed as aliens, urged to make their visits relatively brief and to leave no mark of their passage. There are good reasons for this. Some people, oblivious to any but their own needs, leave a bright spoor; others have a resident's instincts and wish to build corrals and emergency shelters in country they visit regularly. But there is something unsettling in this kind of purity. To banish all evidence of ourselves means the wilderness is to that extent contrived. We are not, in fact, aliens; and Yukon-Charley offers a chance to reconsider this aspect of wilderness, and better determine what we mean by "human disturbance" in such places.

There are a handful of miners in Yukon-Charley. Under the present law they are allowed to stay on and work their claims. Resident subsistence hunters, who have expressed a desire to stay on but whose way of life is seen to be less responsible, less defensible than the gainfully employed miners', have been urged to go. These latter lives are ones of humble scale. To allow them to remain here, to go on working as indigenous guides, as translators of this experience for the rest of us, infrequent visitors, is an attractive thought. Such a life speaks to a need many of us have but few can attend to—long-lived intimacy with a place, being able to speak of it knowledgeably to others.

Such people, of course, would have to kill the local animals to feed themselves, and take fish from the river for their dogs. They would run trap lines. Perhaps it would never do. But if it could be people like the Moore family living on Sam Creek it would be worth it. To let them stay and hunt and speak with us would mean we would have to rid ourselves of some abnegation.

The Moores have been living in Yukon-Charley for almost four years. George traps mink, marten, lynx, and wolverine and trades the furs for staples, clothing, and personal items. He and his wife, Kelly, both in their early thirties, have an infant son, Zachariah. They are congenial, alert, resourceful. They share everything they have with us— food, time, conversation, books. When George says quietly after dinner, "I want for nothing," the self-knowledge, the self-confidence in his words, has a ten-winter ring.

The Moores live a life many have tried and abandoned. They husband the scraps of the animals they hunt and trap, and put them to use in some way or other. From the furs they realize a small cash income. Their world, from what I saw, is not haunted by imaginary enemies. They are not hamstrung by schedules. Like the Athabascans in this country before them, they have a manifestly spiritual relationship with the landscape. Considering the physical labor and the harsh climate, their stamina and their belief in themselves are remarkable. To chance on them was to hear of the subtleties of making a living here, to have a history, to sense one's own visit against other years, other seasons.

In the face of such lives, one feels villainous, a coward, agreeing to a proposed federal order that would one day tell them to leave. It is shameful enough to want to kick them out; what is more distressing is that we have made this mistake before. We told the native peoples of North America that their relationships with the land were worthless, primitive. Now we are a culture that spends millions trying to find this knowledge, trying to reestablish a sense of well-being with the earth.

I would not argue with the need to preserve a wilderness free of human enterprise, but I would not want to be the one, if it came to their leaving, to explain the official reasoning to George Moore.

We broke camp at the mouth of the Charley River late one afternoon and the Yukon drew us north toward Circle. The aluminum canoe revolved slowly in the current. We reached out with a paddle only to fend off trees torn loose in breakup or to negotiate fast water at the foot of a bluff.

This is remote wilderness, not apt to draw many visitors. Mosquitoes swarm in the summer and the days of winter are short, brutally cold. For all its undistinguished square miles, however, it is a good place to have set aside. Migratory birds by the hundreds of thousands find sustenance here, spring and fall. It is a country that allows bears room enough to hunt moose, and it provides fox ptarmigan enough to get through the winter. It casts up mile after mile of small, beautiful stones on its river banks; and its cliffs support endangered peregrines that range to Oklahoma and Mexico. Its air is laced with the sweet odor of balsam and the honk of geese, and its meadows burst in spring with clusters of bluebells and white-headed cotton grass, with the evaporating pinks of primroses, the camellia-like blossoms of bunchberry, the soft purple of wild gentian, and red bundles of the fireweed's blossoms. Its green hills stretch back easily from the river.

It is the sort of ordinary place that shaped many people in rural America. It is straightforward country that drives home two lessons. People who do come here will find, on looking, a mix of color, of smells, of events, that can be found nowhere else in the world. So the country, finally, is exceptional. And the profound elevation of the spirit in a wild place, rejuvenation, does not always require a rush of adrenaline. Sometimes lingering in a country's unpretentious hills and waters offers all one might wish of wisdom.

Somewhere down the Yukon, king salmon were coming. In a month dog salmon will follow. After them, after freeze-up, silver salmon. In some way each species will contrive to see in the dense, brown water, to smell, to clear its gills in order to breathe. They will nourish grizzly bears and subsistence families like the Moores. Wolverines and bald eagles will scavenge their carcasses. Pike will eat their young.

In the low evening light stands of birch become dazzling rivers of white light, pouring down the hills of dark spruce. The bulbous ground shadows of cumulus clouds glide silently over distant slopes.

I remember an erratic wind blowing white blossoms off Kelly Moore's tomato plants; the *pas retiré* of willows; the abalone nacre in a dragonfly's back. And sensing one afternoon at the edge of a thick stand of spruce, drawn by a commotion of ravens, that we were suddenly too close, much too close, to a fresh bear kill. And how that wolverine *had just appeared in camp* while I was pulling on my boots.

The last night on the river we unroll goose-down bags on a tarpaulin thrown down over river sand and small stones. We do not speak of anything we have seen. We each wish in our different ways for some insurance against the disappearance of wild relationships here. These dreams of preservation for the very things that induce a sense of worth in human beings must have been dreamt seven thousand years ago on the Euphrates. They are dreams one hopes are dreamt on the Potomac but suspects may not be, dreams of respectful human participation in a landscape, generation after generation. Dreams of need and fulfillment. Common enough dreams. Poignant, ineffable, indefensible, the winds of an interior landscape. A handful of beautiful damp stones in arctic sunlight, a green duck feather stuck to one finger. Water dripping back to the river. I fumble at some prayer here I have forgotten, utterly forgotten, how to perform. I place the stones back in the river, as carefully as possible, and move inland to sleep.

Content Considerations

1. Summarize the arguments for wilderness.
2. What arguments for wilderness does Lopez support most strongly?
3. In what ways do humans fit into the wilderness?

The Writer's Strategies

1. In paragraph 34, Lopez gathers stones from the river; in the last paragraph he replaces them. Why? How do these two actions reflect his attitude toward wilderness?
2. How does Lopez use his own wilderness experience to shape and answer questions regarding wilderness and landscape?

Writing Possibilities

1. Respond to the conflict presented by the idea that we may "need to preserve a wilderness free of human enterprise" (paragraph 44) but that an attempt to exclude humans "means the wilderness is to that extent contrived" (paragraph 38).
2. Explore the ways places and landscapes shape people and their lives.
3. Enter the argument for or against wilderness areas by evaluating and adding to the arguments Lopez presents.

EDWARD ABBEY

The Great American Desert

This essay concerns the Sonoran, the Chihuahuan, the Mojave, and the Great Basin deserts, areas that collectively make up the North American Desert, and the people who choose to visit them. Before his death in 1989 author Edward Abbey worked as a ranger in several of the area's national parks and monuments. The area is one Abbey loved and sought to protect.

His proclaimed means of protection was radical, at times militant, for he believed no human interest or profit should in any way endanger the ecosystems of desert regions. A good many people argued with his views, and a good many abhorred his suggestions—suggestions culled from events in the novel The Monkeywrench Gang *and allegedly practiced by some environmental groups.*

Before you start the essay, make a list of places you hold so dear that you would protect them at almost any personal cost. Write about one of the places, about the kinds of dangers it could face and what you would do to ward them off. If there is no place you value that highly, write about why that might be.

IN MY CASE it was love at first sight. This desert, all deserts, any desert. No matter where my head and feet may go, my heart and my entrails stay behind, here on the clean, true, comfortable rock, under the black sun of God's forsaken country. When I take on my next incarnation, my bones will remain bleaching nicely in a stone gulch under the rim of some faraway plateau, way out there in the back of beyond. An unrequited and excessive love, inhuman no doubt but painful anyhow, especially when I see my desert under attack. "The one death I cannot bear," said the Sonoran-Arizonan poet Richard Shelton. The kind of love that makes a man selfish, possessive, irritable. If you're thinking of a visit, my natural reaction is like a rattlesnake's—to warn you off. What I want to say goes something like this.

Survival Hint #1: Stay out of there. Don't go. Stay home and read a good book, this one for example. The Great American Desert is an awful place. People get hurt, get sick, get lost out there. Even if you survive, which is not certain, you will have a miserable time. The desert is for movies and God-intoxicated mystics, not for family recreation.

Let me enumerate the hazards. First the Walapai tiger, also known as conenose kissing bug. *Triatoma protracta* is a true bug, black as sin, and it flies through the night quiet as an assassin. It does not attack directly like a mosquito or deerfly, but alights at a discreet distance, undetected, and creeps upon you, its hairy little feet making not the slightest noise. The kissing bug is fond of warmth and like Dracula requires mammalian blood for sustenance. When it reaches you the bug crawls onto your skin so gently, so softly that unless your senses are hyperacute you feel nothing. Selecting a tender point, the bug slips its conical proboscis into your flesh, injecting a poisonous anesthetic. If you are asleep you will feel nothing. If you happen to be awake you may notice the faintest of pinpricks, hardly more than a brief ticklish sensation, which you will probably disregard. But the bug is already at work. Having numbed the nerves near the point of entry the bug proceeds (with a sigh of satisfaction, no doubt) to withdraw blood. When its belly is filled, it pulls out, backs off, and waddles away, so drunk and gorged it cannot fly.

At about this time the victim awakes, scratching at a furious itch. If you recognize the symptoms at once, you can sometimes find the bug in your vicinity and destroy it. But revenge will be your only satisfaction. Your night is ruined. If you are of average sensitivity to a kissing bug's poison, your entire body breaks out in hives, skin aflame from head to toe. Some people become seriously ill, in many cases requiring hospitalization. Others recover fully after five or six hours except for a hard and itchy swelling, which may endure for a week.

After the kissing bug, you should beware of rattlesnakes; we have half a dozen species, all offensive and dangerous, plus centipedes, millipedes, tarantulas, black widows, brown recluses, Gila monsters, the deadly poisonous coral snakes, and giant hairy desert scorpions. Plus an immense variety and near-infinite number of ants, midges, gnats, blood-sucking flies, and blood-guzzling mosquitoes. (You might think the desert would be spared at least mosquitoes? Not so. Peer in any water hole by day: swarming with mosquito larvae. Venture out on a summer's eve: The air vibrates with their mournful keening.) Finally, where the desert meets the sea, as on the coasts of Sonora and Baja California, we have the usual assortment of obnoxious marine life: sandflies, ghost crabs, stingrays, electric jellyfish, spiny sea urchins, man-eating sharks, and other creatures so distasteful one prefers not even to name them.

It has been said, and truly, that everything in the desert either stings, stabs, stinks, or sticks. You will find the flora here as venomous, hooked, barbed, thorny, prickly, needled, saw-toothed, hairy, stickered, mean, bitter, sharp, wiry, and fierce as the animals. Something about the desert inclines all living things to harshness and acerbity. The soft evolve out. Except for sleek and oily growths like the poison ivy—

oh yes, indeed—that flourish in sinister profusion on the dank walls about the quicksand down in those corridors of gloom and labyrinthine monotony that men call canyons.

We come now to the third major hazard, which is sunshine. Too much of a good thing can be fatal. Sunstroke, heatstroke, and dehydration are common misfortunes in the bright American Southwest. If you can avoid the insects, reptiles, and arachnids, the cactus and the ivy, the smog of the southwestern cities, and the lung fungus of the desert valleys (carried by dust in the air), you cannot escape the desert sun. Too much exposure to it eventually causes, quite literally, not merely sunburn but skin cancer.

Much sun, little rain also means an arid climate. Compared with the high humidity of more hospitable regions, the dry heat of the desert seems at first not terribly uncomfortable—sometimes even pleasant. But that sensation of comfort is false, a deception, and therefore all the more dangerous, for it induces overexertion and an insufficient consumption of water, even when water is available. This leads to various internal complications, some immediate—sunstroke, for example—and some not apparent until much later. Mild but prolonged dehydration, continued over a span of months or years, leads to the crystallization of mineral solutions in the urinary tract, that is, to what urologists call urinary calculi or kidney stones. A disability common in all the world's arid regions. Kidney stones, in case you haven't met one, come in many shapes and sizes, from pellets smooth as BB shot to highly irregular calcifications resembling asteroids, Vietcong shrapnel, and crown-of-thorns starfish. Some of these objects may be "passed" naturally; others can be removed only by means of the Davis stone basket or by surgery. Me—I was lucky; I passed mine with only a groan, my forehead pressed against the wall of a pissoir in the rear of a Tucson bar that I cannot recommend.

You may be getting the impression by now that the desert is not the most suitable of environments for human habitation. Correct. Of all the Earth's climatic zones, excepting only the Antarctic, the deserts are the least inhabited, the least "developed," for reasons that should now be clear.

You may wish to ask, Yes, okay, but among North American deserts which is the *worst*? A good question—and I am happy to attempt an answer.

Geographers generally divide the North American desert—what was once termed "the Great American Desert"—into four distinct regions or subdeserts. These are the Sonoran Desert, which comprises southern Arizona, Baja California, and the state of Sonora in Mexico; the Chihuahuan Desert, which includes west Texas, southern New

Mexico, and the states of Chihuahua and Coahuila in Mexico; the Mojave Desert, which includes southeastern California and small portions of Nevada, Utah, and Arizona; and the Great Basin Desert, which includes most of Utah and Nevada, northern Arizona, northwestern New Mexico, and much of Idaho and eastern Oregon.

Privately, I prefer my own categories. Up north in Utah somewhere is the canyon country—places like Zeke's Hole, Death Hollow, Pucker Pass, Buckskin Gulch, Nausea Crick, Wolf Hole, Mollie's Nipple, Dirty Devil River, Horse Canyon, Horseshoe Canyon, Lost Horse Canyon, Horsethief Canyon, and Horseshit Canyon, to name only the more classic places. Down in Arizona and Sonora there's the cactus country; if you have nothing better to do, you might take a look at High Tanks, Salome Creek, Tortilla Flat, Esperero ("Hoper") Canyon, Holy Joe Peak, Depression Canyon, Painted Cave, Hell Hole Canyon, Hell's Half Acre, Iceberg Canyon, Tiburon (Shark) Island, Pinacate Peak, Infernal Valley, Sykes Crater, Montezuma's Head, Gu Oidak, Kuakatch, Pisinimo, and Baboquivari Mountain, for example.

Then there's The Canyon. *The* Canyon. The Grand. That's one world. And North Rim—that's another. And Death Valley, still another, where I lived one winter near Furnace Creek and climbed the Funeral Mountains, tasted Badwater, looked into the Devil's Hole, hollered up Echo Canyon, searched for and never did find Seldom Seen Slim. Looked for *satori* near Vana, Nevada, and found a ghost town named Bonnie Claire. Never made it to Winnemucca. Drove through the Smoke Creek Desert and down through Big Pine and Lone Pine and home across the Panamints to Death Valley again—home sweet home that winter.

And which of these deserts is the worst? I find it hard to judge. They're all bad—not half bad but all bad. In the Sonoran Desert, Phoenix will get you if the sun, snakes, bugs, and arthropods don't. In the Mojave Desert, it's Las Vegas, more sickening by far than the Glauber's salt in the Death Valley sinkholes. Go to Chihuahua and you're liable to get busted in El Paso and sandbagged in Ciudad Juárez—where all old whores go to die. Up north in the Great Basin Desert, on the Plateau Province, in the canyon country, your heart will break, seeing the strip mines open up and the power plants rise where only cowboys and Indians and J. Wesley Powell ever roamed before.

Nevertheless, all is not lost; much remains, and I welcome the prospect of an army of lug-soled hiker's boots on the desert trails. To save what wilderness is left in the American Southwest—and in the American Southwest only the wilderness is worth saving—we are going to need all the recruits we can get. All the hands, heads, bodies, time, money, effort we can find. Presumably—and the Sierra Club, the Wilderness Society, the Friends of the Earth, the Audubon Society, the Defenders of Wildlife operate on this theory—those who learn to love what is spare, rough, wild, undeveloped, and unbroken will be willing to fight for it, will help resist the strip miners, highway builders, land developers, weapons testers, power producers, tree chainers, clear cutters, oil drillers, dam beavers, subdividers—the list goes on and on—before that zinc-hearted, termite-brained, squint-eyed, near-sighted, greedy crew succeeds in completely californicating what still survives of the Great American Desert.

So much for the Good Cause. Now what about desert hiking itself, you may ask. I'm glad you asked that question. I firmly believe that one should never—I repeat *never*—go out into that formidable wasteland of cactus, heat, serpents, rock, scrub, and thorn without careful planning, thorough and cautious preparation, and complete—never mind the expense!—*complete* equipment. My motto is: Be Prepared.

That is my belief and that is my motto. My practice, however, is a little different. I tend to go off in a more or less random direction myself, half-baked, half-assed, half-cocked, and half-ripped. Why? Well, because I have an indolent and melancholy nature and don't care to be bothered getting all those *things* together—all that bloody *gear*—maps, compass, binoculars, poncho, pup tent, shoes, first-aid kit, rope, flashlight, inspirational poetry, water, food—and because anyhow I approach nature with a certain surly ill-will, daring Her to make trouble. Later when I'm deep into Natural Bridges Natural Moneymint or Zion National Parkinglot or say General Shithead National Forest Land of Many Abuses why then, of course, when it's a bit late, then I may wish I had packed that something extra: matches perhaps, to mention one useful item, or maybe a spoon to eat my gruel with.

If I hike with another person it's usually the same; most of my friends have indolent and melancholy natures too. A cursed lot, all of them. I think of my comrade John De Puy, for example, sloping along for mile after mile like a god-damned camel—indefatigable—with those J. C. Penny hightops on his feet and that plastic pack on his back he got with five books of Green Stamps and nothing inside it but a sketchbook, some homemade jerky and a few cans of green chiles. Or Douglas Peacock, ex-Green Beret, just the opposite. Built like a buffalo, he loads a ninety-pound canvas pannier on his back at trailhead, loaded with guns, ammunition, bayonet, pitons and carabiners, cameras, field books, a 150-foot rope, geologist's sledge, rock samples, assay kit, field glasses, two gallons of water in steel canteens, jungle books, a case of C-rations, rope hammock, pharmaceuticals in a pig-iron box, raincoat, overcoat, two-man mountain tent, Dutch oven, hibachi, shovel, ax, inflatable boat, and near the top of the load and distributed through side and back pockets, easily accessible, a case of beer. Not because he enjoys or needs all that weight—he may never get to the bottom of that cargo on a ten-day outing—but simply because Douglas uses his packbag for general storage both at home and on the trail and prefers not to have to rearrange everything from time to time merely for the purposes of a hike. Thus my friends De Puy and Peacock; you may wish to avoid such extremes.

A few tips on desert etiquette:

1. Carry a cooking stove, if you must cook. Do not burn desert wood, which is rare and beautiful and required ages for its creation (an ironwood tree lives for over 1,000 years and juniper almost as long).
2. If you must, out of need, build a fire, then for God's sake allow it to burn itself out before you leave—do not bury it, as Boy Scouts and Campfire Girls do, under a heap of mud or sand. Scatter the ashes; replace any rocks you may have used in constructing a fireplace; do all you can to obliterate the evidence that you camped here. (The Search & Rescue Team may be looking for you.)
3. Do not bury garbage—the wildlife will only dig it up again. Burn what will burn and pack out the rest. The same goes for toilet paper: Don't bury it, *burn it.*
4. Do not bathe in desert pools, natural tanks, *tinajas,* potholes. Drink what water you need, take what you need, and leave the rest for the next hiker and more important for the bees, birds, and animals—bighorn sheep, coyotes, lions, foxes, badgers, deer, wild pigs, wild horses—whose *lives* depend on that water.
5. Always remove and destroy survey stakes, flagging, advertising signboards, mining claim markers, animal traps, poisoned bait, seismic exploration geophones, and other such artifacts of industrialism. The men who put those things there are up to no good and it is our duty to confound them. Keep America Beautiful. Grow a Beard. Take a Bath. Burn a Billboard.

Anyway—why go into the desert? Really, why do it? That sun, roaring at you all day long. The fetid, tepid, vapid little water holes slowly evaporating under a scum of grease, full of cannibal beetles, spotted toads, horsehair worms, liver flukes, and down at the bottom, inevitably, the pale cadaver of a ten-inch centipede. Those pink rattlesnakes down in The Canyon, those diamondback monsters thick as a truck driver's wrist that lurk in shady places along the trail, those unpleasant solpugids and unnecessary Jerusalem crickets that scurry on dirty claws across your face at night. Why? The rain that comes down like lead shot and wrecks the trail, those sudden rockfalls of obscure origin that crash like thunder ten feet behind you in the heart of a deadstill afternoon. The ubiquitous buzzard, so patient—but only so patient. The sullen and hostile Indians, all on welfare. The ragweed, the tumbleweed, the Jimson weed, the snakeweed. The scorpion in your shoe at dawn. The dreary wind that blows all spring, the psychedelic Joshua trees waving their arms at you on moonlight nights. Sand in the soup de jour. Halazone tablets in your canteen. The barren hills that always go up, which is bad, or down, which is worse. Those canyons like catacombs with quicksand lapping at your crotch. Hollow, mummified horses with forelegs casually crossed, dead for ten years, leaning against the corner of a barbed-wire fence. Packhorses at night, iron-shod, clattering over the slickrock through your camp. The last tin of tuna, two flat tires, not enough water and a forty-mile trek to Tule Well. An osprey on a cardón cactus, snatching the head off a living fish—always the best part first. The hawk sailing by at 200 feet, a squirming snake in its talons. Salt in the drinking water. Salt, selenium, arsenic, radon and radium in the water, in the gravel, in your bones. Water so hard it bends light, drills holes in rock and chokes up your radiator. Why go there? Those places with the hardcase names: Starvation Creek, Poverty Knoll, Hungry Valley, Bitter Springs, Last Chance Canyon, Dungeon Canyon, Whipsaw Flat, Dead Horse Point, Scorpion Flat, Dead Man Draw, Stinking Spring, Camino del Diablo, Jornado del Muerto . . . Death Valley.

  Well then, why indeed go walking into the desert, that grim ground, that bleak and lonesome land where, as Genghis Khan said of India, "the heat is bad and the water makes men sick"? 21

  Why the desert, when you could be strolling along the golden beaches of California? Camping by a stream of pure Rocky Mountain spring water in colorful Colorado? Loafing through a laurel slick in the misty hills of North Carolina? Or getting your head mashed in the greasy alley behind the Elysium Bar and Grill in Hoboken, New Jersey? Why the desert, given a world of such splendor and variety? 22

  A friend and I took a walk around the base of a mountain up beyond Coconino County, Arizona. This was a mountain we'd been planning to circumambulate for years. Finally we put on our walking shoes and did it. About halfway around this mountain, on the third or fourth day, we paused for a while—two days—by the side of a stream, which the Navajos call Nasja because of the amber color of the water. (Caused perhaps by juniper roots—the water seems safe enough to drink.) On our second day there I walked down the stream, alone, to look at 23 the canyon beyond. I entered the canyon and followed it for half the afternoon, for three or four miles, maybe, until it became a gorge so deep, narrow and dark, full of water and the inevitable quagmires of quicksand, that I turned around and looked for a way out. A route other than the way I'd come, which was crooked and uncomfortable and buried—I wanted to see what was up on top of this world. I found a sort of chimney flue on the east wall, which looked plausible, and sweated and cursed my way up through that until I reached a point where I could walk upright, like a human being. Another 300 feet of scrambling brought me to the rim of the canyon. No one, I felt certain, had ever before departed Nasja Canyon by that route.

But someone had. Near the summit I found an arrow sign, three feet long, formed of stones and pointing off into the north toward those same old purple vistas, so grand, immense, and mysterious, of more canyons, more mesas and plateaus, more mountains, more cloud-dappled sun-spangled leagues of desert sand and desert rock, under the same old wide and aching sky.

The arrow pointed into the north. But what was it pointing *at*? I looked at the sign closely and saw that those dark, desert-varnished stones had been in place for a long, long time; they rested in compacted dust. They must have been there for a century at least. I followed the direction indicated and came promptly to the rim of another canyon and a drop-off straight down of a good 500 feet. Not that way, surely. Across this canyon was nothing of any unusual interest that I could see—only the familiar sun-blasted sandstone, a few scrubby clumps of blackbrush and prickly pear, a few acres of nothing where only a lizard could graze, surrounded by a few square miles of more nothingness interesting chiefly to horned toads. I returned to the arrow and checked again, this time with field glasses, looking away for as far as my aided eyes could see toward the north, for ten, twenty, forty miles into the distance. I studied the scene with care, looking for an ancient Indian ruin, a significant cairn, perhaps an abandoned mine, a hidden treasure of some inconceivable wealth, the mother of all mother lodes. . . .

But there was nothing out there. Nothing at all. Nothing but the desert. Nothing but the silent world.

That's why.

Content Considerations

1. Summarize the reasons Abbey gives for staying out of the desert. Explain how serious you think he is.
2. After expounding on why one should never visit the desert, the author

enumerates several rules for hikers who do visit. Are any of these spurious?
3. What do the "hardcase" names of the desert features suggest about the way man views them?

The Writer's Strategies

1. How does Abbey's persona, his voice, suggest very early to the reader the content of the essay?
2. Most of the author's caveats are true; the desert dangers are real. How, then, does he convey a sense of irony?
3. What purpose is served by the incongruity of Abbey's obvious love for the desert and his dire warnings to stay out of it? Who is the audience for Abbey's essay?

Writing Possibilities

1. Consider an area you love—the mountains, a beach, a river, or a woodland—and write an essay in which you extol its virtues while warning others to stay out because of its inherent dangers.
2. Write a reply to Abbey's essay in which you support or disagree with his advice.
3. With two or three classmates, speculate on the result of halting development in an area of your choice. In a short essay, use your conclusions to persuade the rest of the class that halting development of the area you selected is a good idea.

SUE HUBBELL

Summer

The essay that follows is from A Country Year: Living the Questions, *in which writer and beekeeper Sue Hubbell presents the natural history of a year on her farm in the Ozarks. The book begins and ends with spring, and through its pages one can learn about the land and people, plants and animals of the area.*

Presenting natural history, though, almost always includes a variety of other topics, for speaking of nature quite often demands side trips into zoology, psychology, literature, geology, biology, or any one of several other disciplines. Hubbell writes here of snakes, of how they behave in the relative wilderness of her land, and her observations lead into folklore and popular belief.

Before you read Hubbell's essay, recall what folklore or tales you have heard about various kinds of snakes. Then write a paragraph or two about how such tales may shape the way people believe and act.

I'VE BEEN OUT IN BACK TODAY checking beehives. When I leaned over one of them to direct a puff of smoke from my bee smoker into the entrance to quiet the bees, a copperhead came wriggling out from under the hive. He had been frightened from his protected spot by the smoke and the commotion I was making, and when he found himself in the open, he panicked and slithered for the nearest hole he could find which was the entrance to the next beehive. I don't know what went on inside, but he came out immediately, wearing a surprised look on his face. I hadn't known that a snake could look surprised, but this one did. Then, after pausing to study the matter more carefully, he glided off to the safety of the woods.

He was a young snake, not even two feet long. Like the other poisonous snakes found in the Ozarks, the cottonmouths, copperheads belong to the genus *Agkistrodon*, which means fish-hook toothed. The copperheads in my part of the Ozarks are the southern variety, *Agkistrodon contortrix contortrix*, which makes them sound very twisty indeed. They are a pinkish coppery color with darker hourglass-patterned markings. They have wide jaws, which give their heads a triangular shape. Like the cottonmouths, they are pit vipers, which means that between eye and nostril they have a sensory organ that helps them aim in striking at warm-blooded prey. They eat other small snakes, mice, lizards and frogs.

The surprising things about copperheads are their mild manners, timidity and fearfulness. They have, after all, a potent defensive weapon in their venom, and yet their dearest wish when they are discovered is to escape. This rocky upland peninsula of land between the river and creek is a lovely habitat for copperheads. I often find them under the beehives, and they are common in the open field. Twice I have had them in the cabin. Every time I come upon copperheads they simply try to get away from me and never offer to strike.

Once I had an old and heavy Irish setter who was badly bitten when he clumsily stepped on a copperhead. To the snake, being walked on by eighty-five pounds of dog represented a direct attack, and so he struck. The dog's leg swelled and he was in obvious pain. Within a few hours his heartbeat was rapid, his breath shallow, and I took him to the vet. Afterwards he watched where he put his feet. I do, too, and wear leather boots when I walk through the field or in the woods, and in the warm months I give decent warning when I turn over a stack of old boards. I have enormous respect for a small animal with venom so potent that it can make a large dog very sick. I weigh more than the dog, and so I might not have such a severe reaction; there is no record of a human death caused by a copperhead bite in Missouri, but I don't want to risk the pain.

I respect copperheads, but I also have another set of feelings toward them, a combination of amazement and sympathy that an animal should be so frightened by me, so eager to escape, so little inclined to use the powerful means that he has to defend himself.

Copperheads contrast oddly with the eastern hognose snakes, sometimes called puff adders, that I also see here sometimes. These are harmless, but put on a tremendous show of ferocity. I came upon a hognose one day in the field, and he raised up the first third of his body and spread his neck wide, hissing horribly, trying to convince me that he was a cobra. I was fooled hardly at all and stood quietly watching him; after some more halfhearted hissing and spreading he gave up the attempt to frighten me, remembered urgent business he had elsewhere and slithered away into the tall grasses.

Apart from copperheads, there are few dangerous snakes here. There are supposed to be rattlers, but in the twelve years I have been walking the woods and river banks I have never met one. Most of the snakes around are harmless or, like the black rat snakes, which eat rodents, beneficial, and I have no sympathy with the local habit of killing every snake in sight. It is an Ozark custom to pack a pistol along with the beer on a float trip. The pistol is for shooting the cottonmouths that are supposed to fill the river and be thick upon its banks.

In point of fact, the river is too cold and swift for cottonmouths, and since I have lived here I have seen only one. He was idling in a warm, shallow pool at the side of the creek that runs along my southern property line. I stopped to look at him from a safe distance. Heavy-bodied and dark, confident and self-assured, he watched me in his turn and did not retreat as a copperhead would have. Instead, he coiled and raised his head, in a defensive posture, ready to strike if I were to advance. He opened his mouth wide, and I could see the white, cottony-looking interior that gives the snake his common name. After he was sure I was not going to come closer, he dropped into the water slowly and with dignity and swam away from me to the bank, where he disappeared under the safe cover of the branches hanging over the pool.

This species of cottonmouth is called *Agkistrodon piscivorus leucostoma*, or white-mouthed *agkistrodon* who eats fish. They are often found in warmish water, and are primarily fish-eaters, but they also feed on other snakes, rodents, frogs and lizards.

The cottonmouth I saw was evidently an old one, for he was big, probably nearly four feet in length; only a slight hint of his cross-banding was visible. Young cottonmouths are lighter, more patterned, and newborn cottonmouths, like copperheads, have yellow-tipped tails. Both cottonmouths and copperheads belong to the evolutionarily advanced group of snakes that do not lay eggs. The young are retained within the mother's body, protected by a saclike membrane, until they are born. Like other snakes, they shed their skins as they grow.

A treasured possession of mine is one of those snakeskins, fragile but perfect, with even the eye scales intact. It is always startling to me to notice people shudder when they see it. There is enough psycho-mythology about humankind's aversion to snakes to reach from here to Muncie, Indiana, some of it entertaining, much of it contradictory. Whatever the reason, many people are irrationally afraid of snakes, and this makes for poor observation. It is hard to tell what a snake is up to if you are running away from it or killing it. This may account for the preponderance of folklore over natural history in conversations about snakes. It may be why Ozarkers have told me that the hognose snake is poisonous, that snakes go blind in August, that the hills are full of the dread hoop snake who holds his tail in his mouth and comes roaring down hillsides after folks to attack them with the horn on his tail—a horn so deadly that if it gets stuck in a tree, the tree will die within a few days.

My favorite folk story about snakes is the one about copperheads, who are said to spit out their venom on a flat rock before taking a drink of water and then, having drunk, suck it back up into their fangs. I always liked it because I thought it was a grand Ozark stretcher. But then I found the exact same story in a Physiologus, a medieval bestiary. It is a snake story at least eight hundred years old, perhaps more. So it turns out that the yarn is a piece of natural history after all. It is just that it has to do with the nature of the human mind, not nature of the snakish kind.

Content Considerations

1. What has been Hubbell's experience with snakes?
2. In what ways can fear interfere with truth?
3. What kinds of knowledge can be gained from folklore? From natural history?

The Writer's Strategies

1. What main idea evolves through this essay? How do snakes serve as its illustration?
2. How do the differences between folklore and natural history shape Hubbell's essay?

Writing Possibilities

1. Reflect upon what this essay suggests about how people coexist with nature, focusing on the practices you support most strongly.
2. What is the nature of the human mind? As you prewrite for this response, consider how fear, superstition, hope, uncertainty, and conviction may sometimes affect how your own mind works.
3. With several classmates, recall folklore stories about animals. In a paper, present the ones you like best and speculate on how they affect people who hear them. What do these stories suggest about the cultures that created them?

CHRISTIAN KALLEN

Eco-tourism: The Light at the End of the Terminal

In the following article, Christian Kallen writes about eco-tourism, a small but growing part of the travel industry. Kallen, coauthor of several adventure books and contributor to several newspapers and national magazines, examines the causes and consequences of eco-tourism in several areas—economic, political, environmental, cultural, and individual.

Travel is the world's largest industry, and it creates for millions of people opportunities not only for pleasure and recreation but for education and awareness. But eco-tourism, which includes a range of travel from touring endangered areas to bird-watching expeditions, may not be without problems.

Before you start Kallen's article, write a few sentences about what those problems might be. Compare your predictions with your classmates', then discuss the consequences of the problems predicted.

IT'S TIME FOR A VACATION, but the thought of a self-indulgent week at a Mexican beach just doesn't have the appeal it once did. Besides, your social and environmental consciousness keeps telling you there must be a better way to spend your time. If this sounds like your scenario, then you're a ripe candidate for an environmental vacation, or "eco-tour."

Eco-tourism is the latest buzzword in organized travel, and it's as hot as a travel trend has ever been. An offshoot of the adventure travel industry that came into its own in the 20 years since the first Earth Day, eco-tourism shares many of the same destinations and itineraries of adventure travel.

Through treks, river rafting, visits to natural history reserves and countless other offerings, adventure travel promises close and meaningful encounters with the natural world, in contrast to more traditional vacations. The objectives of eco-tourism are more conscientiously environmental and "politically correct" still. In fact, many travel companies now encourage visiting destinations simply because they support endangered species, are controversial or are in danger of deforestation.

Even the World Wildlife Fund has recognized the potential impact of eco-tourism, and offers the definition: "tourism to protected natural areas, as a means of economic gain through natural resource preservation." It sounds like the ideal merger of recreation and responsibility, but does merely calling a trip "eco-tourism" make it so?

Eco-tourism opportunities are many, and the companies that offer them are increasing almost daily. You can visit the same troops of mountain gorillas in Rwanda that ethologist Dian Fossey worked so hard to save; you can stay at a jungle lodge in the rainforests along the Rio Napo; you can even float along the coast of Antarctica with one of a handful of cruise ship companies who are following self-imposed guidelines for "low-impact tourism."

Common to all these trips is a tolerance for the inconveniences of remote travel. The nature tourist is more likely to put up with rustic camping accommodations, muddy trails and reduced services than the resort traveler, if the payoff is closer contact with the bird life, native cultures or endangered habitat he or she has come to see.

In some cases, such as the offerings of the Massachusetts-based Earthwatch Expeditions, field work is the point of the tour—more a tour of duty than a form of tourism, a "short-term scientific Peace Corps," as their public affairs director puts it. Other outfitters, such as Sobek Expeditions of California, evidence their commitment to the cause by donating a portion of trip fees to nonprofit groups such as the Rainforest Action Network (RAN). Still others, such as Journeys of Ann Arbor, Michigan and its affiliate, Wildland Journeys of Seattle, Washington have created nonprofit partners—in their case, the Earth Preservation Fund—to oversee local rehabilitation or cleanup projects in their host countries.

Behind all of these approaches is the recognition that travel, whether individual or commercial, may be the only way for many of us to personally experience endangered creatures and ecosystems. Environmental tourism makes its strongest and simplest case by taking people to natural environments, where they can see what stands to be lost to unbridled development. "The goals of eco-tourism have always been my goals," says Vicki Fittsmilgrim, founder of Breakaway Travel of Bedford, Massachusetts. "To take people to places they wouldn't normally go, to see things they wouldn't normally see, to get a sense of the global village we all live in. And I hope the experience jars them a little."

This stance is accepted by environmental activists like RAN's Randy Hayes, who at one time advocated a tourism boycott of Hawaii in protest of several land development schemes. "We have reversed ourselves on the tourism boycott," he now says. "We feel it's important to get people there to see for themselves these unique areas. In Hawaii, we're using eco-tourism to build an army of activists."

Any consideration of the impact of eco-tourism on the environmental movement must begin with the recognition that travel is the largest industry in the world. The World Tourism Organization counted nearly 370 million tourists in 1987, supporting 65 million jobs and 25 percent of world trade. Combined national and international spending on tourism is over $2 *trillion* annually. Adventure travel, which includes environmental tours in its definition, accounts for less than 10 percent of the total tourism figures; but it is the fastest growing segment of the industry, growing nearly 30 percent annually.

A recent Gallup poll indicated that 76 percent of Americans regard themselves as "environmentalists," and that figure is not lost on business planners. International travel is an increasingly popular vacation choice for the baby boomer generation. Eco-tourism seems to offer not only a "politically correct" approach to a global problem, but an on-target marketing ploy to the largest consumer group in the nation—thirty or fortysomething men and women with disposable income, accumulated vacation time, and a desire to see the world.

But just how effective is a vacation on an individual's subsequent commitment to an environmental lifestyle? Above the Clouds Trekking of Worcester, Massachusetts offers tours to Nepal and other prime trekking destinations. When I asked owner Steve Conlon if he found that his clients became more aware of environmental issues after their trip, his reply was swift: "Definitely. People's awareness of the fragility of the environment just takes a giant leap forward. I think tourism has the potential to have a tremendous impact upon the relations between countries and the whole direction the planet has to move in."

Whether or not Conlon's idealism is justified, supporters of environmental tourism say it can have a beneficial effect on a host nation's economy, providing an alternative to exploitative economic growth industries such as logging and mining. According to Elizabeth Boo of the World Wildlife Fund (WWF) the nature traveler spends more money in a country than the recreational traveler. In her recent study, *Eco-tourism: The Potentials and Pitfalls,* people who cited parks and protected regions as their main reason for visiting certain Latin American countries spent over $1,000 more in two weeks than other tourists. This is the "economic gain" referred to in the WWF definition of eco-tourism.

Such evidence is sure to carry weight with other debt-burdened countries struggling to manage their resources. Nations such as Costa Rica, Ecuador, Belize, Rwanda and Kenya have recognized this, and have set aside significant percentages of their land for protection, with an eye toward courting tourist revenues as an alternative to industrial exploitation. Kenya recently acknowledged that, with a tourist industry of 735,000 visitors bringing in $250 million annually, conservation is far more lucrative than development. Some see this as the primary benefit of eco-tourism: its positive impact on local economies.

There is a limit to how far tourism can go toward rescuing an economy, however. Despite the attention given to such plans as the "debt-for-nature" swaps proposed to help relieve Latin America's $3 trillion foreign debt in exchange for protecting their national reserves, some areas have more than enough visitors already. The obvious case is the Galapagos, the desert islands off Ecuador's coast known for their diversity of wildlife. The Galapagos saw only 4,500 visitors in 1970; currently they are burdened with over 30,000 per year. Despite this onslaught, the permit quota was recently raised to 87,000 annually. The sheer number of visitors to such delicate environments is sure to have a negative impact.

Similarly, large numbers of visitors to wildlife preserves in East Africa have been shown to impact both the population and behavior of the animals there—the number of cheetahs in Kenya's Amboseli Reserve dropped from 15 to 5 in 1987 alone. And it was a passenger-carrying Argentine supply ship that ran aground in Antarctica in early 1989, spilling some 600 metric tons of diesel fuel along the coastline. The days of "low impact tourism" are hardly upon us.

The problem clearly goes beyond litter lining the trekking trails of Nepal to other considerations such as sanitation, public health, depletion of fish and wildlife, and the cultural impacts on the inhabitants of many remote areas. While some indigenous people seem more immune to Western influences than others (travel writer Tim Cahill notes, "The Balinese aren't interested in bargaining for your Levis—they don't understand why you don't wear a sarong."), other cultures seem to evaporate under the onslaught of portable radios, digital watches, slogan-bearing T-shirts (such as those supplied to "clothe the natives" by missionary programs) and motorized vehicles.

Even politics can be affected by tourism. In recent events in Kathmandu, government troops opened fire and killed some 50 demonstrators. Though hesitant to assign full responsibility to tourism, Steve Conlon of Above the Clouds Trekking remarked, "It may have been a contributing factor. Certainly the presence of Westerners was a reminder to people that they weren't simply born at the wrong time in the wrong place, but that someone was actively keeping them down." As well as the obvious economic effects of tourism—which can be controlled or encouraged by governments—intangible effects that cannot be controlled are almost inevitable.

"We don't want to deny the fruits of the modern world to local people—change is inevitable," said Conlon. "But we can determine whether our impact is positive or negative." By way of example, Conlon started the Society for the Preservation of Traditional Cultures to balance his for-profit trekking company, and solicits funds from his trekking clients for such operations as the Himalayan Earthquake Relief Operation (HERO) to deal with crises in his host country.

Similarly, Journeys founder Will Weber started the nonprofit Earth Preservation Fund (EPF), financed by proceeds from commercial trips. EPF has many cultural and environmental programs in the countries where Journeys does business, such as restoring monasteries in Ladakh, and reforestation and trail clean-up campaigns in the Himalayas. The Seattle-based affiliate Wildland Journeys bills many of its programs as "environmentally conscientious," and its president Kurt Kutay has become an outspoken advocate of eco-tourism.

There's yet another side to the issue: that taking an environmental tour may be mistaken for taking an environmental stance, that tourism itself may become a substitute for activism. After all, an environmental vacation is still a vacation, a "feel good" solution to a pressing global problem. Earthwatch Expeditions is quick to differentiate itself from eco-tourism, notes Public Affairs Director Blue Magruder. Begun in 1971 by educators and scientists, Earthwatch offers people the chance to spend between two and six weeks time assisting research scientists doing field work. "We've been very careful not to call what we do 'tourism.' We're supplying people power to research professors to facilitate their fieldwork." In Earthwatch's case, since they are entirely nonprofit and funded in part by foundation grants, much of the fee paid to join a project goes directly to the field project itself.

Magruder is not alone in questioning the motives of eco-tourists. She sees personal education as a by-product of Earthwatch projects. As with service programs, gratification is long term, associated with the completion of a project. Others are less critical of tourism as a solution. "I don't know if the motive really matters if the end result is good," says Tensie Whelan, Vice President for Conservation Information at the National Audubon Society. Currently working on a book on eco-tourism and its role in developing countries based on her two years in Costa Rica, Whelan is generally positive about the environmental tourism movement, whatever its origins. "While some sites in Costa Rica are overused, the overall impact is to raise awareness within the country of the importance of their own natural areas. It has caused the government to start a ministry of the environment, to put aside more than 25 percent of their land. Eco-tourism does make a difference."

For many within the travel industry itself, personal education and environmental enlightenment are the primary motivations behind their commitment to eco-tourism. Leslie Jarvie came to the position of Director of Environmental Programs for Sobek Expeditions indirectly, while the adventure travel outfitter was serving R. J. Reynolds on their Camel Trophy Four-Wheel Drive Rally in Brazil in 1988. Pushing tobacco products and an aggressively masculine image in an endangered landscape proved troubling for Jarvie, and a meeting with an outspoken Brazilian journalist, Juan de Onis, convinced her she had to assume some responsibility for her work. Back in San Francisco she looked up Randy Hayes.

"We handed out RAN press packets to the 250 journalists from all over the world covering the rally," she said. "Everyone was looking for something that would give their article more substance than just a bunch of macho men tearing through the rainforest in 4-wheel drive." Following the rally, Jarvie easily convinced Sobek founders John Yost and Richard Bangs that the company had to take a more positive environmental stance, and talked herself into the job of directing the program. "We decided to use some of our tours to communicate our concerns about the environment, increase awareness and—we hope—to raise consciousness," she said. The company introduced five eco-tours in 1990.

Their Amazon Basin itinerary includes stays at isolated rainforest lodges. And educational visits to mining and gold rush towns along the Amazon and the frantic trading center of Manaus broaden their experience with the region. Meetings with local environmental activists are also built into the itineraries, and between $250 and $400 of the fees for each booking go to Rainforest Action as a tax-deductible contribution.

Of course, it's not necessary to join an organized environmental tour to make a commitment to the environment when you travel, and many people find it preferable to travel alone or with friends, staying in locally-owned hotels and guest houses, eating in the market, packing out all their trash and batteries. Still, many problems arise due to independent trekkers who haven't done their homework about local customs and who argue over already-low accommodation, food and crafts prices.

Experience and professional responsibility can make the impact of a well-organized tour group more positive than that of a single traveler. Within the travel industry, consumer demands can have an almost immediate effect on how trips are conducted. An individual's complaint about a hotel's waste disposal practices is likely to produce a shrug; a commercial trip operator who books 300 clients a year in that hotel might receive more solicitous attention.

Though eco-tourism is clearly not a win-win situation, many such as Audubon's Tensie Whelan regard it as a more responsible solution than strip mining or clearcutting. Nearly all agree that, to make eco-tourism a more powerful force in the environmental movement—and to assure its integrity as a solution—standards should be set within the travel industry that define what an "eco-tour" is, how it should be run and what it should accomplish. A recent Eco-tourism and Resource Conservation Project seminar at the Environmental Law Institute in Washington, D.C. worked on this issue. The common goal of the participating nature tour companies, government agencies and conservation groups was to develop something akin to the Valdez Principles for tourism—guidelines that, if followed, would assure the safe and sound establishment of tourism to culturally and environmentally sensitive areas. Several agencies have their own principles on tourism, including National Audubon, the Center for Responsible Tourism, and the Ecumenical Coalition on Third World Tourism.

However faddish eco-tourism may be now, it's important to recall the motivating rationale for it, expressed in terms of "changing consciousness" by Leslie Jarvie, "raising awareness within the [host] country" by Tensie Whelan, or its "potential to have a tremendous impact" by Steve Conlon. Everyone seems to agree with Randy Hayes's bold goal for eco-tourism: to create "an army of activists" from the 30,000 that visit the Galapagos yearly, the 735,000 who go to Kenya, the 55,000 trekkers in Nepal, or even the half million that run rivers in North America each year. If eco-tourism works even a fraction as well as its supporters hope, it will create that army of activists, primed and committed to save the rainforest, the llamas and the elephants, and poised to take a stand in the place where they live.

28

Content Considerations

1. What does it mean for an idea, action, or opinion to be "politically correct"?
2. What goals do promoters of eco-tourism claim?
3. What are some of the environmental, cultural, economic, and political benefits of eco-tourism? How might eco-tourism cause harm?

The Writer's Strategies

1. What is Kallen's attitude toward eco-tourism? Where do you find it expressed?
2. With how much authority does Kallen speak? How is that authority established?

Writing Possibilities

1. Which of the effects of eco-tourism do you find most worrisome? Write about them as detrimental effects and propose any solutions you think of.
2. With a classmate, suppose you were contemplating an eco-tour to a remote region. Discuss your questions and concerns, then draft them in letter form to a tour organizer.
3. Several of the people Kallen interviewed believe that eco-tours will increase environmental awareness. What do you think?

LORI NELSON

The Dolphins of Monkey Mia

Monkey Mia, an area on the coast of western Australia in Shark Bay, has drawn dolphins to its beaches for more than twenty years. Because there are dolphins, there are people—people who come to watch and pet and play. Writer Lori Nelson was one of those people; in the following article she tells of her visit, of the tourists and researchers she met, and of the dolphins.

The dolphins, she discovered, were in danger.

That danger focuses on how wild the wilderness must remain, on how much development or even human contact can occur before harm is done. Nelson, whose articles on marine mammal and ocean conservation issues have been included in national and international publications, presents observations and interviews as she explains what is known about the ways in which the dolphins are threatened.

Before you read Nelson's article, take a few minutes to write about how well-meaning human beings might be causing harm.

"THEY'RE HERE!"

WITHIN MINUTES OF THE CRY, I crawl from my tent and join a sleepy crowd at the beach of the campgrounds, where we watch three dorsal fins gliding shoreward through the dawn-streaked water. Children race into the shallows, followed by parents readying cameras, grandmothers gamely hoisting their skirts, and young backpackers in neon t-shirts. Surging into the group, the dolphins weave through the forest of knees, arching their bodies just enough so that outstretched hands can stroke them along the sides. The dolphins pass back and forth, tossing their heads, and soon bring everyone together to witness their antics.

Watching the encounter between species of land and sea, I am struck by the animals' easy familiarity. The wild dolphins of Monkey Mia in Shark Bay, a remote area of Australia's western coast, have been hosting such get-togethers almost daily for the past 20 years. Today's visitors have traveled thousands of miles—and in my case halfway around the world—to meet these sea celebrities.

Kelly Waples, a graduate in marine biology from the University of Santa Cruz who is spending the summer studying the dolphins with researchers from the University of Michigan, joins me at the water's edge. "That's Holey Fin, she's about 24 years old," Kelly says, pointing to a speckled dolphin with a hole in her dorsal fin. "She's been coming in since the beginning, more than 20 years. The one with the jagged fin is her daughter, Nicky, and the other one is Crooked Fin, another grandma." She scans the horizon, looking for dorsal fins, then shakes her head. "It's a shame you missed the others."

Kelly and Richard Connor, a graduate student at the University of Michigan, had greeted me with news when I arrived the night before: Several weeks earlier a newborn calf had been found dead and then, during the next 18 days, six more of the shore-visiting dolphins disappeared from the area. I had come to Monkey Mia to learn about the research and found everyone in the community, from researchers to rangers who monitor the tourist center, to the small group of people who live in the caravan park to be near the dolphins, gripped by the loss and the mystery that surrounded the dolphins' fate.

The tragedy began January 23, 1989, when the rangers discovered Holey Fin's six-week-old calf, Koorda Fin, dead by her side in the shallows. Over the next 18 days, Nicky's and Puck's newborn calves disappeared, the mother appearing inshore without them. Then Holly, a juvenile female whose serene disposition made her nearly everybody's favorite, disappeared, followed—one by one—by three male dolphins, Sickle Fin, Snubby, and BB, members of a "coalition," or group of males who spend most of their time together.

While the death of a newborn calf is not uncommon and the dolphins often leave the area for a few days, especially during mating season, the combined disappearances were alarming. The three males, in particular, had never been absent for more than four days at a time; in 1988, they had visited Monkey Mia 345 days. Worried, Connor began a daily search of the area by boat. No bodies were recovered, but both Puck and Holey Fin were observed mating offshore. Because female dolphins go into estrus following the loss of a calf, the researchers were forced to conclude that the calves, if not all of the missing dolphins, had indeed died.

"Who's here today?" asks Richard, appearing nearby. As he studies the turquoise bay through binoculars, the rangers arrive with buckets of fish and wade into the water, beginning to distribute them to visitors eager to feed the three females. I watch Holey Fin patiently submit to the clumsy pats of a group of young, excited children. Her forbearance is remarkable, that a woman her age might assume with her grandchildren. But these dolphins are far from tame, as Nicky demonstrates with a powerful flick of her flukes at one man who unwisely tries to stroke her on the melon, or forehead, a most sensitive spot.

The visitors cannot know the enormity of the dolphin loss and, as I watch them enjoying this moment, I think of how especially difficult this must be for Richard. In 1982, he and Rachel Smolker, undergraduates at the University of California at Santa Cruz, saw a photograph in *National Geographic* of the Monkey Mia dolphins and came to Australia to investigate. They never left the beach that summer, so enthralled were they by the daily lives of the dolphins unfolding before them inshore. Funding from the National Geographic Society, the National Science Foundation, and private contributions enabled them to set up a base camp on the beach and purchase boats and equipment for tracking the dolphins across the bay. They have spent several months of every year since at Monkey Mia amassing invaluable data on both the inshore and offshore dolphins, and have identified, by dorsal fin markings, a local population of nearly 200.

PRIMATE PARALLELS

Richard's research focuses on the adult male dolphins, and his painstakingly accumulated data on the activities both inshore and offshore of Sickle Fin, Snubby, and BB, have yielded exciting new information about male roles in dolphin society, about which little had been known before. His work suggests provocative parallels between dolphin behavior and the social systems of the higher primates. Richard's observations have intrigued other scientists including primatologists Barbara Smuts and Richard Wrangham, who are now embarking on their own Monkey Mia research. "Monkey Mia may prove to be as fruitful for dolphin research as the Gombe Stream Reserve has been for Jane Goodall's primate studies," says Harvard primatologist Irven Devore.

Of the 30 or so species making up the dolphin family (Delphinidae), 19 inhabit Australian waters. The dolphins of Monkey Mia are bottlenose dolphins *(Tursiops truncatus)*, the performing stars of marine parks around the world, as well as the species most frequently sighted in Australia. Bottlenose dolphins may grow to 4 meters and are the most studied species both in the wild and in captivity.

These dolphins have proved difficult to study in the wild, as rigorously monitoring their behavior in the open sea by boat is almost impossible. Yet, only by observing them in their natural habitat can scientists collect accurate information. Forming a perfect natural laboratory, Shark Bay is the dolphin researcher's paradise. In a shallow basin of 5,000 square miles, the bay's sheltered waters teem with life of such variety and uniqueness that the government is moving toward placing it on the UNESCO World Heritage List, where it will join 300 other sites, such as the Grand Canyon and the Taj Mahal, that are protected from change because of their historical, geographical, and cultural value. Seagrasses cover nearly 2,500 square miles of the seafloor and provide a nursery bed and home for a wealth of fishes and crustaceans, including whiting, mullet, bream, snapper, giant tuna, marlin, manta rays, lobsters, prawns, and scallops.

Though hunted to near extinction in the tropical Indo-Pacific range, the dugong *(Dugong dugon)*, or sea cow, a completely vegetarian sea mammal with a population numbering only about 1,000, thrives on Shark Bay's underwater seagrasses. Other endangered species, the green turtle *(Chelonia mydas)* and loggerhead turtle *(Caretta caretta)* have also found unmolested nesting areas along Shark Bay's beaches. Kangaroos share the dunes with 96 species of reptiles, including 8 found only in Shark Bay.

Of course, there are also the sharks. In 1699, William Dampier, an English sea captain, sailed midway into the bay during his famous journey along the coast of what was then known as New Holland and wrote in his journal, "There are an abundance of them in this particular Sound that I therefore give it the name of Shark's Bay."

BLOOD-RED SANDS

A strip of sandy beach punctuated by a fishing jetty and backed by miles of blood-red sand dunes and desert scrub, Monkey Mia is situated on a finger of land, Denham Peninsula, that crooks into the bay roughly 400 miles north of Perth, the capital of Western Australia. The nearest settlement, Denham, is a fishing village with a population of 400, comprised mostly of the descendants of the Malay, Filipino, Chinese, and aboriginal workers imported to man the pearling camps that formed the area's sole industry from the mid-1800s until 1940, when fishing became more lucrative. In 1986, when the daunting 25-mile road of rock and sand from Denham to Monkey Mia was paved, droves of tourists suddenly swept through the peninsula, and Denhamites are still reeling from an onslaught that shows no signs of abating. In August 1984, 10,000 tourists, conservationists, scientists, and journalists passed through Monkey Mia. In 1988 more than 100,000 visited.

The event that launched Monkey Mia's fame is local legend. One hot, summer night in 1964, a fisherman and his wife anchored off the jetty and unwittingly created a moment of history. Unable to sleep, "Ninny" Watts watched a dolphin splashing near the boat in the moonlight and tossed it a fish from the icebox. The next night Charlie, as the couple later named it, returned, but this time it took the fish from her hand. Within a short while, it was bringing along its friends and accepting fish from them on the beach. Other local fishermen picked up the habit, and the herd of well-fed dolphins became mascots of the Monkey Mia fishing community.

By the time Wilf and Hazel Mason took over and modernized the primitive campgrounds in 1974, Charlie had died, but members of the original group, speckled grandmothers like Holey Fin and Crooked Fin, along with their offspring, Nicky, Puck, Holly, and the males were still coming in to shore. While the dolphins, knowing a good deal when they've got one, may be motivated by free fish, it's equally apparent that they enjoy their contacts with Monkey Mia's visitors, which have been a part of the dolphins' daily agenda for three generations.

Of all the marine environments in the world, why was Monkey Mia singled out for this phenomenon? Hec Goodall, one of Australia's marine mammal authorities, believes a rare combination of factors created the unique situation. "First of all," says Goodall, "you have a remote and isolated area, a unique environment because of the low shelving beach, an abundance of fish, and perhaps most importantly, a group of gentle, retired people who were willing to spend time coaxing the dolphins inshore."

With the burgeoning tourist traffic, the dolphins have become an integral part of the region's economy. Their disappearance attracted a considerable amount of national media attention and, shortly after my arrival, government agencies stepped in. The Department of Conservation and Land Management (CALM) requested a full investigation from the Environmental Protection Authority (EPA). The EPA's findings were horrifying. After extensively testing the pollution levels of the shore water, the EPA determined that high levels of sewage contamination had occurred just before the dolphins disappeared. The Monkey Mia rangers were informed that the toilets installed at the tourist center near the beach were leaching fecal matter directly into the water habitually visited by the dolphins. A few weeks before the dolphins' disappearance, the EPA reported, a national holiday had brought a drastically higher number of visitors. At the same time, tide and water movement had been lower than usual. The combination of factors could indeed have proved fatal for the dolphins, though, without a necropsy or post mortem exam of the bodies, the report could not definitively conclude this was the cause of death.

A Terrible Irony

"It's the only explanation that fits in with the facts," said Connor, when told of the results. He explained to me that the calves would have been most susceptible to contamination and the three adult males had been at the peak of mating season, when their testosterone levels were high and their immune systems depressed. As males fight among themselves during this period, they earn plenty of cuts and scratches where infections can take hold.

No one in the community missed the terrible irony: The tourist center installed to protect the dolphins had most likely caused their deaths. The toilets were closed immediately, and a reassessment of the situation at Monkey Mia began. Authorities examined the impact on the dolphins of other previously suspected dangers, such as pesticide use around the center, sunscreens and lotions used by visitors in the water, and types of fish being fed to the dolphins. By making changes now, the rangers hoped to protect what was left of their fragile treasure. As Connor pointed out, "There are quite a few dolphins here that come into shore, but don't allow themselves to be fed or petted. They could easily be coaxed in to replace the others, with a little effort, but we're not going to do a thing until we're absolutely certain they won't be endangered."

Meanwhile, other troubling changes were in the works at the caravan park. Even as the EPA released its final study, the Masons were in the process of selling the caravan park to a large resort hotel conglomerate that intends to capitalize on the dolphin phenomenon by turning the campground into a deluxe resort complete with swimming pools, tennis courts, and a cabana bar on the beach. Rumors flew that regularly scheduled air flights into Denham were also planned to bring unprecedented numbers of tourists. Just how much added stress would the remaining dolphins be able to tolerate, those of us anxiously watching the bulldozers wondered.

The newspapers reported that other cities on the coast such as Bunbury and Canarvon had hired dolphin trainers to lure their own populations of wild dolphins to recreate the Monkey Mia gold mine. An entirely new concept in marine park was being invented in Australia, one that freed the dolphins from the cement tanks of the oceanaria, but raised issues never considered before. Who would oversee and ensure the safety of the dolphins, especially once a profit motive was the bottom line in their welfare? Once the dolphins were habituated and dependent on shore-feedings, were they really free?

For the remaining shore dolphins of Monkey Mia—Holey Fin, Crooked Fin, Nicky, and beautiful Puck—life continues as it always has, with fish from the rangers and the eager attention of the visitors.

It is the human beings of Denham Peninsula, feeling the full weight of responsibility for the tragedy, who must decide how best to protect one of the greatest natural resources in the world. There is much we can learn from the dolphins, but they also take back with them into their watery world a lasting impression of us. Like any animal, if threatened, they will simply go away. Head Ranger Sharon Gosper told me, "When dolphins and people meet, they always seem to bring out the best in each other," but whether that remark will hold up, only people can decide.

Content Considerations

1. What is the most likely cause of the death or disappearance of the dolphins in Shark Bay?
2. What changes have been made or are planned to accommodate tourists visiting the area? What might be the consequences of these changes?
3. What attracts people to dolphins?

The Writer's Strategies

1. Lori Nelson uses the first-person point of view in this article, readily including her own observations and reflections. How does this point of view shape your response to the article?
2. What is the central question or conflict presented in this article? How does the author build toward it?

Writing Possibilities

1. Determine what limits you think should be set for people in wilderness areas, then argue for those limits.
2. Explore the reasons people are lured to remote regions in order to observe wildlife in natural surroundings. Include a part about your own desire for such experiences.
3. With a classmate, brainstorm a list of problems that might arise when people flock to the wilderness. Then brainstorm solutions. Together, write your conclusions in the form of a proposal.

BROOKS ATKINSON

The Great Swamp

"If you believe that," one guy tells another, "then I've got some great swampland for you at a good price." So goes an old joke or a piece of American lore that implies the value most people find in a swamp. Swampland is worthless and watery, not firm enough for a house or hotel; it crawls with foul and dangerous creatures that would need to be exterminated to make it fit for human habitation.

In their natural state swamps seem to offer little promise to people. To become valuable, they must be drained and developed—then, of course, they are no longer swamps. In the following essay author Brooks Atkinson explores the past of a wetland in New Jersey called Great Swamp.

The essay is from Atkinson's This Bright Land, *a book published in the early 1970s during America's first round of ecological awareness. Looking at America then, Atkinson saw squalor, waste, too much development planned too poorly, a deplorable loss of wild places. Before you read this essay, write for a few minutes about the environmental changes you think America has experienced in the last twenty years.*

IT IS JANUARY. As in the past ten or eleven thousand years, another winter has seized Great Swamp, in Morris County, New Jersey. The trees are bare, the snow is crusted, the marshes are frozen, and little slivers of ice cling to the edges of brooks. In comparison with the life of the other seasons, there is not much to see except tree sparrows in the bushes or among the weeds, titmice, a mockingbird puffed out against the cold, a pheasant, Canada geese, a barred owl, and a few indolent white-tailed deer. Winter has locked up this ancient reservoir of life.

Although the difference between ten and eleven thousand years is stupendous by human standards, it is trifling in geological terms. The exact date when the Wisconsin ice sheet melted into the north and left the physical structure of Great Swamp is not a matter of much consequence. All that concerns the residents of the region today is that the retreating glacier left a wild swamp that has remained autonomous to this day, and puts a solid foundation under modern life.

It is a natural masterwork, only thirty miles west of Times Square. When the air is clear you can see the peak of the Empire State Building from the ridges of the geological till that surrounds the swamp. The contrasts are dramatic. Thirty million people live in big and small houses in the surrounding terrain; automobiles and trucks choke the cement highways that sweep around the swamp; railroads cut by on the north and the south. Great Swamp is so centrally located to business and industrial institutions that for years the Port of New York Authority wanted to tear it apart to build a jet airport, and real-estate developers have long looked at it in the hope of finding some way to make something profitable out of something unused.

But, for many thousands of years, Great Swamp has retained its independence, and preserves in the midst of megalopolis a living patch—seven miles long and three miles wide—of primitive America. Progress has stopped where Great Swamp begins. In New Jersey the density of the population is 833 people to one square mile. But no one lives in the interior of Great Swamp. Even today, some parts of it are impenetrable because of quicksand and thick vegetation.

There was a time, not more than one half century ago, when the word "swamp" was the equivalent of "waste" or "danger." Swamps occupied wasteland that most people thought should be "improved." Swamps harbored snakes and vermin and mosquitoes. Swamps held stagnant water that was suspected of breeding agues and diseases. Escaped slaves hid in swamps; escaped prisoners disappeared there. In most respects swamps were regarded as blights on the community.

But swamps are now recognized as essential links in the chain of life. When in 1959 the Port Authority first proposed to degrade Great Swamp into an airport, people in the surrounding communities organized themselves into local units and raised $1.5 million to buy three thousand acres, which they turned over to the U. S. Fish & Wildlife Service to be managed as the Great Swamp National Wildlife Refuge. The basic motive was no doubt to spare the community the bedlam of an airport; and the non-professional campaign to raise money was widely recognized by public-spirited people as an admirable instance of voluntary civic action.

But exclusion of an airport was not the only motive. Ever since the Morris County Park Commission built a nature center for classes and lectures, and the Fish & Wildlife Service built a boardwalk into one part of the swamp, and the Garden Clubs of Summit and Somerset Hills built an observation booth in another wild area, thousands of visitors have been going to Great Swamp to see the splendor of the laurel blooms in the spring, the calla lilies, marsh marigolds, and pink lady's-slippers, the wood ducks, the herons, and the bitterns, as well as deer and foxes.

For many years, professional and amateur naturalists have realized that Great Swamp teems with multitudinous forms of life. It has become a classroom for children. For students at nearby Drew University, at the College of St. Elizabeth, and at Fairleigh Dickinson, it serves as a field laboratory. James W. Hand, Jr., of Green Village, a former president of the Summit Nature Club, compiled a list of 178 species of birds seen in Great Swamp from September 1949 to March 1964. More than half of them nest in the swamp. Woodcock, snipe, grouse, and pheasant are abundant, either as migrants or residents. Dr. Robert K. Zuck and Mrs. Zuck, of the botany department of Drew University, are collecting the plants of Great Swamp. In previous lists the species had been estimated at about eighty-five. The Zucks have already collected more than six hundred and expect to collect more than one thousand before they have finished.

The deer population is thought to be about two thousand. Other mammals include muskrats, weasels, mink, raccoons, striped skunks, red and gray foxes, opossums, woodchucks, and cottontails. Their tracks are everywhere. In Great Brook and the Passaic River, which gets about half its water from Great Swamp, bass, white perch, catfish, sunfish, and carp can be taken. Primrose Brook, which is a branch of Great Brook, has been stocked with trout.

Every swamp is unique. But those who explore Great Swamp regard it as unique because it contains both northern and southern species of trees, and because little islands throughout the swamp retain trees that are virgin growth. In addition to sweet gum, sour gum, tulip trees, and black and red oak, the swamp contains some white oaks of prodigious size. One of them stands ninety-five feet high and has a diameter of four feet nine inches. Some of the beeches are also enormous; one has a trunk fourteen feet in circumference. Some of these massive trees are from three hundred to five hundred years old, in other words dating from colonial times or possibly from several years before Columbus sighted San Salvador Island. An oblong bowl of lowlands that used to be regarded as worthless is now recognized as a museum of thousands of years of natural history and a rich storehouse of contemporary wildlife.

Although planners and engineers can destroy Great Swamp, no organization of human beings and inhuman machines could have built it. The structure is too vast in scale. It began 185 million years ago, when the red shale on which Great Swamp rests was laid down. Ten million years later the Watchung Mountains, now covered with suburban houses, boiled out of the ground in the form of volcanic lava. Fifty thousand years ago (recent history in geological terms) the Wisconsin glacier started creeping south; it reached its southern limit about twelve thousand years ago. When it started to melt, it left walls of gravel and sediment that contained a lake 160 to 240 feet deep, 30 miles long, and 8-10 miles wide. The lake existed for about one thousand years.

People who lived in new houses on the hills surrounding Great Swamp can blame their steep driveways on the Wisconsin glacier. It left a terminal moraine not well suited to level lawns or roads. The trouble caused by icy driveways in winter began thousands of years ago. Traces of the sediment left by the glacier can still be found on the slopes of the Watchung Mountains and Long Hill, which were islands originally. Before Lake Passaic drained out through the Little Falls gap, near Paterson, it lasted long enough to lay down a solid clay deposit sixty to eighty feet deep that today keeps the ground water flowing through a gravel-and-sand aquifer beneath. Surface water hardly penetrates the clay bottom of Great Swamp. Surface water nourishes the vegetation and transpires into the atmosphere; and since the swamp slopes seventeen feet, surface water slowly moves through the marshes, and three quarters of it eventually flows into the Passaic River. Between the time of the advance of the glacier and the time it retreated, about four thousand years elapsed. That is a tremendous investment of time in the shallow bowl of forest and marshes that now constitutes a unique entity. Modern bulldozers, draglines, and dump trucks could reshape it in the form of an airfield or industrial community in three or four years.

Man has had a long history in the swamp. In 1965 a member of the Archaeological Society of New Jersey found a stone javelin estimated to be about nine thousand years old. That was the Paleo-Indian period of nomadic big-game hunters. Members of the chapter who have dug on the perimeter of the swamp have discovered thirty-seven Indian sites of much more recent origin—three thousand years, perhaps. In modern history, the records of deeds show that William Penn and two of his sons owned six thousand acres of Great Swamp in 1667. Men have been picking away at it ever since. During the American Revolution, trees were cut to make charcoal for smelting iron. Brick and potash industries were begun. In the nineteenth century the forests were ransacked for ships' timbers and railroad ties. A factory that made fruit baskets depended on the swamp for raw materials. Swamp meadows still produce what is termed "foul" hay. There have been farming and grazing projects on fields that stand above the waterline.

People still live as close to the swamp as possible, paying for their proximity with wet cellars and roads that are occasionally flooded and impassable. Attempts have been made to drain the swamp. But up to now Great Swamp has been essentially impregnable. The energy of every attempt to tame the swamp has eventually succumbed to the swamp's indomitable independence. It cannot be imposed upon. Some kinds of compromise can be made with it, but, so far, it has not been absorbed into the fabric of civilization and it contributes nothing to the Gross National Product.

To people in love with the life and lore of the swamp and its sequestered beauty, the dynamic signs of the Fish & Wildlife Service are gratifying. The design of the flying goose posted everywhere means that Great Swamp is now under professional management. It means that the wildlife will be protected in the midst of a massively populated area and that the public is invited in, but disciplined. It means that hunting will be supervised and controlled.

Apart from the wood ducks, blue-winged teal, and other waterfowl that nest in the swamp, millions of land and water birds migrate twice a year through the swamp, which is only twenty-five miles from the mouth of the Hudson River and is part of the great Atlantic flyway. At present, the refuge produces about 275 newly hatched ducks a year. The Fish & Wildlife Service expects to increase production to fifteen hundred young ducks and one hundred fifty young geese every season. By building low dikes that will stabilize the level of the water, the service will also make the swamp more useful to migrating birds. It has planted about three hundred acres of the small grains that water birds feed on. Eventually the agency hopes to make the refuge the sort of habitat it was a century ago, before thousands of dwelling houses sprang up all around it.

No man is an island, as John Donne declared; and in ecology no species, including man, can destroy the environment and survive. Great Swamp is a spiritual resource. In the midst of a tangle of people, facilities, and factories, it provides an enclave of about ten thousand acres where people can return to the solitude that was once available to everyone. Anyone who travels from Hoboken to Morris County can see how urgent it is to conserve an open retreat from the bedlam and ugliness of an industrial jungle. The Jersey marshes are composed of concentrated hideousness. It offends the spirit of the people who cross it; it has permanently suffocated the spirit of the people who no longer notice it. Apathetic people have become one of the depressed species. On the Jersey marshes the landscape looks frantic. The wild scramble of bridges, tracks, elevated highways, and power lines; the hodgepodge of billboards, storage tanks, and junkyards are not merely desperate. They devastate the landscape. They symbolize the shrill, nervous disorder that has destroyed much of the beauty of America and much of the joy of being alive. The pace is fast but the blight is permanent.

Since the climate has not been uniform during the past ten or eleven thousand years, we cannot assume that all the winters at Great Swamp have been identical with this one. Some have been colder; some have been warmer. But this shallow bowl of water, marsh, meadow, and forest preserves in the midst of a harsh metropolitan area a harmonious part of the America that has always been. In Great Swamp nothing essential has been changed since the last glacier receded. The rhythms of life through the seasons and through the ages continue their mysterious dance. In the winter the woods are virtually silent. There are no sounds except the detached tap of a woodpecker or the chance scream of a bluejay. Under the pale sky the marsh looks gray, with occasional flecks of gold where the sun touches it.

On all sides of Great Swamp property values rise rapidly because the population is increasing and the supply of land remains the same. In Great Swamp the property values are low because the land is good for nothing except life, knowledge, peace, and hope.

Content Considerations

1. Why did the people in surrounding communities buy 3,000 acres of Great Swamp?
2. What problems does Great Swamp cause its nearby residents?
3. According to Atkinson, why do people need a "spiritual resource" such as Great Swamp?

The Writer's Strategies

1. How does Atkinson use the idea of time in this essay?
2. What is the author's attitude toward wilderness? How does he value land?

Writing Possibilities

1. Evaluate how the communities about which Atkinson writes have balanced the demands of civilization with the need of and for wilderness.
2. Describe a wilderness area with which you are familiar. What is its purpose and value?
3. With a classmate, consider the urge "to make something profitable out of something unused" (paragraph 3). Explore the effects of this urge and speculate upon how it has transformed an area you know.

Additional Writing and Research

Connections

1. Compare what Atkinson, Hubbell, and Quindlen have to say about how people may live in or with wilderness.
2. The pieces by Abbey, Kallen, Lopez, and Nelson concern human journeys through wild areas. What concerns do these writers share?
3. Using information from any of the readings, write about how people may seek both adventure and solitude from the wilderness.
4. Write about how wilderness is affected by the numbers of people who use or visit it; consider the pieces by Abbey, Atkinson, Kallen, Nelson, and Quindlen.
5. Reflect upon nature as a spiritual resource; use your own beliefs and experiences as well as those presented by Abbey, Atkinson, Hubbell, Lopez, and Quindlen.
6. Using any three of the readings as well as your own observations, discuss whether America seems to be trying to dominate wilderness or support it and whether the current attitude is a departure from past ones.

Researches

1. Find information about how the wildlife refuge or national preserve systems obtain land and how their governing agencies operate them.
2. From magazines, books, advertisements, travel agencies, and interviews, investigate the eco-tours now available. You may want to focus on operators' claims or stated purpose, the way tours are conducted, or the kinds of people who take such tours.
3. Explore the impact of eco-tourism upon a certain area.
4. Determine what kinds of problems wilderness or wildlife creates where you live; interview residents as well as government agency workers or local specialists.
5. Through bookstores, newspapers, and libraries, discover natural history writers in your area. Read their work and, if possible, arrange an interview. Focus on the relationship of communities to wilderness or how the sense of landscape forms their perspective.
6. For any area you choose, find out how people have lived off the land.

7

Destroying the Forests

LT. JOEL HEINEY WAS A NAVAL FLIGHT OFFICER aboard four-engined Orion antisubmarine planes based in Jacksonville, Florida. Six months of each year, however, his squadron settled down in Reykjavik, Iceland, to keep an airborne eye on the northern sea-lanes. It was bitterly cold, the fliers were far from their families, and the long missions over the North Atlantic were rigorous and dangerous. But for Heiney and others in his squadron, the single worst aspect of duty in Iceland was the absence of a familiar landscape feature.

Trees.

Trees—forests—have been such an integral part of America and the New World that we have often taken them for granted. Sailors to seventeenth-century New England knew when they were within two hundred miles of the coast because they could smell the pines. Over the next two centuries the American settlers followed their urge to tame the wilderness by clearing millions of acres of trees, changing forever the character of the country. Many areas claimed to be the logging capitals of the world; countless trees were sawn into lumber for the needs of burgeoning industries, businesses, and communities. Everywhere, it seemed, trees were an endless resource to be used, a limitless opportunity to be seized.

Today much of the world's environmental attention is riveted on the rain forest, which is being razed at an alarming rate—eighty acres a minute, by some accounts. Developed nations, including the United States, are harshly critical of the entrepreneurs and governmental policies of Brazil and other nations that contribute to the destruction of the rain forest. And it is easy to be critical: According to many ecologists, elimination of the tropical rain forest would mean an increase in global warming, a decrease in rainfall and oxygen production, and extinction for many species found only in the unique environment of the lush tropics. But to many who profit from the clearing of the rain forest, such concern is hypocritical. Americans and Europeans have no right, they say, to speak out against the felling of trees after the way forests were clear-cut across America, after trees were devastated across medieval Europe in the name of civilization. Such interest in the internal affairs of other nations is sometimes interpreted as economic colonialism, as efforts by wealthy nations to prevent the growth and enrichment of poorer ones.

But the battle for forests also rages in America's own Pacific Northwest. In some places the logging industry has been forced to halt, at least temporarily, because of the northern spotted owl, which biologists say can only live in old-growth forest. Even while many lumber companies plant new trees to replace those harvested, a cry is being raised that trees are more than board feet of lumber and that silviculture—the systematic, controlled cutting and planting of trees—does little more than create stands of two-by-fours. A forest, many environmentalists insist, is not only trees but an interdependent and diverse environment, one that is essential to the continued existence of many species of plants and animals. Biological diversity is a key to continued survival of all species, they say, and this diversity is threatened by clear-cutting anywhere or by elimination of old-growth forests. But loggers insist that clear-cutting and planting actually increase habitat for wildlife and that trees must be cut to supply wood, a natural product in continuous demand.

Intentional destruction of forests is not the whole story. While logging versus wildlife is hotly debated in some regions of the United States, elsewhere, particularly in the Northeast and across Europe, acid rain is killing forests and aquatic life. The effects of acid rain on forests are visible and widespread: thousands of square miles of dead trees in Germany, Switzerland, Sweden, and other European nations; deterioration of high-elevation conifers in the eastern United States; dead sugar maples in large areas of Canada. But though the effects are visible and unarguable, the causes are not. The United States has at the time of this writing still not agreed to steps to eliminate or curtail the burning of fossil fuels, which many scientists cite as the major cause of acid rain.

In cities and suburbs across the country, construction practice is to bulldoze the land, cutting down every tree in a tract, and then to go back and plant saplings so the neighborhood will one day again have the shade and charm of a wooded area. More recently, as concern for trees has taken root, developers are leaving the oldest and largest trees and building around them, a practice applauded by many but sometimes cited as a cause for higher construction costs because crews must wrestle with tangles of roots and low-reaching branches.

In the following essays you will meet writers who see trees as more than standing lumber. Doug Stewart describes the engineering marvel that is a sequoia, the largest living thing on earth—a masterpiece of hydro-engineering, of adaptation, of resistance. John Hay paints a picture of trees as habitat, as virtual cities for legions of creatures large and small. Julie Sloan Denslow traces the consequences of destruction of the rain forest, not only to many plants and animals unique to that environment but to humans also. As you read these and the other essayists here, consider the posting of Lieutenant Heiney and others who have experienced land without trees. What would our world be like without them?

JOHN HAY

Cove and Forest

Writer, naturalist, and conservationist John Hay spends his life on Cape Cod and in Maine; both places, as well as others, appear in his work. What follows is from In Defense of Nature, *published in 1969. In the excerpt Hay describes a forest, one filled with animal and plant life and connected to the larger life of its surrounding world.*

If one carefully observes a spot of nature—a tide pool, a patch of grass, the bank of a stream—one begins to see movement and color. Creatures no longer blend with the background but become distinct; the background itself reveals its character. Such observation is one of the marks of the naturalist, the person who watches to see how nature functions, connects observations and facts, and arrives at conclusions, speculations, and possibilities.

John Hay takes just such a look in the next piece; moving from a cove to the forest, he sees life abundant and linked, the trees serving animals and earth and humans. Before you read, write for a few minutes on the ideas of community and regeneration. What might each idea mean when the forest is at its center?

> Self-centered beings, we civilizados tend to assume that all life proceeds on the same plane as our own, that of the horizontal, whereas in the forest most life struggles up toward the sun and at death drops away from it. It is the rhythm of living.
> —ADRIEN COWELL, *The Heart of the Forest*

THE SOIL IS NO LESS SHIFTING and often precarious in its nature than the water, and when you think how thin this surface mantle is you can bless the grass and trees for holding it down. Like the tidewaters in the cove, it is marked by sensitivities, degrees of awareness, sometimes explosive reactions, which no one skinning off ten acres of woodland and calling it "dirt" could possibly imagine. Just at the high tide mark, flicking over the surface of the pools, there are often small masses of blue-gray fleas, one of the few marine insects, and known as *Annurida maritima*. One day in spring there were a great many of them showing on the slanting surface of rocks exposed between the water's surface and the belt of grass and forest land above them. This was a zone marked by a thin white line of salt at the top of the rock. There were two principal masses, one where the rock slanted in slightly, and the other below, where it shelved out, and there was a constant exchange between them. At their closest range, a matter of inches, the fleas were bouncing back and forth in a frenzied way. There was intense activity, almost a boiling made by the jumping and twitching of these thousands of animals. Each single one was in a state of nervous agitation, and the nearer they came to each other the more they seemed inspired to their crazy leaps. There was a magnetic, raining exchange between the two groups—I find no other way to describe it—and never before had I been so aware of what power of affinity there is in the communities of life. If this happens in the cool and rocky state of Maine, how will it be in Brooklyn?

Above the tide line, above the narrow belt of grasses and plants adapted to periodic flooding and wetting by salt water, the forest starts. Its growth has been slashed, cut over, burned, reduced to fringes between one city and another, between one attempt on its life and the next. The part of it where I go in is more properly called a wood, since there is a field beyond it and its boundaries are constantly being chewed back by the chain saw; in fact, given a little less restraint on the part of its owners it would not take more than a few days to have it a mass of stumps and chewed-up ground. Still this area belongs with the great northern forest, and its trees insist on an appropriate longevity in spite of everything, a growth and a stature which has no obligations to any needs but its own.

We have relatively narrow means by which to approach a tree. It may be in the way, or it may have ornamental value. For those who deal in lumber it will have another; and most people do not know its name. We very rarely assume that any such silent, faithful, available plant would be able to draw any more out of us than a tacit acceptance. On the other hand, trees breathe, a slow, quiet, tireless breath, exchanging gases with the outer air. They drink, through the elaborate network of roots that thread the soil. They apparently communicate impulses from cell to cell through the thin film of cambium just under the bark. They manifest the most sensitive and elaborate connections between the earth, the air and the sunlight. They are not only a life environment for native birds and squirrels and the thousands that thread the leaf mold through their roots, as well as a passive shelter for all travelers, but their connections with animate and inanimate things might just as well be called personal. They share with the gall wasp that stings a leaf to make its gall, with the caterpillar that eats leaves or needles, and with the oriole that plucks fibers from bark for its pendulous nest.

I do not, or cannot, go very deep into what a cove or the trees above it may mean in their complexities, but the depth waits. From the centipede that makes its way through particles of soil to the tips of the highest branches the wood stretches from the roots, breathing, groaning, roaring if our ears could make its inner processes their own. Birch, oak, spruce, arborvitae, pine and hemlock, each in their way vulnerable because they have to stand and take it, unable to move like animals in search of better conditions of light, moisture and freedom from enemies, stretch next to each other, continually in exchange with their surroundings. Trees may not have "passion" attributed to them, but they endure as much competitive ardor, and disaster and tense exactitude as the rest of us. That the slaughter of forests has done great things for civilization is no reason not to respect the life of a tree.

So I walked up above an explosion of water fleas at the tide's rim into the self-made darkness of the trees, and the sensitivities, the great springs inside, remained. Modern civilization has become insulated against the trees. Occasionally, passing some northern wood where the trunks of the trees stand out in thick, solid strength, their branches rough and splintered, the undergrowth tangled and covered with briars, some residual understanding comes to me of the way the undominated land used to feel in America. I see why our greed was so much encouraged, why we are so addicted to size, and why Americans have found it more important to win than to lose. But the impetus given us by this original conflict with a wilderness appears to have lost its base. The new world is just as afflicted with raw possessiveness as the old, but it can deal with what is in its way with more dispatch, less close acquaintance.

Still and forever, a wood breathes. Also, its scene, like any other, its particular arrangements, oblige you to it in a way that is the earth's possessive way, insisting on its aptitudes, insisting on unique responses. I passed by the remnants of a woodpecker's wing, a casualty of the winter, and then the foot of a snowshoe rabbit. How does the rabbit survive the weasel, or the woodpecker the great horned owl? The trees were full of the shadows of a hazard of one kind or another, full of the meaning of endurance and escape, of vibrant existence in the measured place where it was held.

As I walked through an open space between the trees that was covered with fallen logs, vines and ferns, I flushed a white-crowned sparrow. It skipped quickly and silently ahead of me from ground level and perched in a nearby tree. Then another appeared, and both sounded their alarm calls. A little search and I found their nest, a small cup of grass next to a fallen log, surrounded by spruce seedlings and ferns. And in this green, dappled place, the nest held two eggs, of a light, flecked, tawny color like immature grapes.

There were patches of spring flowers farther on, fringed milkwort, a little pink flower with extended petals like wings, and starflowers, crisp, white and neat, and white wood anenomes, each of their small flowers having the bravery of a single stem. Out of the will of beauty they lighted their own ground, each defined petal, each thin-veined perfect leaf.

A golden crowned kinglet perched nearby, its feathers blown by the wind so that they were lightly parted on its cloud-gray breast. It had a repeated tiny, low cry: "tseet, tseet," and a faint song: "tee-tee-tee-tee," with a trill at the end. Then its relative the black-capped chickadee passed through, unconcerned, as usual, by human presence. A robin whickered loudly, and an ovenbird sounded. I heard warblers with high-pitched, whispery notes, and after some searching caught sight of two of them flitting in tall spruce and white pines, one a Blackburnian, the other a black-throated green, very trim, as colorful as flowers and as exceptional. They were elusive, bright migratory notes showing that northern trees are touched by the tropics.

The wood was a matter of survival, of adaptation to the fine edges of necessity, so that the red squirrels I heard chattering in the evergreens were in the same frame as the snowshoe rabbit. Each species used it as their home range, the rabbits with trails, the squirrels with tunnels at the foot of the trees and a knowledge of them, from crown to root, which was a part of every twitch and spring they made. Hairbreadth survival depends on exceptionally clear senses finely tuned to the world they travel in.

The bright-eyed red squirrels added a magic spontaneity to the dipping arms of the trees. I remember as a boy sleeping out one night and being woken up by a red squirrel scolding from the top of a white pine for all it was worth, shaking and chattering as the dawn spread across the sky. Now on this day so many years later, I sat still under a few giant white pines, resinous of scent in the warm sun, with long, thin silvery-green needles shimmering. After a while a whole pack of red squirrels came by, chasing each other excitedly over the needles of the forest floor. One of them had a pinkish-red cast to the top of its back and tail but was otherwise white. They dashed over the needles, they ran up the trunks of the trees, skipped along the branches, ran down again and after each other, making little nasal, grunting noises as they went.

Later I moved to a spot about a hundred yards away and met the albino again in the middle of a gay chase with two other squirrels, and since I stood very still behind a tree, one of them came right up between my feet without seeing me. It stopped spasmodically, splayed its hind legs wide, twitched and shook with alarm, then gave a little shriek and scampered off.

The trees held these spontaneous and sensitive lives the year around, or like other trees great distances away, received them when they came, being at once part of the substance of a local year and of forces thousands of miles in extent. This northern forest environment, of rabbits, squirrels, fishers and wildcats, crows, warblers in season, minks, weasels or porcupines, all using it as they could, was manifest regeneration. The young spruce seedlings were starting again, as they did around the world, crowding the slashed areas where their progenitors had come down, to battle for light and space, to take on the sun and the snow. In them was the latent power of all nature to reestablish itself. In a minute or over unseizeable periods of time, that force, sometimes totally hidden, sometimes blotted out, shows itself again, demanding its primal right again in a seed that is to fill a desert, or a needle that claims the light.

The tidewaters rose and lapped beyond the scaly trunks of the trees, whose branches lifted and fell in the rhythmic swing of the air, sounding sure and deep. An occasional strong gust would make them sound with a long "whaughh" and then would die down again, while a herring gull wheeled high and free out in the sky with a trumpeting laugh, and a green heron crouched on the rim of the tide, stealthily moving forward like a thief.

Content Considerations

1. Make a list of the ways in which trees interact with their environment in nearly animal-like ways.
2. Why does Hay say that "modern civilization has become insulated against the trees"?
3. What is the difference to Hay between a "forest" and "a wood"?

The Writer's Strategies

1. Hay writes a very personal essay, putting himself into the text. How does this contribute to a sense of voice, of intimacy among writer/subject/reader?
2. The author focuses on life processes in a single wood, then broadens his view to a philosophical discourse on life on a larger stage. How is this specific-to-general approach effective?

Writing Possibilities

1. Hay says "the slaughter of trees has done great things for civilization." Consider ways in which your life would be altered without trees or wood products.
2. Do you agree or disagree with Hay's assertion (paragraph 5) that trees have helped form an American national character?

LAURA TANGLEY

The Last Stand?

Forests located in places that experience a dry season—months with no rain—are called dry forests. The tropical dry forest has dwindled; since the 1500s, about 98 percent of it has disappeared, mostly converted to agricultural uses, and the remaining 2 percent lies so scattered and isolated that some of its plant and animal species are threatened with extinction.

In the following article writer Laura Tangley explains restoration efforts being undertaken in the dry forests of Costa Rica, a Central American country located between Nicaragua and Panama. These efforts focus on holding human destruction at bay and on intervening in natural processes so that the forests will have time to regenerate.

Before you read her explanation, which appeared in Earthwatch *during the summer of 1990, work with classmates to brainstorm ways such restorations may be carried out. Then, as you read, keep track of all you already knew about rebuilding the forests.*

"THERE'S NOTHING OUT HERE this time of year," says biologist Daniel Janzen as he pops into a plastic collecting bag the sixth caterpillar he's found on an early January walk through Costa Rica's Santa Rosa National Park. "It's starvation city for anything that eats insects," he says later, after bagging a cluster of bright yellow-and-brown bugs of the Coreidae family. Overhead, a noisy magpie jay complains from the branches of a bare tree, while a woodcreeper silently circles the trunk, probing its bark for insects. Near a dry streambed, a herd of collared peccaries drinks from a water hole. When the startled animals thunder off, inspection of their tracks reveals that a tapir also has visited the spot.

Even in the midst of a half-year-long drought, a tropical dry forest teems with life. This five- to seven-month dry season is what distinguishes these forests from the far more familiar tropical rain forests. Unlike rain forests, dry forests are deciduous. In January, a visitor flying down from New England would find the trees—except those in flower—just as bare as the trees back home. Without a canopy of leaves to shield them, dry forests turn scorchingly hot and "as dry as a desert for this half of the year," says Janzen, a University of Pennsylvania professor who has studied these forests in northwestern Costa Rica for more than two decades. "But when the rains return in the spring," he adds, "they are as rainy as, or rainier than, rain forests for the next six months."

Dry forests once covered more of the tropics than rain forests ever did. But whereas primary rain forests still cover substantial portions of Latin America, primary dry forests have been reduced to a few small patches. Despite their endangered status, says Janzen, they have been largely ignored by the conservation community, which has focused instead on rain forests. To offset the oversight, Janzen since 1985 has spearheaded an effort to regrow devastated dry forests.

Janzen believes conservationists focus on rain forests "because in obvious ways they are more spectacular. They are wet, dark, and mysterious." But people may also ignore dry forests, he adds, because there are too few of them left to even notice that they're disappearing.

When the Spanish conquistadors arrived in the New World in the sixteenth century, 550,000 square kilometers of tropical dry forest—an area the size of France, or five times the size of Guatemala—blanketed the Pacific slopes of Mesoamerica from Mexico to Panama. Today, just 2 percent of those original forests are considered intact, and most of them are broken up into small, isolated fragments. Now, even these last precious remnants are vanishing. According to Janzen, only 0.09 percent of what's left of the Western Hemisphere's tropical dry forests—about 480 square kilometers—lies within protected areas such as national parks.

Why have the dry forests fared even worse than rain forests? According to Janzen, when human beings began farming some 5,000 years ago, "the dry-forest sides of continents and isthmuses like Central America were more hospitable than the wet-forest sides." Unlike most rain forests, dry forests often lie on rich soils. Farms located here produce more crops and livestock than do farms carved out of rain forest. The dry forest is a healthier place to live, says Janzen, because the long, rain-free season keeps many pathogens and pests in check. ("It also keeps your clothes from rotting off your back.") And low humidity for half the year means that crops can be stored without spoiling.

Human preference for the dry slopes of the New World tropics probably preceded settled agriculture, says Janzen. He cites evidence showing that Pleistocene hunters extinguished most of the region's large mammals as far back as 9,000 years ago. More species of edible plants grew in dry forests, the "parental climate" for such major crops as rice, corn, beans, and sorghum. "From the Indians' standpoint, you started over here and not in the rain forest," says Janzen. "I think it's fair to say that the indigenous groups of the rain forest were outcasts who then became specialists at living in a crummier habitat." When Europeans arrived, he added, they also preferred to settle in dry forests.

Dry forests were—and still are—extremely easy to clear and to keep clear. During the dry season, even the smallest fire can quickly turn into a raging giant and gut a forest if not put out immediately. Forests cleared by fire are soon invaded by exotic grasses—primarily jaraguá, which was imported from Africa in the nineteenth century to feed cattle. Turning yellow and straw-like at the end of the rainy season, jaraguá creates ideal conditions for even larger fires and the further spread of grasslands the following year. Today, in a region that had no natural savanna, jaraguá grasslands dominate the lands that once supported tropical dry forests.

Despite the near-complete obliteration of their habitat, a surprising number of dry-forest species have, at least so far, managed to cling to survival. Janzen attributes this to the necessary hardiness of organisms that have adapted to annual fluctuations between two climate extremes. Plants and animals have evolved strategies to escape or remain dormant during long periods of hot, dry weather. Some of the caterpillars Janzen collected, for example, were inside of rolled-up leaves, where they remain dormant until the rainy season.

Other dry-forest species do more than simply survive the dry season; they thrive in it. At least half of Costa Rica's dry-forest trees flower, produce fruit, and disperse their seeds during the dry season. Many trees replace their leaves before the rains return. How the trees obtain enough water for these demanding physiological activities has kept biologists puzzled. The poro-poro tree *(Cochlospermum vitifolium)*, for instance, sheds all its leaves in December and January, then begins producing brilliant yellow flowers about a month later. Six to eight weeks after the flowers are pollinated by bees, the tree bears fruit. Then in May, after all flowering and fruiting is over (but before the rains begin), the poro-poro produces new leaves. For the many animals that feed on the flowers and fruit of trees like the poro-poro, says Janzen, "the dry season is the bountiful time of year and the rainy season, inimical."

Although the ability to survive half a year of rough weather has made dry-forest species better able than rainforest organisms to survive adverse conditions, time is running out for even the toughest survivors, says Janzen. Today's scattered forest fragments are simply too small to support most of their inhabitants over time. As an example, Janzen points out that the guapinol *(Hymenaea courbaril)*, a dominant canopy tree in many lowland dry forests, depends entirely on agoutis to move its seeds to appropriate germination sites (though now-extinct large mammals almost certainly served as seed dispersers in the past). Although guapinols are often left standing in forest patches and even pastures, the trees, when isolated from forests large enough to support agoutis, cannot produce viable offspring. "They may flower and bear fruit each year, but they don't ever produce any kids," says Janzen. He calls such trees "the living dead." "They may be physiologically alive, but are as dead as if they were already lying in the litter."

Here, adds Janzen, lies the irony of a dry-forest species' ability to hang on through the destruction of its habitat. Although many of the last dry-forest patches may look like living, thriving forests, some are as extinct as a rain forest that has been burned to the ground. When the surviving species age and die off—as is beginning to happen throughout the fragments—nothing will remain of these forests and their inhabitants. Although Janzen says he does not know of a single dry-forest species that has become extinct during the past 100 years, he cautions that "if we keep on behaving as we are now, over the next 20 to 25 years we'll lose a quarter to half of them."

To prevent this mass extinction, Jansen says that protecting existing dry forests will not be enough. We must also restore some of the lands that once supported them. "A traditional conservation battle for tropical dry forests would have to have been fought in 1900," he writes. Janzen says that if lands between forest patches—no matter how degraded—are protected from fire, restoring a forest is relatively easy. The forest will practically restore itself. After the fires are gone, he says, tree seeds carried in by the wind—once killed by the flames—will finally be able to take root. When the saplings grow tall enough, they will kill the grass and thus pave the way for animals—and the tree seeds they disperse—to colonize the new forest.

For the past five years, Janzen has been coordinating efforts to restore tropical dry forests in the vicinity of Santa Rosa National Park. With the participation of the nation's conservation-minded citizens—and financial assistance from other countries and international organizations—the private Fundación Neotropica has purchased a number of degraded forest lands and private ranches for this effort and donated them to the government. In July of last year, then-president Oscar Arias united these lands with existing protected areas in the region, including Santa Rosa, to create a single, 100,000-hectare protected area called the Guanacaste Regional Conservation Unit.

At several locations throughout this new protected area, Janzen and his Costa Rican colleagues are immersed in the process of restoring dry forest. Their activities include establishing fire patrols, cutting fire lanes, planting trees, and stopping hunters from killing important seed dispersers. In addition to this ecological restoration work, Janzen is also spearheading an effort he calls "biocultural restoration." Its purpose, he says, is to reestablish the cultural ties local people once had with the natural world, through activities that bring people to the park—field trips for school children, for example. Children who grow up understanding and feeling connected to nature will not only have richer lives, says Janzen, they will also make better neighbors for Costa Rica's vanishing dry forests.

Already, large areas in and around Santa Rosa are in the early stages of forest restoration. The Guanacaste Regional Conservation Unit has tripled the area of protected dry forest. Although these successes are certainly worth celebrating, Janzen cautions that the amount of protected forest remains trivial compared with the vast areas these forests once covered. Without more efforts to protect and restore tropical dry forests, he says, we will soon extinguish an ecosystem that has hung on through 9,000 years of ever-intensifying human disturbance. "Either we act very soon," says Janzen, "or we will witness the elimination of species that have persisted through many seemingly more severe perturbations than the contemporary, innocuous-looking clearing of the last hedgerows."

Content Considerations

1. Summarize the reasons dry forests have drastically declined with comparatively little notice.
2. Why have individual dry forest species survived despite the widespread destruction of habitat?
3. What is being done to conserve, and perhaps expand, existing dry forests?

The Writer's Strategies

1. The writer quotes only one authority in her essay. In what other ways does she establish the case for preserving dry forests?
2. This essay begins with a narrative about a walk through a dry forest in Costa Rica. How does the author attempt to make the reader care about dry forests?

Writing Possibilities

1. What would your life be like if trees disappeared from your region?
2. Can humans be concerned with every species of plant and animal life? With a classmate, prepare an argument for or against such a proposition.
3. Talk to a local forester or conservation agent about the effects of trees on plant and animal life in your region.

DOUG STEWART

Green Giants

In the next few pages writer Doug Stewart focuses on the sequoia (also called redwood), those trees that grow hundreds of feet high and have lived, so far, as long as 3,200 years. Stewart explains the natural history of two species and discusses the implications of that history for the trees in their current environments.

A species survives over long periods of time by adapting to ecological changes. The sequoia adapts slowly, and like so many other plant and animal species, it may not be able to withstand the increasingly quick and extreme environmental changes caused by technological advance and practice. Flooding, erosion, and air and water pollution pose problems for these trees that began to grow as early as 1200 B.C.

Stewart's article appeared in the April 1990 issue of Discover; *before you begin reading it, write a few paragraphs on the value of such senior members of the biosphere.*

TURN-OF-THE-CENTURY LOGGING killed off many of them, and pollution threatens the rest. But for now, at least, the towering sequoia survives as nature's most fabulous anomaly. Other threatened species may have their own remarkable traits, but none inspires the same awe as these cloud-piercing trees.

Naturalist John Muir was one of the first to rhapsodize on the grandeur of the mammoth trees. In his book *Our National Parks* he wrote: "Who of all the dwellers of the plains and prairies and fertile home forest of round-headed oak and maple, hickory and elm, ever dreamed that earth could bear such growths, trees that the familiar pines and firs seem to know nothing about—lonely, silent, serene, with a physiognomy almost godlike . . . ?"

Muir was writing about the giant sequoia *(Sequoiadendron giganteum)*. This tree is often confused with its kin, the taller but more slender coast redwood *(Sequoia sempervirens)*. In botanical usage the term *sequoia* refers to both, but many people prefer to group the two under the simple heading *redwoods*. No matter what they're called, the majestic trees belie the notion that plants are too static, too quiet to be as thrilling as stampeding elephants or tail-pounding dinosaurs.

Consider their superlatives: The coast redwood is the world's tallest living thing. It's tough to determine the trees' exact heights while they are standing, but the tallest coast redwood ever measured—with surveying instruments, in Redwood National Park, in the northwest corner of California—was 368 feet high, which is (take your pick) just 22 feet shorter than a dead-center-field home run at Fenway Park, 35 feet taller than a Saturn V rocket, or the same length as 1,766 standard-size toothpicks laid end to end. A 20-year-old sapling, a mere sprig of a thing, may already be 50 feet high.

The tallest giant sequoias are not quite as high as the tallest coast redwoods, rarely exceeding 300 feet. But what the giant sequoias lack in height, they make up in bulk and mass. The General Sherman Tree in California's Giant Forest is thought to weigh 1,250 tons (the same as 13 space shuttles or 20 million boxes of toothpicks) and to be 103 feet around at its base. Oh, and it's still growing.

Neither of the giant trees can claim to be Earth's oldest living organism. That honor goes to the bristlecone pine, a gnarled, low-lying conifer of the High Sierras that reaches 25 feet on average. Some bristlecones are thought to be close to 5,000 years old, meaning that the trees that still grow today were already 500 years old when the pyramids were built. The oldest giant sequoias are a comparatively callow 3,200 years old; the oldest coast redwoods are just 2,200. Of course, in sequoia country, the usual notions of age must go out the window. In its seventies—sunset years for a human being—a sequoia is still a teenager, having just begun to bear its first tiny seeds (each of which may weigh just an eight-thousandth of an ounce, or a hundred-billionth the weight of a full-size tree); until three centuries have passed, the tree is still considered a youngster.

As a sequoia begins to grow, it doesn't behave dramatically differently from other trees. The secret of the trees' success is a handful of adaptive tricks that let it live long enough to grow taller and taller, and stay strong enough to support all that height and weight.

Trick number 1: *Grow in the dark.* Sequoias—coast redwoods in particular—are among the plant world's most efficient practitioners of photosynthesis. A coast redwood can grow up in a spot so shadowy that only one percent of the incoming sunlight ever makes it down to the leaves. Without this adaptation infant sequoias would never have a chance to survive at ground level in the dark stillness of a "cathedral grove" of mature sequoias.

Trick number 2: *Keep your weight centered.* Even among conifers, sequoias are unusually thick (the bark can be more than two feet thick) and pole-straight. They are also branchless for much of their height, thus achieving the no-nonsense, cantilever-free simplicity of the Washington Monument. So much bulk in the trunk and so little branching gives the trees great stability, even after a wet snowfall.

Trick number 3: *Anchor yourself.* If the roots of any tree are its structural base, the roots of a sequoia are a full-fledged launchpad. The matlike network of roots in a mature coast redwood fans out for at least 50 feet on all sides, and can stretch two or three times that distance; a giant sequoia's roots can cover more than an acre. Despite their enormous area, these subterranean networks remain close to the surface: a coast redwood's roots are usually only six feet deep; a giant sequoia's roots only three to five feet.

It may seem impossible that such a shallow base could allow a sequoia to withstand the wind—for comparison, the foundation of an office building of the same height may use dozens of feet of structural steel anchored in concrete. Of course, trees are not subject to the restraints that govern buildings: sequoias can safely sway with the wind, to a degree that would make human dwellers in a similarly swaying building decidedly unhappy. But perhaps most important in keeping a sequoia upright is that as the roots fan farther and farther out, they intertwine with the roots of their sequoia neighbors and any other vegetation in the vicinity; the trees essentially lash themselves into the ground.

Trick number 4: *When pressed, grow a buttress.* When sequoias do develop a lean, the tree ring on the side that's settling responds with a growth spurt. Over 100 years—brief to a sequoia—this can lead to a supporting ridge of "compression wood" several feet thick and up to ten feet high on one side of the trunk.

Trick number 5: *Be unappealing to pests.* Sequoia wood, especially in giant sequoias, is rich in acidic tannins. Most parasitic insects and fungi find tannin-saturated wood distinctly unappetizing and tend to avoid it. In 1964, when a well-preserved remnant from an old sequoia log was carbon-dated, it was revealed to have died some 2,100 years earlier. Bugs had sidestepped it since the beginning of the Roman Empire.

Standing tall and saying no to pests, however, is not enough to ensure a sequoia's survival. The trees also need to take in water—quite a bit of it. The average sequoia requires at least 300 gallons of water in a 24-hour period. Pumping all that fluid—about 2,500 pounds' worth—from the roots to the leaves requires the tree to do about 750,000 foot-pounds of weight lifting every day. This is roughly the muscle that would be needed to loft a can of soda to a low Earth orbit. Nature writer Ferdinand Lane once called a mature sequoia "a veritable fountain of ascending moisture, a vegetable geyser."

The key to any tree's hydraulic system is the strong affinity that water molecules have for one another, clinging together as though they were sticky. A sequoia simply makes the most of this property. As water in its needle-shaped leaves evaporates, moisture in the leaves' narrow central tube is drawn out to replace what's lost. Other molecules farther down the line move up in turn, all the way down to the roots. The vertical capillaries through which all this water is drawn are fantastically narrow and fantastically numerous. Hundreds of tiny tubes run through every square inch of sapwood, each tube hundreds of feet long but only a few cells wide.

If the soil in sequoia country were moister than it is, this elaborate pumping system would be more than adequate to quench the tree's thirst; the volume of soil gripped by a coast redwood's roots is capable of holding more than 130,000 gallons of water. But the West Coast is prone to long periods of drought, and the soil on which the trees rely often turns bone-dry. The coast redwoods in the southern portion of the trees' range would probably have died out long ago were it not for a phenomenon known as fog drip. During the dry summer months sea fog condenses on the trees' foliage and is absorbed steadily through the leaves; without a rain cloud in the sky, the tree often receives as good a dousing as it would get during a sudden thunderstorm.

Despite all its adaptive features, a sequoia is by no means invulnerable. Forester Douglas Piirto of California Polytechnic State University has spent years studying the fallen remains of giant sequoias to determine the species' most common causes of death. Piirto's autopsies have revealed that fire damage, insect infestation, and fungus infections are the most frequent problems spotted in the roots or trunks of fallen trees. The rich tannin content of the wood helps resist attack by many pests and fungi, but tannin is not an all-purpose repellent, and some types of invasive life do get a toehold. And although the thick, heat-resistant bark helps protect the trees against fires, the wood is certainly combustible. Floods and soil erosion also take their toll.

Twenty-five million years ago sequoias were common in what is now Texas and Pennsylvania and in areas that are now France and Japan. As the world's climate cooled and dried, however, the range of the sequoias shrank. Today coast redwoods are confined to a 30-mile-wide strip of rugged land that runs some 500 miles along the Pacific from the Oregon border to south of Big Sur. The thickest, tallest forests are in California's Humboldt and Del Norte counties, where whole groves average 300 feet.

The giant sequoias are even more isolated. These trees grow not in a continuous strip of forest, but in 75 mostly small groves. All are found along the western slopes of the Sierra Nevada Mountains of east central California. Despite their relatively small numbers, giant sequoias seem to be more tolerant of environmental extremes than their coastal cousins. The trees can withstand 20-foot snowfalls and zero-degree temperatures.

Some individual sequoias have survived even worse conditions. In Sequoia and Kings Canyon National Parks, most of what's left of a tree known as the Black Chamber is a hollow, fire-blackened stump 18 feet across. Yet the charred ruin has managed to sprout new branchlets and leaves and so continues to add new rings. More dramatic (to the delight of tourists who can't afford a trip to Pisa) is the Grizzly Giant at Yosemite's Mariposa Grove, which leans a full 17 degrees without falling. Botanists believe that nearby ground erosion is responsible for the list, causing the tree to tilt to one side faster than it could develop a buttress; when it finally did develop one, the new wood could not correct the list, it could only arrest it.

An increasingly important question is whether continuing climatic change, accelerated by human activity, is likely to doom the sequoias. Ground-level ozone in Sequoia and Kings Canyon National Parks now exceeds the federal Clean Air Act limits, thanks to San Francisco Bay—area traffic several hundred miles upwind; the result is tissue damage to the sequoias' leaves that renders the trees vulnerable to diseases or infestations. Additionally, increasing temperatures and decreasing rainfall caused by the greenhouse effect could cause further drying of the sequoias' often parched soil.

Although the giant sequoia and the coast redwood have both succeeded in adapting to major changes in their habitat since their earliest days on Earth, researchers worry that pollution-linked changes could be altering the habitats so quickly that the trees will not be able to adjust. You must add to the sequoias' list of superlatives the caveat that they may be the slowest adapting species on the planet. To an organism that may not reproduce for three-quarters of a century, a change in the atmospheric mix brought about by the introduction of the internal-combustion engine is an extremely sudden one.

El Palo Alto (literally, "the tall stick"), the tree for which the California city was named, offers a chilling look at what could be the future of sequoias caught in an increasingly dirty world. The tree is an elderly coast redwood close to 200 feet tall that now finds itself surrounded by cars, pavement, and train tracks—a drier, hotter, windier climate than it experienced in its medieval youth. To simulate sea fog, the city ran a pipe with spray nozzles up El Palo Alto's trunk and above its leaves, where a steady mist keeps the tree quenched. Without its synthetic fog, the aged tree would dry up and die. City foresters know that the solution is not elegant, nor would it be cost-effective for an entire redwood forest. But the tree does stubbornly keep adding rings—board foot by reddish brown board foot.

Content Considerations

1. What is the difference between a giant sequoia and a coast redwood?
2. How do sequoias adapt to their environment?
3. How are human activities now affecting the sequoias?

The Writer's Strategies

1. Stewart writes with a light, humorous touch. How does the tone affect your attitude toward his subject?
2. The author makes comparisons several times between a sequoia and how many toothpicks it might make. How does this trivialization contribute to his theme?

Writing Possibilities

1. With a classmate, consider a serious topic and then write an ironic essay about it, such as how many lint balls would be lost if . . .
2. Stewart closes this article with a reference to growth of the sequoia "board foot by board foot." This suggests several ways of viewing trees—either as potential lumber or as something else. Write an essay in which you explore other possibilities.

MICHAEL FROME

Forestry: Only God Can Make a Tree, But . . .

Taken from Michael Frome's Conscience of a Conservationist *(1989), the next selection explores the causes and consequences of clearcutting, the practice of mowing down and uprooting all plant life in an area where timber is being harvested. Frome's report centers on the debate over this forestry practice that occurred in the early 1970s. Almost twenty years later, the debate continues, as does clearcutting.*

One of the primary arguments against clearcutting is that it destroys entire ecosystems—trees, animals, plants, microorganisms, soil. Such an ecosystem is difficult to rebuild. The planting of new trees may indeed supply more timber, critics say, but the land mechanism requires entire ecosystems to retain its balance—a balance that no one fully understands. Proponents of clearcutting argue its economy and efficiency, assigning priority to the benefits it yields civilization.

Before you read Frome's essay, take a few minutes to write about the ways that land and its species interconnect. Then write about how people depend on resources such as a forest.

THE PRACTICE OF CLEARCUTTING TIMBER on public and private forests has become the subject of heated debate, both nationally and in various regions of the country—as well it should, considering the high stakes in economics, ecology, and esthetics. Controversies have arisen over the management of the California redwoods, the Douglas fir forests of the Pacific Northwest, spruce forests of southeast Alaska, lodgepole and ponderosa pine forests of the Rocky Mountains, the hardwood forests of the Northeast, and the mixed hardwood-pine forests of Appalachia and the South. Within the past three years, hearings on clearcutting have been conducted before committees of both houses of Congress. Reports and studies have been made by the Council on Environmental Quality, the United States Forest Service, deans of forestry schools, and concerned citizen groups. Yet there is still no resolution of the debate.

What is clearcutting? In an editorial appearing in its April 1972 issue, the magazine *Field & Stream* presented this blunt appraisal of the practice:

"It is a method of harvesting trees which causes complete devastation. It is more harmful than a forest fire. The land is churned up over vast areas by big machinery. *Every* tree is cut down and most of the surface plants are killed. Until grasses and shrubs can get started again, the land is wide open to extreme erosion. Timber companies prefer this method of harvest because it is cheaper than cutting selected mature trees and leaving the remainder unharmed and because having once destroyed a mixture of kinds of trees on a certain tract they can plant one type of tree, which they prefer. These trees are planted in neat little rows, all standing at the same height and all reaching maturity at the same time for another cutting. But it is no longer a forest any more than an orchard is a forest. There are no open grassy glens, no bushes, no aspen or alders. Everything is crowded and shaded out by the trees, and for the major part of the growth years of the trees there is little food for animals or birds. It is a sterile sort of forest designed by a computer."

Quite different descriptions of clearcutting are offered by its advocates, who include, in addition to most of the forestry profession, large corporations, holding investments in timber lands or in mills.

"It is efficient, economic, and in general produces forest products and resources useful to man," declared Dr. Kenneth P. Davis of Yale University, president of the Society of American Foresters, while testifying before a Senate committee in 1970. The immediate consideration was a proposed moratorium on clearcutting in the national forests, as urged by citizen conservationists. To halt clearcutting, he warned, "would place an unwarranted and disruptive restriction on using a proper and, in many situations, necessary method of managing forest lands."

Edward P. Cliff, Chief of the Forest Service, addressing the National Council of State Garden Clubs in May 1965, declared that the practice "is something like an urban renewal project, a necessary violent prelude to a new housing development. When we harvest overmature, defective timber that would otherwise be wasted, there is bound to be a temporary loss of natural beauty. But there is also the promise of what is to come: a thrifty new forest replacing the old. The point is that there often must be a drastic, even violent upheaval to create new forests. It can come naturally—and wastefully—without rhyme or reason as it has in the past, through fires, hurricanes, insects, and other destructive agents. Or it can take place on a planned, purposeful, and productive basis."

Mr. Cliff served as Chief Forester from 1962 to 1972. Under his aegis clearcutting came of age; he defended and promoted it with fervor. "For the young 'citified,' articulate part of our citizenry," he declared before the Pacific Logging Congress of 1966, "it is especially easy and natural to get stirred up about outdoor beauty, recreation, wilderness, vanishing wildlife species and environmental pollution. It is not likely that very many know or even particularly care much about how timber is grown, harvested and used to meet their needs." The Chief likened accelerated timber cutting through modern technology to gardening, or farming of field crops. "Wild old stands have pristine beauty which is instantly felt and appreciated," he wrote in the 1967 *Yearbook of Agriculture*. "But a newer forest, man-planned and managed and coming up sturdily where century-old giants formerly stood, also has its brand of beauty—similar in its way to the terraced contours and the orderly vegetative growth upon well-managed farmlands."

The clearcutting concept, as enunciated by Dr. Davis and Mr. Cliff, plainly emphasizes the anthropocentric—the design of nature for the use of man; it rejects the notion that resource managers must "think in ecosystems"—that they must relate every decision and every action to the entire complex picture rather than to an isolated component of the ecosystem, let alone to considerations of expediency or short-term economic returns. It denies the principles evoked by Aldo Leopold, forester of another generation and yet a pioneer of today's ecological movement, who wrote: "The land is one organism. Its parts, like our own parts, compete with each other and cooperate with each other. The competitions are as much a part of the inner workings as the cooperations."

Clearcutting's bias toward commodity production in the short run, rather than toward protection of the resource in all its aspects for the long run, is often the main basis of attack in the media. Strong criticisms along these lines in *Field & Stream* and other publications (notably the *New York Times, Reader's Digest, Atlantic Monthly, Des Moines Register and Tribune* and *Montana Daily Missoulian*) have been summarily dismissed by spokesmen for the timber industry, the forestry profession, and the Forest Service with such epithets as "sensationalism," "hit-and-run reporting," and "yellow journalism." In a speech on "The Nature of Public Reaction to Clearcutting," delivered in February 1972, an official of the Forest Service, John R. Castles, declared: "Probably the most frustrating and insidious form of pressure is that generated by irresponsible or ill-informed news media people who seize on unsubstantiated reports, half-truth rumors, misinformation, or outright distortions without checking them further."

But the evidence, even from timber-forestry witnesses, appears to substantiate charges that clearcutting is environmentally pollutive and ecologically disruptive, as well as designed principally for immediate profit. In an article in the February 1972 issue of the *Journal of Forestry*, for instance, Dr. David M. Smith, professor of silviculture at Yale, described the emergence of the new synthetic forest: "Combinations of herbicides, prescribed burning, and powerful site-preparation machinery made it possible, almost for the first time, to start new stands entirely free of the competition of preestablished vegetation. In some localities, it has become possible to contemplate deliberate efforts to eliminate natural populations and replace them with the planted products of conscious genetic selections. . . . It takes no great wit to see that within this frame of reference, the optimum cutting practice will be that which removes nearly everything that will pay its way out of the woods. The future benefits which might be derived from growth on reserved merchantable trees are quite intangible from this point of view."

Professor Smith went further, joining the critics in their concern over damage done by heavy machinery used in logging and site preparation. "The vegetation can be swiftly repaired," he wrote, "but it may take centuries or millenia to repair the kind of damage to the soil that can result from deep gouging or scraping action. It is time there was more concern for adapting the machinery to the silviculture and less resignation to the idea that soil damage is an inevitable consequence of practical forest operation."

John McGuire, who succeeded Mr. Cliff as Chief of the Forest Service, also conceded, in an interview published in *American Forests* magazine of October 1972, that: "Roads have been cut where they shouldn't have been permitted. Erosions have followed that make it impossible to get a forest of quality, or even any forest, in that area again." But instead of talking about eliminating roads in order to regain protection of the natural resource, he proposed construction of an additional 100,000 to 150,000 miles of highway in the national forests, thus tending to give credence to charges that the Forest Service is the handmaiden of special economic interests. Or, as Justice William O. Douglas declared in his dissent in the *Mineral King* case (Sierra Club v. Morton), issued April 19, 1972: "The Forest Service—one of the federal agencies behind the scheme to despoil Mineral King—has been notorious for its alignment with lumber companies, although its mandate from Congress directs it to consider the various aspects of multiple use in its supervision of the national forests." In the same message, Justice Douglas likened clearcutting to strip mining.

The most dangerous kinds of chemical poisons, poured into the soil and seeping into streams, are implicit tools of clearcutting. Entomologists warn that a pure stand of timber forms an ideal situation for damage from insects and disease; infection is rapid from tree to tree, and if one species is destroyed, there is nothing left. A monoculturally managed forest, therefore, creates the need for pesticides and herbicides. Ultimately these chemicals do more harm than good, for the biotic diversity is destroyed. Nevertheless, the Forest Service has poisoned millions of acres of public land, encouraged the use of ecologically crude poisons on millions of additional acres, and ignored pleas and protests.

14 The general fear among environmentalists is that more wood is being cut than grown on both public and private forests. The timber industry called for increased logging of the national forests—even though three-quarters of commercial forest land is in private hands—and the Forest Service responded by trebling its cutting of timber in the period from 1950 to 1970. Still, the industry wants the Forest Service to increase production by accelerated cutting of old-growth forests and by ecologically questionable programs of thinning and fertilizing. It fought enactment of the Wilderness Act of 1964 and now opposes establishment of additional inviolate wilderness, although such areas stabilize soil and watersheds, provide habitat for a variety of wildlife species, and are cherished for recreational pursuits.

15 Justification of clearcutting is repeatedly attempted on grounds that it produces more game. "Actually, this is the strongest argument for clearcutting, because artificial openings in the forest are a boon to wildlife," wrote William E. Towell, executive vice-president of the American Forestry Association.

16 It is true that clearcuts produce quail habitat, often where nonexistent before, and that an abundance of deer browse is produced on many clearcut areas. Biologists note that these benefits are temporary, however; before many years, quail habitat and deer browse decline. Within ten years following planting the pine canopy can be expected to close; until thinning, this clearcut is of use only as cover to wildlife. With increasingly short cutting rotations, it is difficult to anticipate how "mast" (foods such as berries and nuts) will be provided in the future for turkeys, squirrels, and deer. Removal of mast trees and cover is now destroying prime squirrel and turkey habitat, and lack of mast may reduce the carrying capacity for deer after a relatively few years. In the sequence of events, the clearing and conversion to pine in natural pine-hardwood areas, or hardwood areas with high deer populations, sometimes induces destruction of planted pines by deer, with the accompanying demand for hunters to "bring the deer population into balance."

17 Even though logging may improve deer habitat, serious disturbance eliminates such species as spotted owls and pine martens, which require old growth conifers for survival. Birds actually furnish the most efficient and least costly form of insect control in the forest. It is their definitive function in providing balance to the ecosystem. A single woodpecker, for example, has been estimated to be able to consume the larvae of 13,675 highly destructive wood-boring beetles per year. It is fair to generalize that the more numerous and varied the bird population of a forest, the broader the spectrum of natural insect control. John

Smail, executive director of the Point Reyes Bird Observatory, a California research organization focusing on the ecology of nongame species, has reported on an analysis of nine breeding-bird censuses in coniferous forests in California, Colorado, and South Dakota. The analysis showed that 25 percent of the total number of birds using these forests are of species that nest in holes. These hole-nesters require older trees with some decayed portions in order to breed successfully (and feed large broods of young on destructive insects), although they forage on trees of various ages. "Any forestry practice producing solid stands of trees the same age reduces the diversity of bird species able to breed, and this in turn severely reduces possible insect control," according to Mr. Smail. "Clearcutting is the most drastic example."

The South is perhaps being hit harder by clearcutting than any other section of the country. Vast areas that once supported mixed forests have been reduced to even-aged stands of pine, like apple orchards or orange groves. The sole purpose in transforming forests into farm lots is to provide pulp and paper for an affluent, throw-away society. "For a paper company, the obvious objective of pine management is to produce the largest volume of usable wood fibre per acre," wrote Henry Clepper in the August 1971 issue of *American Forests*, in describing the operations of International Paper Co., a timberland giant which owns 8 million acres in the United States and Canada—including 5 million acres of the South—and controls an additional 15 million acres under long-term lease from the Canadian government. "To attain this goal, foresters must control the site, which is to say the forest environment."

The *Louisiana Conservationist* in 1971 described how this is achieved. Under a headline, "Flourishing Forests Threaten Wildlife," this state publication noted: "When the stand reaches the desired stage of maturity the entire timber crop will be cut and the whole process repeated. To complete this cycle anywhere from 15 to 80 years may be required, depending upon the wood products desired. Already thousands of acres in blocks, ranging from 160 to well over 1,000 acres, have been stripped of existing timber, bulldozed, chopped, or burned clean, and then seeded or planted with pine. The small stream bottoms which have historically supported hardwoods are now the main targets. They provide a last and most critical retreat for game within the great sea of pine."

The industry's design to transform the rural South into a man-dominated forest, or massive pine factory, is embodied in a highly publicized campaign called "The Third Forest." According to the industry's report, titled *The South's Third Forest—How It Can Meet Future Needs*, there are now 24 billion cubic feet of "cull" trees—unwanted hardwoods—which take space needed for "better" trees. The removal of cull trees in order to provide for future growing stock would constitute the bulk of timber stand improvement on no less than 90 million acres. Dr. George Cornwell, professor of wildlife ecology at the University of Florida, commented on this proposal as follows:

"As wildlife habitat, cull trees usually mean food (mast) and housing (den and nest). In terms of natural beauty, most culls would be more highly valued by the forest recreationist than the 'better' trees planted in their places. Imagine my disappointment on learning that, after several decades of wildlife managers' pleading with forest managers to retain these den, nest, and mast-producing trees for wildlife use, a major Southern forest policy plan would call for their removal throughout the Third Forest. This approach to the cull tree is symptomatic of the recommended silvicultural practices in the South's Third Forest and would appear to reflect a nearly total contempt for non-timber values."

Until the upsurge of recent years, clearcutting had been accepted as a silvicultural practice only in certain short-lived forest types which reproduce easily, such as aspen, jack pine, lodgepole pine, and some southern pines. But it always had been applied in small patches, so that surrounding trees deterred hot, dry winds from desiccating the forest soil, and were close enough to supply the openings with seed for regeneration while providing shade cover for young seedlings. The more prevalent system of silviculture was selective logging, or "selection-cutting." Essentially, this system is designed to follow and fit into nature's pattern of growth, maturity, and decline by selecting individual trees, or very small groups of trees, in order to favor species tolerant of shade, or larger groups up to quarter-acre clearings to favor species intolerant of shade.

With the advent of large machinery, however, clearcutting became a habit. It began in the Pacific Northwest, on the basis of assertions that Douglas fir, the most profitable—and hence most desirable—species, reproduces only in full sunlight. Since then clearcutting has spread to cover nearly all forest types.

What alternatives are there to clearcutting? Dr. Leon Minckler, professor of forestry at Syracuse University, who spent twenty years in research for the Forest Service, insists that clearcutting is not the way to go, that Eastern hardwood trees do regenerate better through other techniques. Other forest technicians are now challenging the idea that Douglas fir must be cut in large blocks. In an article in the *Journal of Forestry* for January 1972 Dr. Minckler wrote:

"For integrated uses (such as timber, wildlife, watershed, recreation, and esthetics), management should aim toward maximum diversity and minimum damage to the environment. This can be accomplished by single tree selection, group selection, small patch cutting of a few acres, or a combination of these. Clearcutting, on the other hand, tends to minimize diversity and makes it almost impossible to avoid damage to the site, to streams, and to esthetic qualities. Most of all, it eliminates the forested character of a particular area for a long time. Ecologically it is a major disturbance. When harvesting mature stands, clearcutting is a cheap and effective way of extracting timber, but the sacrifice of other values may be a poor trade-off for cheap timber harvesting. In immature or partially mature stands, clearcutting may not even be the cheapest way of harvesting timber."

According to Gordon Robinson, a veteran California forester and consultant to the Sierra Club, good forestry consists of growing timber on long rotations, generally from 100 to 200 years, depending on the species and quality of the soil, but invariably allowing trees to reach full maturity before being cut. "It is not enough to have orderly fields of young trees varying in age from patch to patch," he declares. "In looking at a well-managed forest one will observe that the land is growing all the timber it can and that most of the growth consists of high-quality, highly valuable material in the lower portions of the large older trees. It will be evident that no erosion is taking place."

In short, while the corporate forester or timberman may insist that trees can be harvested and cultivated like any farm crop, in a genuine balanced-use forest immediate values must be integrated with long-range protection of soil, water, wildlife, wilderness, and scenery, and with assurances that harvested areas will grow more trees for future timber needs.

Business has a natural and understandable tendency to stress economics rather than ecology when thinking about resources. But land is an integral part of all life, and its resources remain part of the environment. In dealing with them, business needs to blend ecology and economics in its thinking. No landowner, large or small, should be able to control land use entirely for his own benefit without regard for what his actions do to others. Ownership is a trust which must be exercised in the interest of all—and one of the prime ingredients of that interest is the quality of the environment.

Dealing as we do with a complex earth mechanism which we only partially understand, we should be cautious in tampering with natural forces. Clearcutting has been subject to so many challenges and criticisms, and may do such serious long-range damage to soils and streams of the nation, that it needs to be curbed at once and restricted to experimental uses only, until answers are fully known.

Certainly any system of conservation based solely on commodity production or economic self-interest is hopelessly lopsided. It tends to ignore, and thus to eliminate, elements in the life-community of the land that lack commercial value, but which are essential to its well-being; if the land mechanism as a whole is good, then every part is good, whether we understand it or not. Perhaps the first rule to guide those who use and administer the land should be that economic parts of the biotic clock will not function without the uneconomic parts. Once that rule is learned and applied, then and only then can we sustain healthy, productive forests for the long-term future.

Content Considerations

1. What are the main arguments in favor of clearcutting? Against? Who articulates those arguments?
2. What is silviculture?
3. Describe the Third Forest concept.

The Writer's Strategies

1. Frome quotes many specialists and experts in forestry on both sides of the argument. What does the order in which he presents these quotations suggest about his own view?
2. How does the title affect the way you approach this essay?

Writing Possibilities

1. Visit a replanted forest or view a series of photographs of an area that has been clearcut and write about your impressions.
2. Make a list of the negative aspects of clearcutting as described by Frome. Which of these would have the greatest impact on your life? Write about it.
3. With a classmate, consider alternatives to clearcutting that would meet the demand of future generations for wood products. Together, prepare a report urging adoption of your plan.

JOHN NIELSEN

Expanses of Trees Fall Sick and Die

In this article from the March/April 1990 issue of International Wildlife, *John Nielsen presents a brief history of the earth's boreal and temperate woodlands and an even briefer history of their demise. These forests of the Northern Hemisphere are the forests of industrialized nations and have been decimated to feed initial and subsequent development.*

Although logging in some areas has slowed, environmental distress causes additional decimation through acid rain, loss of biodiversity, air pollution, and disease. The nature of forests has changed; many people and programs are now devoted to restoring forests or, at least, making it possible for forests to restore themselves.

Write for a few minutes about the myths and legends of forests—the folk tales and children's stories that your culture embraces. Then, as you read Nielsen's article, consider how these stories have shaped your attitude toward the value of woodlands.

WHEN ADRIAN DORST FIRST HEARD THE NEWS that the giant Sitka spruce trees on Canada's Meares Island might be logged, he built a trail through the pristine forest and started leading tours. "I wanted people to understand what an ancient forest is," explains the amateur naturalist and photographer. "If enough people got interested, I thought it might slow things down."

Dorst, who lives on the western side of Canada's Vancouver Island in sight of tiny Meares, is hopeful that this forest can be saved, but he knows that history is not on his side. Eight thousand years ago, wild woodlands blanketed much of the Northern Hemisphere. No longer. From the beginnings of civilization, such forests have been leveled to make way for crops and grazing animals, chopped down for fuel or systematically logged to build homes.

The sad fact is that in Europe, North America, the Soviet Union and Asia, the temperate and boreal forests of the northern world have been transformed. By some estimates, more than one-third of the woodlands that once covered most of the Northern Hemisphere are gone, along with much of the wildlife they sheltered. Although the total acreage of forest is no longer declining today, this seeming equilibrium actually represents an era of furious change in which the last remaining "virgin" forests are threatened.

"In the case of tropical forests, intense commercial and subsistence pressures are the agents of devastation," says Peter Hazlewood, a forest specialist with CARE (Cooperative for American Relief Everywhere). "But the forests of the developed world have another set of problems."

In addition to the continuing loss of old, magnificent trees like those on Meares Island, these problems include acid rain, loss of diversity through conversion to "managed" forests and the continuing inroads of development and suburban sprawl. "We need to worry about degradation as well as disappearance," warns Elliott Norse, senior ecologist with the Wilderness Society.

The problems are all the more poignant to Westerners because temperate forests often have special meanings. These are the forests of Western myth, religion and culture, harboring everything from Goldilocks to Big Foot, from truffles to grizzly bears.

Roughly speaking, they begin at the southern edge of the tundra of the far north, where the land thaws slightly, and mosses, lichens and shrubs give way to a globe-girdling ring of stark, boreal woodlands. Here, spruce and fir predominate. Further south, as the climate warms, the forests increase in diversity and complexity, forming a gigantic mosaic that is astounding in its majesty and range. These are the huge, aged oaks of Sherwood Forest in Britain; the towering sequoias of northern California; the intricate beeches of northern Japan; the lush Scotch pines of Scandinavia; and the endless swamp-bound larches of Siberia.

From these temperate and boreal woods comes a staggering amount of lumber, most of it used for homebuilding, papermaking and other industrial purposes. The lion's share comes from the world's 7.4 billion acres of coniferous forest—roughly 90 percent of which are in North America or the Soviet Union. The developed world also contains most of the 624 million acres of the wood plantations owned and operated by timber companies. Major exporters of timber include Canada, the United States and Scandinavia. Major importers include Japan, Great Britain and parts of Europe.

As early as the ninth century B.C., logging in Europe was proceeding fast enough to impress Homer. The poet described a battle so fierce that it matched "the din of woodcutters in the glades of a mountain." Five centuries later, the pace of logging sickened the philosopher Plato, who described a formerly wooded spot as "the skeleton of a sick man."

That was how it went. In China and Japan, huge stretches of natural forest were long ago converted to farmland. In the lands around the Mediterranean Sea, what was once a carpet of oak, pine and laurel was cut to build everything from fires to Roman warships. Today, that area is covered with a shrubby growth known as maqui. The region is heavily grazed, which prevents new forests from taking root and confines the famous stately cedars of Lebanon to tiny patches of land.

By the turn of the twentieth century, the attack on the world's temperate woodlands reached a level of ferocity that had never been matched before or since, even when compared with today's destruction of tropical rain forests. In some areas, the destruction was so complete that observers at the time worried that the loss of these forests would soon bring on a "timber famine" and cause widespread economic havoc.

Those times are gone. In the last several decades, the pace of logging has slowed considerably. Steel, iron and concrete have replaced wood for industrial and many construction needs. The rush to convert forests to farmland has virtually stopped, and in some places, trees have grown back. New England, for example, went from predominately farm country to more than 80 percent forest between 1830 and 1950. Germans refined a type of forest management known as "silviculture," in which woodlands are harvested and replanted in cycles. In the United States, Theodore Roosevelt helped establish a Forest Service, although as things turned out, this government agency is actually responsible for many of the problems in U.S. forests today, some critics say.

Forest growth has been further bolstered by extensive replanting campaigns. In Scandinavia, Nepal and the United Kingdom, government reforestation programs greatly increased the woodland totals. During the 1980s, China planted more trees each year than the rest of the world combined, although only about 30 percent have survived.

Whether or not the recovery will continue remains an open question. Although people speak less frequently of timber famines, they do speak of *Waldsterben,* the German word for an ominous and mysterious forest death.

During the last two decades, more than 14.6 million acres of forest in West Germany, Austria, Czechoslovakia, Poland and Yugoslavia have succumbed to what appears to be a lethal combination of air pollution, weather patterns and pathogens. In Europe, virtually all common species of forest trees are affected, including white fir, Scotch pine and larch. In the United States, growth rates of many trees are slowing, while red spruce on Eastern ridges are dying.

Another debate has been spawned by genetic advances that have enabled lumber companies to squeeze even more from rapidly-expanding "plantations," most of them consisting of neatly-spaced conifers and little else. While some describe these single-species designer forests as economically essential, critics such as the Wilderness Society's Norse see them as virtual biological deserts. "Some of what we now call 'forested lands' are little more than parking lots for trees," he says.

By all accounts, centuries of relentless logging have irrevocably changed the nature of temperate forests around the world. In Europe, China and Japan, "natural forests" are minuscule, more like theme parks than wilderness areas. They are hardly the magical, eerie places that harbored Scandinavian gods or Spenser's Fairie Queen.

That loss is exactly what saddens Adrian Dorst, who sees forests as more than potential lumber. In recent years, more people have come to share Dorst's attitude. There is a growing movement in places like Western Canada, the U.S. Pacific Northwest and Alaska to preserve millions of acres of unlogged, ancient forest in the region and creatures, such as the northern spotted owl, which live only in these old-growth woodlands. However, powerful timber companies still dominate the economy and politics of these regions and have therefore historically prevailed.

Dorst doesn't know yet if he has won his battle to save the trees on Meares Island. The forest's fate is now tied up in Canadian courts, which must decide if trees can be cut. But Dorst keeps fighting, leading tours to give people one final glimpse of the majesty of one of the last-remaining original temperate forests.

Content Considerations

1. What are the causes of problems besetting temperate forests in developed nations, according to the author?
2. How do these problems differ from those in underdeveloped countries?
3. If the total forest acreage has stabilized around the globe, why does the author perceive a problem?

The Writer's Strategies

1. Nielsen provides a perspective of today's forest by writing thumbnail sketches of ancient logging practices. How does this affect you as you read his essay?
2. Why does the author start and end with the threatened extinction of one forest on one island, and then broaden into other times and cultures? Does this circular construction seem to work?

Writing Possibilities

1. Nielsen quotes an ecologist who says that efforts to replace logged forests result in "parking lots for trees." Write an essay in which you consider how a forest differs from a field of trees in rows.
2. With a classmate, think about tales from mythology, legend, or religion that center around forests. Together, write a paper on how forests have affected our culture.
3. Think about what your response might be if your favorite grove of trees were going to be logged. What arguments could you raise to prevent the trees from being cut?

Don Hinrichsen

Acid Rain and Forest Decline

The author of the next article explains that acid rain forms when emissions of sulphur and nitrogen, and sometimes hydrogen chloride, mix with "water vapour, sunlight, and oxygen in the atmosphere." As moisture or dry particles, this substance returns to the earth and settles within the land's ecosystems.

Increased acidity causes damage in many ways—trees and fish cannot survive in habitats where soil and water chemistry is so altered. Airborne acids damage such monuments of civilization as the Parthenon, the Taj Mahal, and the Statue of Liberty. Other structures—homes, railroads, buildings of all types—become less sound as acid eats away at them.

Don Hinrichsen, former editor of Ambio *and of* The World Resources Report, *explains the causes of acid rain, its effects, and the movements toward solving the problems it creates in this article from* The Earth Report (1988). *Before you begin it, write for a few minutes on the role of rainfall in the biosphere, and then consider what might happen if that rainfall contains dangerous elements.*

AFTER NEARLY TWO DECADES of intensive scientific research regarding the acidification of aquatic and terrestrial ecosystems, coupled to equally intensive squabbling over the results and implications, it is clear that the acidification of the environment remains one of the industrialized world's most intractable problems.

Despite the collective efforts of a host of scientific bodies and wellmeaning governments, the scourge of acid rain continues to plague the industrialized regions of the northern hemisphere. And it is now spreading to the rapidly industrializing areas of the south as well.

Acid rain spares nothing. What has taken humankind decades to build and nature millenia to evolve, is being impoverished and destroyed in a matter of a few years—a mere blink in geologic time.

In Poland, for example, acid deposition (which includes both wet and dry acidic substances) erodes iron railway tracks in the highly polluted Upper Silesian industrial area, limiting the speed of trains to 40 km per hour. In nearby Krakow, the ornate facades of historic buildings slowly disintegrate from this domestic form of "chemical warfare." Acid rain, along with other airborne pollutants, is dissolving Greece's classic past, eating into the marble of such priceless monuments as the Parthenon in Athens. It is also responsible, in part, for the sad state of the Cologne Cathedral, which is literally falling apart in a hail of masonry. Even the exquisite Taj Mahal in India is under assault from airborne acids.

Nature fares little better. In Scandinavia—one of the world's most "acidified" regions—20,000 of Sweden's 90,000 lakes are acidified to one degree or another, and 4,000 of these are said to be totally devoid of fish life.

The situation is even worse in Norway where 80 per cent of the lakes and streams in the southern half of the country are either technically "dead" or on the critical list. Norwegian authorities say that fish have been wiped out in more than 13,000 square kilometres of lakes.

Researchers in the Federal Republic of Germany have identified acid rain as one of the prime suspects in the *Waldsterben* (tree death) syndrome currently affecting 38,000 square kilometres, or 52 per cent of the country's forests. Acid deposition is thought to be one of the principal culprits in the decline and death of Swiss forests as well, particularly those in the central alpine region, where 43 per cent of the conifers are dead or severely damaged. Acid rain is under investigation in southern Sweden where about ten percent of the conifers are beginning to show *Waldsterben*-like symptoms of decline.

On the other side of the Atlantic the picture is not much different. Acid rain, in combination with other airborne pollutants, has stricken the Canadian Parliament building in Ottawa with a creeping blight that is turning the facade black. Meanwhile, officials report that more than 300 lakes in the province of Ontario are estimated to have pH values below 5, with an additional 48,000 lakes (roughly three per cent of the total number) designated as "acid sensitive." Similarly, in Nova Scotia, nine rivers have average pH values below 4.7 and are no longer capable of supporting salmon or trout reproduction.

United States scientists claim that thousands of lakes on the eastern seaboard, especially in the Adirondack Mountains, are so acidic as to be virtual "fish graveyards." The U.S. National Surface Water Survey discovered that at least ten per cent of the lakes in the Adirondack region have pH values below 5.

Airborne acids have also been implicated in the deterioration of eastern North America's higher elevation coniferous forests, stretching up the spine of the Appalachian Mountains from Georgia to New England. And Canada's sugar maples are dying out over wide areas; due in large measure to the increasing acidification of soils in eastern Canada.

While the causes of acid rain are more or less understood, its effects are still hotly contested. Certainly acid deposition is responsible for the death of fish in thousands of lakes and streams across substantial areas of northern Europe and North America. However, its effect on forests and croplands is a grey area that has spawned scientific controversy. And its economic consequences are even more difficult to pin down.

It is generally agreed that acid deposition is caused mainly by sulphur and nitrogen emissions from the burning of fossil fuels, such as coal and oil in power plants, and from various industrial processes such as metal smelting. Nitrogen oxide emissions also originate about equally from motor vehicle exhausts.

Ironically, one of the reasons acid rain is so widespread is due to the policy "mind-set," prevalent in the 1960s and 1970s, which proclaimed that "the solution to pollution is dilution." Consequently, over the past three decades power plant and industrial smokestacks were built much higher, so emissions wouldn't pollute the immediate environment. Instead, sulphur dioxide and oxides of nitrogen are carried by prevailing winds over considerably greater distances. There are documented cases of sulphur dioxide compounds covering 1,000–2,000 kilometres in a period of three to five days. The nickel and copper smelting plant at Sudbury, Ontario, makes a good example. It has the dubious distinction of being the largest single source of sulphur dioxide pollution in the world. One 400-metre high stack belches out more than 650,000 tonnes of sulphur dioxide a year, acidifying lakes and forests hundreds of kilometers away. By comparison, Sweden's total yearly sulphur dioxide emissions from all sources amount to around 300,000 tonnes.

When these high-flying pollutants combine with water vapour, sunlight, and oxygen in the atmosphere, they create a diluted "soup" of sulphuric and nitric acids. In some heavily industrialized regions, hydrogen chloride gases are mixed up in this atmospheric soup kitchen to produce hydrochloric acid, which can also be an ingredient in acid rain.

It has been estimated that, in the northeastern U.S., 65 per cent of the acid rain is due to sulphuric acid, 30 per cent to nitric acid and five per cent to hydrochloric acid.

"There are two phases involved in the formation of acids—dry and wet," writes John McCormick in *Acid Earth*. "In both, sulphur dioxide and nitrogen oxides are converted to sulphate and nitrate. Dry or 'gas phase' conversion predominates in the vicinity of emission sources, whereas wet or 'aqueous phase' conversion involving reactions within water droplets is more predominate at a greater distance. Dry phase sulphur dioxide, nitrogen oxides, sulphates and nitrates return to earth by direct deposition on surfaces in the form of gases and particles, especially in the vicinity of the sources (i.e., within 300km). This is known as dry deposition. The rest of the oxides, converted to acids by wet or 'aqueous phase' reactions, eventually return to earth as acid rain, hail, snow, sleet, or fog. This is known as wet deposition."

After this witches' brew settles to earth—washed out of the atmosphere by rain, encapsulated in snow crystals or in the form of dry particles—it increases the acidity of freshwater lakes and streams (and in some cases terrestrial ecosystems) by *decreasing* the pH values. The pH scale is used to express the extent of acidity or alkalinity of a solution, and is based on a solution's concentration of hydrogen ions.

Some scientists define acid rain as any precipitation with a pH value below 5.6, others push it down to pH 5.

Acid rain is not a new environmental phenomenon brought to us by "progress." It is, in fact, as old as the Earth itself and can be triggered by volcanic eruptions and forest fires, among other natural processes. Nevertheless, allowing for a certain amount of acidity from natural causes, this does not account for today's apparent creeping acidification of large chunks of the northern hemisphere.

Nature's own "doses" of sulphur and nitrogen oxides are dwarfed by man's industrial-based pollution. Every year, somewhere between 110 and 115 million tonnes of sulphur dioxide are spilled across Europe and North America, while the countries in the Organization for Economic Co-operation and Development (OECD) as a whole generate about 37 million tonnes of nitrogen oxides.

Scientists agree that on a global basis probably 50 per cent of the atmospheric sulphur is from natural sources, but in industrialized regions, like Europe and eastern North America, more than 90 per cent of the deposited sulphur is from man-made emissions.

Data gathered by the European Monitoring and Evaluation Programme (EMEP), which now consists of 82 monitoring stations in 23 countries, show that the average pH levels in central Europe are 4.2 or below. And according to the OECD, polluted areas in Scandinavia, Japan, central Europe, and eastern North America have annual pH values ranging from 3.5 to 5.5. Furthermore, the sulphate content of rainfall in these same regions varies from one to 12 milligrammes per litre, while nitrate concentrations average 0.5–6 milligrammes per litre.

Nitrate deposition in Europe has doubled since the 1950s. And this rise in deposition corresponds to increasing emissions of nitrogen oxides. Currently, annual nitrogen deposition in central Europe is 30–40 kilogrammes per hectare, levels at which forest ecosystems can expect to suffer from nitrogen saturation. Bengt Nihlgard, a plant ecologist at the University of Lund in Sweden, has surmised that "nitrogen-saturated forests would begin to appear after 20 to 25 years if the nitrogen deposition rate was in the range of 30 kilogrammes per hectare per year."

Ironically, it was an observant English chemist named Robert Angus Smith who first discovered a relationship between the increasingly sooty skies of industrial Manchester and the acidity he found in precipitation. The year was 1852! One of the earliest modern accounts of the effects of acid rain on fish dates from 1926 and was reported by the Inspector for Freshwater Fisheries in Norway. He noted that the rather sudden death of alarming numbers of newly hatched salmon fry was linked to water acidity. However, these early warnings went largely unheeded.

In the late 1950s acid rain was detected in Belgium, the Netherlands and Luxembourg. A decade later it was showing up in West Germany, France, Great Britain and southern Scandinavia, while at the same time silently spreading over the entire eastern half of the USA and Canada.

Today, the effects of acid rain (and dry deposition) are impossible to ignore. Last year Norway experienced rainfall that was so acidic it might just as well have been lemon juice. And the southern part of the country has suffered through acidic snowstorms that deposited a sickly black film instead of the usual white powder. In North America, precipitation as acid as vinegar has fallen on Kane, Pennsylvania, and "rainfall" a notch away from battery acid once poured on Wheeling, West Virginia.

If such staggering amounts of sulphur dioxide and nitrogen oxides are being pumped into the atmosphere every year, then why aren't more lakes and rivers in North America and Europe acidified? And how does acid rain actually kill fish and damage soils?

In the first place, every lake and river is different, and soil types vary considerably from region to region. Just as nature induces acid rains, it also provides natural "buffers" for some regions. Alkaline soils, like those covering most of the US midwest, can tolerate higher amounts of acid fallout. So, too, can lakes that are cradled over beds of limestone and sandstone (for example, those found in southern England, parts of France, and in the Allegheny Mountains, USA).

On the other hand, areas where lakes and soils sit atop thin glacial tills or thick slabs of granitic rock—as in most of Scandinavia, Scotland, and central Europe—the buffering capacity is greatly reduced. Nature's defenses soon break down. And it is these sensitive areas that are hardest hit by acid deposition.

When researchers finally found the "smoking gun"—linking biological effects with acid deposition—they discovered that acid rain is not the real killer. Something else pulls the trigger. After seven years of detective work, Norway's largest natural science research project "Acid Precipitation: Effects on Forests and Fish" (known as the SNSF project), concluded that no fish have probably died as a *direct* result of acid rain. Instead, sensitive fish like salmon, trout, roach, minnows, and arctic char—which begin to die out with just slight decreases in pH—succumb to the lethal water chemistry that acid rain fosters. Investigators have determined that low pH levels are associated with elevated concentrations of heavy metals such as mercury, aluminum, manganese, lead, zinc, and even cadmium. However, it is aluminum—the most common metal found in soils—leached into lakes and streams that delivers the final *coup de grace*. Aluminum toxicity depends on water pH and appears to have a maximum around pH 5, where it is most lethal to fish because it precipitates in their gills as aluminum hydroxide, reducing the oxygen content in the blood and causing internal salt imbalances. As to be expected, the problem is most acute where the organic content in the water is low, a state characteristic of many acid-sensitive lakes. Experiments carried out in the United States demonstrated that brook trout could tolerate water with a pH of 4.9 for some days, but as soon as minute quantities of aluminum were introduced, 50 per cent of the exposed fish died within two days.

It was further noted that the overall toxicity of heavy metals increases in acidified waters. Hence, at a pH of 5 or below, nearly all of the game fish and other top predators will be extinct. And this, in turn, gives rise to a drastic reduction in species composition and diversity. Such changes in a lake's ecology impoverishes the entire ecosystem. In the end the only things that seem to survive are some species of water beetles and old, resistant eels. Most everything else may be wiped out.

In addition to the long-term, chronic effects of acid deposition on freshwater ecosystems is the sudden and deadly effects of "pulses" of acidity. These can result from precipitation with a very low pH, or from the rapid melting of snow during spring thaws. "In either case," states *World Resources, 1986,* "the volume and rapidity of the pulse can overwhelm the catchment's ability to neutralize or absorb the acid input so that the depressed pH values generate elevated levels of toxic metals. When the acid pulse is from spring snowmelt, it often occurs at the most vulnerable time in fish life cycles. Snow normally contains more nitrate than do summer rains, so the spring shock can exceed a system's ability to absorb nitrogen. In this way, nitrogen deposition can be more important in some fish kills than is suggested by its relative contribution to total acidity."

Until recently, the effects of acid deposition on soils and forests was a grey area of research. Within the last three years, however, scientists investigating *Waldsterben*—which has now claimed more than 70,000 square kilometres of forests in 15 European countries—have made some important discoveries. Not only does acid rain mobilize heavy metals in poorly buffered soils, increasing their toxicity, but it also leaches calcium, magnesium, and potassium from the soil, depriving trees and other vegetation of these essential nutrients for growth.

Scientists investigating *Waldsterben* have discerned three ways in which acid rain affects forests: through the foliar leaching of nutrients; from the leaching of nutrients out of the root zone; and through the mobilization of phytotoxic concentrations of aluminum in the soil. These "agents of destruction" can work alone or in combination.

A number of studies in North America have shown that acid rain in the range of pH 2.3–5.0 leached potassium, calcium, and magnesium from the leaves and needles of sugar maples, yellow birch, and white spruce. When considering the fact that many forest ecosystems in North America and Europe regularly receive up to 30 times more acidity than is deposited in "pristine" environments, it is not surprising that nutrient loss occurs.

Hand-in-hand with leaching from the canopy also comes nutrient leaching from the root zones. Research on 194 forest sites in the Federal Republic of Germany confirmed that trees attempt to compensate for the loss of nutrients in their leaves or needles by taking up more nutrients from the soil. If sufficient stocks of soil nutrients are not available—due perhaps to the effects of acid deposition—then the trees are rendered more susceptible to climatic stresses like frost and winter damage.

Dr. Bernhard Ulrich at the University of Göttingen, West Germany contends that acid deposition accelerates normal soil acidification processes. Dr. Ulrich's work on the Solling Plateau in West Germany has conclusively demonstrated that acid deposition does leach essential nutrients like calcium from the soil, while at the same time mobilizing toxic quantities of aluminum, which then damage the tree's fine feeder roots. Obviously, thin soils with more restricted root zones are more vulnerable to the effects of acid rain than richer, deeper soils containing higher concentrations of buffering agents such as potassium and calcium.

Just as freshwater lakes and streams experience acid "pulses" during spring snow melts, so do soils. During such episodes huge quantities of aluminum and other heavy metals are liberated at the same time, soaking into soils and leaching into water courses. Dr. Ulrich found that aluminum concentrations of as little as one to two milligrammes per litre in soil solutions was enough to damage root systems. At Solling he found six milligrammes of aluminum per litre under beech forests and 15 milligrammes of aluminum per litre under spruce forests—well above his threshold levels for injury. Dr. Ulrich's work raises some warning flags concerning soil acidification and aluminum toxicity, which appear to be serious problems in parts of Europe, particularly those areas characterized by mineral soils with poor buffering capacities.

It is also known that soil acidity has a profound effect on the mobilization and utilization of lead, as well as other heavy metals. Controlled studies carried out in the USA and Europe show that soil microorganisms exhibit marked metal toxicity as the soil pH is lowered. When heavy metals interfere with soil microbial processes, beneficial fungi and bacteria are inhibited in their efforts to break down organic matter into the nutrients trees need, thus altering normal biogeochemical cycling patterns.

Mounting evidence from Eastern Europe suggests that alterations in soil chemistry are taking place on a wide scale. In Czechoslovakia, large portions of the Erzgebirge Mountains northwest of Prague now resemble a wasteland. Ecosystem studies carried out by Czech geochemist Tomas Paces found that acidification of this mountainous region has altered the soil's ability to even support a forest. He found that losses of magnesium and calcium from the forest soils averaged, respectively, 6.8 and 7.5 times greater than from an undamaged forest area used as a control. The runoff of aluminum, which normally remains bound to soil minerals, was found to be 32 times greater than from the undamaged "control" forest.

Meanwhile, a European-wide forest damage survey compiled in 1987 by the UN's Cooperative Programme for the Monitoring and Evaluation of Long-Range Transmission of Air Pollutants in Europe, highlighted some startling results. Involving 15 European countries, the report flagged the United Kingdom as having the highest percentage of damaged forests in Europe! In all, some 67 per cent of Britain's conifers exhibit slight to severe damage, while 28.9 per cent have "moderate to severe" damage. This puts the UK well ahead of countries like West Germany (52.7 per cent), Switzerland (52 per cent), Czechoslovakia (49.2 per cent), the Netherlands (59.7 per cent), and Yugoslavia at 38.8 per cent—countries thought to suffer far more forest damage from air pollution and acid rain than the United Kingdom. The announcement came as an embarrassment for the government and the Forestry Commission, which were insisting that *Waldsterben*-like forest decline symptoms had not yet begun to show up in Britain.

Another alarming trend is the increasing acidification of groundwater. Acid precipitation, percolating through the soil, can leach heavy metals into groundwater reservoirs. It is already a serious problem in Belgium, Finland, the Netherlands, Norway, Sweden, and the Federal Republic of Germany. In Holland, for example, the worst affected areas have acidified groundwater down to a depth of ten metres. Because of the over-use and misuse of nitrogenous fertilizers, much Dutch soil is saturated with nitrogen and there are lakes with extremely high ammonium levels.

In the province of Bohuslän on Sweden's west coast, 49 per cent of the water wells tested had pH values below 5.5, and a number of people complained about corroded water pipes. Scientists from the Swedish Water and Air Pollution Research Institute observed that the tap water in many houses of the area contained high concentrations of copper, zinc and, in some cases, cadmium. This finding only confirmed what the residents had already experienced—the corrosion of aluminum cooking pots, diarrhoea in young children, foul-tasting water, and in several cases, blond hair that turned green from washing in water with high amounts of copper. Chicken farmers also reported that hens exposed to acid groundwater laid thin-shelled eggs. A large scale survey carried out during 1986 revealed that the acidification of groundwater in Sweden is much more serious than previously thought: 15 to 20 per cent of all private wells dug or drilled in granitic rock contain acidified groundwater. At pH values below six, zinc, copper, and lead begin to corrode. Nearly 70,000 of these 100,000 wells supply people with water all year round (the rest are used by summer homes or temporary dwellings).

Many environmental scientists fear that this is only the beginning—much more serious long-term effects, as yet undetected, may be at work. And it is these subtle effects, especially on soils, crops, and forests, that require substantial research efforts. Indeed, the World Commission on Environment and Development, chaired by Gro Harlem Brundtland, the Prime Minister of Norway, concluded in its final report *Our Common Future*, "Europe may be experiencing an immense change to irreversible acidification, the remedial costs of which could be beyond economic reach."

Damage to metals, building exteriors, and painted surfaces alone cost the 24 member countries of the OECD some $20,000 million a year. If the costs of dead and dying forests, acidified lakes and streams, and crop losses were factored in to the equation, the price might very well prove astronomical.

In the meantime, a number of emergency measures have been launched to save acidified lakes, rivers, and forests. Spreading lime on affected lakes and forests appears to be one method that is getting results, however temporary. By 1984 some 3,000 Swedish lakes had been limed at a total cost of $25 million. Liming programmes have also been carried out in Norway, the Federal Republic of Germany, Austria, Czechoslovakia, Poland, and the Soviet Union. Unfortunately, liming damaged ecosystems is at best a stopgap measure. In the long term, only the imposition of stringent limitations on the amounts of sulfur dioxide and nitrogen oxides discharged into the atmosphere will have the desired effect—that of drastically reducing damage from acid pollution.

Similarly, West German scientists have discovered that fertilizing damaged conifers with appropriate amounts of calcium, magnesium, potassium, zinc, and manganese, reversed the decline syndrome. Sick trees dosed with these chemicals recovered in a matter of weeks. However, this method of treatment works only for those trees suffering from acute nutrient deficiencies. It certainly does not work in all those forests that have been afflicted with *Waldsterben*.

Since Scandinavia was the first region to document ecosystem damage from acid deposition, it is not surprising that Norway and Sweden spearheaded international efforts to bring the "precursors" of acid rain—sulphur and nitrogen oxides—under the heel of international regulation. Since Sweden estimates that 80 per cent of all the sulphur dioxide falling on the country every year comes from abroad (and Norway pushes that figure up to 90 per cent), it is obvious that national control policies in both cases would have merely cosmetic effects. Only concerted international action will reduce the emission and deposition of acidifying substances in Europe and elsewhere. Sweden and Norway were instrumental in persuading the member countries of the UN's Economic Commission for Europe (ECE) to sit down at the conference table. The end result of a tiring process was the signing of the Convention on Long-Range Transboundary Air Pollution (LRTAP) in Geneva in November 1979. Still, the convention did not come into force until March 1983, when the requisite number of signatories ratified it. Dismayed with the slow progress shown by most parties to the convention in carrying out its terms, Norway and Sweden were also instrumental in forming the "30 per cent Club," launched in Ottawa, Canada in 1984. Initially consisting of ten countries—those more polluted than polluting—the "club" called for the reduction of sulphur dioxide levels by at least 30 per cent by 1993 (using 1980 emissions as the baseline). By April 1985, 19 countries had agreed to lower their emissions of sulphur dioxide by 30 per cent, along with unspecified reductions in other pollutants, mainly nitrogen oxides.

When the members of LRTAP met in Helsinki in July 1985 to further discuss the implementation of the convention, another important milestone was reached. The convention itself was given some real teeth. A protocol mandating sulphur dioxide reductions of 30 per cent by 1993—under the same terms as the "30 per cent Club"—was opened for signature. Not surprisingly, it was immediately endorsed by the 21 members (representing 19 countries) of the "30 per cent Club," and was finally ratified by sufficient countries and came into force in September, 1987.

However, two of the largest emitters of both sulphur and nitrogen oxides—the United States and the United Kingdom—refused to sign the protocol on the grounds that scientific uncertainty about the benefits made further reductions problematic.

The European Community (EC) is also moving to reduce pollution loads in member countries. In 1983, the Commission of the European Community proposed a Council Directive calling for significant cuts in three categories of emissions by 1995—60 per cent reduction for sulphur dioxide, 40 per cent for nitrogen oxides, and 40 per cent for dust. The main targets were large fossil-fuel-fired power plants.

At the same time, the commission also recommended that all members of the EC introduce unleaded gasoline by 1989, and that catalytic converters be fitted to all new cars by 1995.

In any case, it is not going to be an easy road to travel. An OECD survey did show that Belgium, Denmark, West Germany, France, and the United Kingdom are net exporters of pollution, while countries like Norway, Sweden, Finland, and Switzerland are net importers. Nevertheless, knowing where the pollution comes from is one thing, quite another is cajoling the exporters into reducing it, especially since control technologies are expensive.

No doubt acid rain will continue to be one of this decade's most troublesome international environmental problems—one that is bound to worsen before it gets better. How, for example, will compensation schemes be worked out? Who will pay what to whom and how much?

Meanwhile, as scientists debate decimal points and administrators are paralyzed by "inconclusive" data, a rain of acid continues to spread across the Northern Hemisphere and perhaps elsewhere. It is not a very comforting notion that, depending on where you live, your next rainfall might have the pH of lemon juice, vinegar, or something stronger.

Content Considerations

1. What evidence suggests that modern use of fossil fuels contributes to acid rain?
2. Why are some areas harder hit than others when equally exposed to acid rain?
3. Why can't the problems of acid rain be solved without international cooperation?

The Writer's Strategies

1. Hinrichsen uses a technical, scientific approach to the topic of acid rain. Is this more or less effective than a tone of advocacy? Why?

2. The author goes into considerable detail when citing evidence or explaining the mechanism of destruction by acid rain. Does this serve to convince you or merely bore you?

Writing Possibilities

1. Using Hinrichsen's article, summarize the ways in which acid rain affects plant and aquatic animal life.
2. With several classmates, consider ways in which pollution-exporting nations should help pay for acid rain damage in pollution-importing countries.
3. Write a letter to your congressman in which you urge action on acid rain. Cite evidence from Hinrichsen or other sources to bolster your argument.

JULIE SLOAN DENSLOW

The Tropical Rain-Forest Setting

It may at first seem unlikely that practices half a world away can affect the environment and quality of life of people living in middle America or along the Mediterranean Sea. But the world shares air and water and species, all of which are harmed when vast tracts of forest are cleared, cut, or burned—damage made evident by the way developed countries contributed to global pollution as they became industrialized.

The next selection, reprinted from People of the Tropical Rain Forest, *discusses some of the issues concerning loss of tropical forest areas. Julie Sloan Denslow discusses what such forests add to the biosphere and what can be lost through their destruction, but she also draws some comparisons between the rain forest and the woodlands of North America, and her discussion never veers far from her focus on the people who populate rain forest land and must survive there.*

As you read her article, think actively about how the needs of humans and the needs of the Earth might be balanced or whether there are alternatives to achieving such a balance.

AROUND THE WORLD, the tropical rain forests are being cut at a staggering rate. One estimate puts the loss rate at ten million hectares (twenty-seven million acres) a year or 1.2 percent of the approximately sixteen hundred million hectares (four billion acres) remaining. Another suggests that deforestation is occurring at a much lower rate, or around 7.5 million hectares (18.5 million acres) a year. All define tropical rain forest and deforestation differently, and all depend on data from often remote forests in Third World countries with inadequate facilities or labor to monitor the fates of their forests. Reliable figures for rates of deforestation are understandably difficult to obtain and have generated much discussion.

There is little disagreement, however, on the importance of such deforestation. It has been linked to changes in global climatic patterns and rising sea levels, to the economic and political instability that accompany declining living standards of the rural poor, to the tragic loss of cultural diversity as rain-forest peoples and their lands come under pressure of national development goals, and to the extinction of species that will dwarf the great natural extinctions of geologic history.

The immediate causes and rates of deforestation vary among regions. Clearing for cattle pasture is a major source of forest loss in the Amazon basin and Central America but is not a significant contributor in Africa and tropical Asia. In the Amazon basin clearing for pasture and, indirectly, land speculation is affecting primarily the southern, western, and eastern edges with relatively little impact in the vast, remote central region. The overall rate of deforestation in the Amazon basin is therefore relatively low (0.33 percent a year). In Central America, however, the forest is more accessible and more vulnerable; it is being cleared at the staggering rate of 3.2 percent a year, primarily for pasture to supply low-grade beef to the U.S. fast-food industry. The rain forests of Asia are being intensively exploited by logging companies for timber and by shifting cultivators who follow logging roads into the newly accessible forest. Deforestation in western Africa primarily is due to timbering and slash-and-burn (swidden) agriculture, but inaccessibility has protected the rain forest of the Congo basin.

Forty percent of the original extent of the world's tropical forests have been destroyed, although in some areas deforestation is even further advanced. Current statistics project that outside of reserves, rain forest will be completely gone from Peninsular Malaysia by the early 1990s and from Central America by the year 2000. They have disappeared now from Haiti, on the island Columbus once described to Ferdinand and Isabella:

> Its lands are high; there are in it very many sierras and very lofty mountains. . . . All are most beautiful, of a thousand shapes; all are accessible and filled with trees of a thousand kinds and tall, so that they seem to touch the sky. I am told that they never lose their foliage, and this I can believe, for I saw them as green and lovely as they are in Spain in May, and some of them were flowering, some bearing fruit, and some at another stage, according to their nature.

We seem to be watching a reenactment of the opening of the great eastern deciduous forest of North America during the eighteenth and nineteenth centuries. Settlement of North America extended pasture and farmland at rates comparable to those now occurring in many parts of the tropical lowlands. Like the present-day tropics much of the clearing was carried out by loggers and homesteaders.

Moreover, the impact of such massive forest destruction in North America does not seem to have been permanent. We can trace few species extinctions to habitat loss in those early years (although several were lost to hunting pressure). Today much of that land has returned to young forest. Economically marginal farms on stoney forest soils of New England were abandoned as the deep prairie soils to the west were brought into production. In America north of the Rio Grande today there is more land in forest than when the pilgrims landed.

Is it alarmist then that we decry the disappearance of the tropical rain forest? Is the tropical forest so very different from the North American deciduous forest? Can't its productivity be channeled for the human good? Isn't the exuberance of its growth evidence of an innate resilience to human disturbance? The answers to these questions lie in an understanding not only of the things that set the tropical rain forest apart from its temperate counterparts but also in an understanding of the role of human activities in its survival.

The rain forest and the people who make their living from it are inextricably interwoven. Not only do the activities of hunter-gatherers, small farmers, plantation owners, and loggers have strikingly different consequences for the survival of the forest, the health of the forest likewise affects their own well being.

From country to country and from people to people, the resources of the forest are used and managed for different purposes and to different advantages. Plantations of rubber trees and oil palms have helped move Malaysia into the ranks of developed nations, where elsewhere plantations have met chronic insect and disease problems. Smallholders in Brazil, Zaire, and Thailand farm a wide diversity of crops on rain-forest soils in such different ways that shifting cultivation almost defies definition.

There remain, however, important biological similarities among rain-forest ecosystems that constrain and influence those who would exploit its resources. Of all the many facets of the tropical rain forest perhaps the most difficult to grasp and the most threatened by deforestation is the diversity of its species. With only 6 percent of the world's land area the tropical moist forests are thought to house almost half its species. Although Costa Rica is smaller than West Virginia, it supports more than 12,000 species of vascular plants, 150 species of reptiles and amphibians, 237 species of mammals, 850 species of birds, and 543 species of butterflies in 3 families alone. That is more species of birds than in the United States and Canada combined. Madagascar has more than 2,000 tree species in comparison with only about 400 in all of temperate North America. Peninsular Malaysia has 7,900 plant species compared to Great Britain with 1,430 in twice the area. The Amazon and its tributaries hold more than 2,000 species of fish. E. O. Wilson counted 43 species in 26 genera of ants in a single tree in Peru's Tambopata Reserve, "about equal to the entire ant fauna of the British Isles."

It is not surprising therefore that large numbers of species are threatened as rain-forest habitats are altered or destroyed, but tropical species are especially vulnerable for other reasons as well. In contrast to their temperate counterparts, tropical species are often highly localized in their distributions. Islands, mountaintops, valleys, drainage systems, watersheds, and local pockets of high rainfall are characterized by high numbers of endemic species: plants and animals occurring nowhere else. For example, almost half the 708 bird species in Papua New Guinea are endemic. In part this is because very few species are common and most are rare. In addition many tropical species do not spread very easily. Many tropical trees have large seeds that are not carried far from the parent tree, and some bird species of the rain-forest understory will not fly across open fields or large rivers. Consequently, the populations of these species tend to be highly localized.

Biological diversity is the true wealth of the tropical forest, but a wealth we are too slowly beginning to appreciate. Many crops that feed the world came from the tropics, and the tropics still house their wild and domestic relations: corn, potatoes, manioc, sweet potatoes, and tomatoes originated in Latin America and rice, bananas, coconut, and yams in tropical Asia. To these should be added important industrial crops such as sugarcane, tobacco, oil palm, coffee, jute, rubber, and cacao. Forty percent of the food-crop production of North America is dependent on crops that originated in Latin America, although not all from rain-forest habitats. The magnitude of our dependence on the genetic diversity of the tropics was highlighted by the corn blight that in 1970 spread to national epidemic proportions in our genetically uniform fields and again eight years later when a previously unknown species of perennial corn was discovered near a Mexican cornfield. Genes from that species are being used to protect the U.S. hybrid-corn crop from fungus, and there is hope that a perennial variety with commercial potential may soon be developed.

Out of the rain forests have also come drugs that changed the course of civilization: quinine from the bark of the cinchona tree is used in the treatment and prevention of malaria; steroids from a Mexican yam were central to the development and early wide dissemination of birth-control pills; curare from a woody vine is used as a muscle relaxant during surgery; vincristine and vinblastine from the Madagascar periwinkle are true miracle drugs for the treatment of childhood leukemia. Tragically the expense of putting new drugs into production and risks of dependence on wild plants have discouraged commercial drug companies from investing in plant exploration. In the United States today most such exploration for potential medicinal plants is in the hands of a few large herbaria such as the New York Botanical Garden.

The rain forests lie between the tropics of Cancer and Capricorn, which are the northern- and southern-most limits to the track of the sun. In fact, the word *tropics* derives from the Greek *tropos,* "a turning."

Between these latitudes, the sun is directly overhead twice a year and all year long its rays strike the ground almost perpendicularly rather than obliquely as in higher latitudes. There are major areas of these complex forests in the Amazon basin of South America, Congo basin of Africa, and islands of Sumatra, Borneo, and New Guinea.

This high influx of solar energy has several important consequences. First and most obviously, the tropics are constantly warm. Near the equator the temperature varies more in a day (five to seven degrees centigrade) than the monthly average temperature varies in a year (less than one degree centigrade). There is no cold month, and where rainfall is abundant, no season of dormancy for plants and animals. Biological activity continues year round. Three crop rotations a year (five, if plantings overlap) can be obtained under such conditions.

Much of temperate North American weather is a product of frontal air masses that move east in response to prevailing westerly winds. Local weather reflects the relative strengths of the cold, dry arctic air masses and warm, moist air masses from the Gulf of Mexico. Much tropical weather, in contrast, is generated locally. Warm air, rising over land heated by the overhead sun, is heavy with moisture transpired from the forest below and evaporated from adjacent warm tropical oceans. As the rising air cools it drops its moisture load as rain, often in late afternoon storms. Brazilian meteorologists estimate that more than half the water falling as rain in the Amazon basin is recycled from the adjacent forest—evaporated and transpired from the large mass of foliage. As a consequence, extensive deforestation has been implicated in changes in local rainfall patterns.

Rain forests also lie at the heart of the earth's heat pump. The warm air generated in the tropics is carried poleward, distributing tropical warmth to the higher latitudes. Any large-scale destruction of this forest would seem to precipitate global changes in climate, although the effects of tropical deforestation on global weather patterns are still highly speculative.

Equally speculative are the effects of deforestation on the global carbon-dioxide balance. Increases in atmospheric carbon dioxide are closely linked to global warming trends and rising sea levels. Scientists estimate that the amount of carbon in the biomass of the world's forests exceeds that in the atmosphere by at least three times. Oxidation of organic matter in vegetation and underlying soils could thus potentially alter the concentration of carbon dioxide in the atmosphere. Some researchers suggest that atmospheric carbon input from the clearing of tropical forests is second only to that from the burning of fossil fuels. Although reliable estimates of the actual amounts of carbon dioxide released by the conversion of tropical forest to agriculture are scarce, the potential for major climatic consequences of widespread deforestation remains.

With the exception of notably rich soils, most old tropical soils are poor agricultural risks for a multitude of reasons, all associated with their great age. Most are poor in such nutrients as phosphorus (the only new sources of which are soil minerals and bedrock), potassium (which is only very weakly held in the soil matrix), calcium, and magnesium. Bacteria, often associated with roots of plants in the legume family, convert nitrogen from the atmosphere to ammonium, which other microbes convert to nitrates that can be taken up by plant roots. Nitrogen is thus replenished by the living vegetation, but the rapid decay of organic matter means that nitrogen is quickly depleted when the forest is cleared. Millennia of weathering have left behind the oxides of iron and aluminum, which give warm-climate soils their red and yellow colors and make tropical soils extremely acidic, with high concentrations of aluminum that are toxic to many crops.

Much of the phosphorus in tropical soils is firmly bound to the clays of the highly weathered soils. Phosphates thus do not reach roots in the dilute soup that flows down through the soil column. Roots must seek it out, and the floors of some tropical forests are carpeted by a thick mat of fine roots at the soil surface. Phosphorus uptake is also facilitated by their association with symbiotic fungi (called mycorrhizae), which grow closely associated with root cells. The fine network of the fungus more quickly and thoroughly penetrates the soil and freshly fallen litter than do plant roots. Through these mycorrhizae, plants are able to obtain sufficient phosphate for growth in otherwise nutrient-poor soils.

Under an intact forest nutrients released by decaying litter are thus likely to be quickly reabsorbed into the living vegetation. Very little is lost into the groundwater, but at the same time little is stored in the soil itself. Tropical soils under many rain forests are thus deceptively poor. The lush vegetation suggested to early settlers that these were regions of great untapped potential that would yield abundant harvests under enlightened modern agricultural techniques. Repeated trials have shown, however, that these soils are very fragile and, unless carefully fertilized, apt to be productive for only a few short years.

Cleared of vegetation the soil is exposed to the full force of the tropical sun and rain. With the root and fungus network no longer in place to capture nutrients released from decaying vegetation, leaching and erosion deplete the soil of its few nutrients. If the trunks and branches are burned, as they are in most forms of swidden agriculture, nutrients in the litter are converted to ash fertilizer for the crops. The ash also lowers the high acidity of soil, improving the availability of phosphorus. The first crops are improved under this scheme, but as the ash is quickly depleted and the remaining trunks and branches decay, soil fertility returns to its original poor state. Experiments in Amazonian Peru show that some crops develop nitrogen and potassium deficiencies in the first eight months, phosphorus and magnesium deficiencies within two years, and deficiencies in micronutrients such as calcium, zinc, and manganese within the next several years. Yields decline sufficiently from weed invasion and decreased soil fertility that small farmers generally abandon their fields after only two years. Pastures last a bit longer, but similar factors generally force their abandonment within ten years.

The use of perennial crops solves many of these problems. The canopies of such crops as coffee, teak, oil palm, cacao, and bananas protect the soils from the impact of rain and dessication of the sun. Litter is often allowed to decay in place or is burned to provide more ash fertilizer. Planting of many tropical crops requires only minimal disturbance of the soil surface; manioc is planted and replanted by inserting a short section of stem in the ground. Yams and plantains are handled similarly. Weeding is accomplished by machete and rarely by tillage of the soil.

Most swidden fields are a mélange of crops of different species, sizes, and growth forms. The mixtures of crop species may prevent population outbreaks of insect pests that might otherwise explode on monocultures of single species. Devastating pest loads are a more serious threat to crops in the tropics than in temperate climates where freezing temperatures annually reduce pest populations. In the absence of chemical controls, tropical farmers rotate crops, intermix species and varieties, and abandon their fields for a long fallow period in an effort to minimize losses. They also mimic many aspects of the natural succession in an old field. Such annual crops as upland (dry) rice and corn are planted at the same time as such longer-lived species as beans, manioc, squash, and yams. Economically important timber species or favored fruit trees may be allowed to stand or are planted in with the short-lived crops. By the time the early crops are harvested, the later species are beginning to spread their crowns over the soil. In this way a plot of land is in continuous production and rarely laid bare to the elements. Even after the growth of weeds and declining soil fertility significantly diminish yields, farmers may return to their old fields to harvest fruits of palms and other trees planted when the field was first opened.

Fields are allowed to lie fallow for varying numbers of years depending on soil characteristics, plant requirements, and local farming practices. In the western Amazon basin fourteen to twenty fallow years seem necessary to sufficiently improve soil fertility. In Peninsular Malaysia a shorter, bush fallow (in which the land is again cleared before trees become large) is common. Differences depend on the quality of the soil, crop requirements, and labor invested in weeding and soil management. Reestablishment of the forest during this period improves the structure and fertility of the soil and diminishes the weed and insect populations. After the fallow period the forest may again be cleared and burned and the land farmed for a year or two in a diverse mixture of tropical root, grain, and fruit crops.

These cropping schemes vary. A single family may have several different fields in various stages of production, and the composition and management of successive fields also differ depending on a family's requirements for crops, the production from other fields, and vicissitudes of local markets. The nutrient requirements of crops importantly influence their management. Corn and upland (dry) rice require relatively high-nutrient availability; they are planted on the best soils or following long fallow periods or soon after the slash has been burned. Manioc is tolerant of very poor soils and will reliably produce tubers for several years following forest clearing. These variations have at least one important characteristic in common: high population densities cannot be supported on slash-and-burn systems that rely on a long forest fallow to renew the soil.

Swidden agriculture is ecologically sound and functional over a large part of the tropics, wherever population density is low. Where forests are extensive and cash and transportation scarce, it is an economically viable method of ensuring subsistence and even producing something for market. Increasing populations and reduced access to forest, however, dangerously reduce the fallow period in many parts of the tropics.

Moreover, national policies, loan programs, and tax incentives often encourage the clearing of large parcels of land. In Amazonia and Central America cattle ranchers and land speculators consolidate the land of smallholders, delaying the return of the forest for the temporary establishment of pasture. Marginal and inadequate soils are often cleared of their forests in large-scale Indonesian and Amazonian resettlement schemes, and under the stimulus of government tax incentives and their own indebtedness, settlers may clear more land than they are able to manage adequately.

Where large areas are cleared or the soil is continuously disturbed (as under pasture), the reestablishment of forest is much delayed. The seeds of most tree species, except fast growing, weedy species, are large, short lived, and poorly dispersed; they are thus unlikely to reach slash-and-burn fields or pasture unless seed trees are close by. Mycorrhizal fungi, which can only survive in association with living roots, also decline in large clearings; seedlings of many forest trees are unable to grow in the absence of their phosphorus-gathering mycorrhizae. Those seedlings that do get started are often those of aggressive tropical grasses and other weeds that strangle crops and tree seedlings. The intense tropical sun and soil compaction, like nutrient depletion, also impede the establishment of young seedlings because their roots are not able to penetrate deeply enough to reach a reliable water supply during the dry season. Great expanses of rain-forest lands, originally cleared for slash-and-burn plots and then converted to pasture, have been severely degraded into scrubland, unusable for either cattle or agriculture and unable to support a potentially productive forest. The forest has been, in effect, mined for a few short years of productivity.

The impact of heavy tropical rains, cattle, and machinery compact the soils, collapsing the fine earthworm tunnels and old root channels. Rainwater is no longer absorbed into the soil but runs over the surface carrying topsoils, silts, and clays into the streams and rivers. Fluctuations in water levels of major rivers become increasingly chaotic; at low water, rivers become unnavigable and floods are higher and more destructive. On slopes, landslides expose the poor subsoils and destroy villages, roads, and bridges. The silt from unprotected watersheds fills reservoirs quickly so that the life expectancy of dams in tropical forests is drastically shortened. In many cases the electricity generated during the brief life of the reservoir falls far short of paying for the construction of the dam.

There is good reason to mourn the loss of the tropical rain forest. The great diversity of its plant and animal life makes the tropical rain forest a resource of special value to humankind, unmatched by any other ecosystem on earth. For all our well-meaning attempts to preserve this diversity in zoos, botanical gardens, seed banks, and germ-plasm preserves, our greatest efforts can only safeguard for a limited time a miniscule number of species and varieties. This diversity and the fragility of most tropical soils make the rain forests especially vulnerable under the heavy hand of large-scale development, endangering both the wealth of the forest and productivity of the land on which it stands. Even more vulnerable is the accumulated knowledge of forest ecology and resources among rain-forest peoples. We will not preserve what we do not know and understand. Our best hope is in the reasoned development of some rain-forest land and resources for the sustained benefit of its people, in the preservation of other forests with all the intricacies of their structure and interactions intact, and in the conservation of the cultural heritage of the people who live close to the forest.

Content Considerations

1. Summarize reasons for the destruction of rain forests in various regions of the world.
2. How is the present destruction of rain forests any different from earlier clearing of trees in North America?
3. List the reasons Denslow enumerates for preserving rain forests.

The Writer's Strategies

1. Denslow first presents an observable fact: rain forests are disappearing; she then asserts that the disappearance is a problem. How does she back up her contention?
2. What is the tone of this essay? How does it contribute to your reception of the information?

Writing Possibilities

1. Compare or contrast the drive across America in the nineteenth century, with its accompanying clearing of forests, to the current deforestation in Central and South America.
2. Consider positive steps that developed nations could take to ensure the continued survival of the rain forest. Write a letter to your newspaper outlining your plan.
3. If the U.S. fast food industry and its need for beef are contributing to the loss of rain forests, what are some actions you can take if you oppose the loss?

Additional Writing and Research

Connections

1. The essays by Tangley, Frome, Nielsen, Hinrichsen, and Denslow all treat the topic of destruction of forests by various means. Summarize these essays and attempt to draw conclusions about the destruction.
2. Hay, Tangley, and Denslow describe forests as habitat. In what ways is the cutting of forests more than the loss of trees?
3. The relationship of trees to humans forms part of the theme of the essays by Stewart and Hay. How do these essays differ? How are they similar?
4. Frome and Tangley both deal with large losses of forested areas because of practices by modern society. What practices seem to be most harmful to forests? Write about those practices and the alternatives the authors suggest, arguing for the alternatives you think are most sound.
5. Using any three of the essays, explain the current state of forests in one region of the world.
6. Discuss the forest restoration and preservation efforts mentioned in five of the essays. How successful do you believe these will be?

Researches

1. Many forests around the world are cut for use in industry. Find out how industry uses various kinds of wood.
2. Trace the history of the New England forests, or another tract, focusing on the recovery of wooded areas.
3. How are companies responding to the public's growing concern about forest decimation? Investigate how some companies are changing their packaging practices in response to this concern as well as the concern with trash disposal.
4. Discover the source of the wood found in a paper product you normally use. Trace its production from the forest to you, focusing on the environmental effects of its manufacture.
5. Explore the biodiversity of one forest region, including how cutting might affect or has affected species in that habitat.
6. What are the political, social, and economic reasons for rain forest clearing? What international efforts are being made to halt the destruction? Include in your research an interview or correspondence with an organization involved in such efforts.

8

Attitudes and Practices

IN MARCH 1845 Henry David Thoreau began constructing what might be the most famous shack in American history. He scrupulously recorded his expenses, and when he completed his house on Walden Pond—on land owned by Ralph Waldo Emerson—he showed a total investment of 28.12\frac{1}{2}$. He resided there for two years and two months, living nearly exclusively on what he grew and gathered. He experimented often, usually out of necessity, and existed on less and less—eventually even leaving yeast out of his bread when he deduced that it wasn't really needed. He honed to a fine art his dictum "simplify, simplify."

Thoreau would have been horrified at today's throwaway world: pens, razors, even cameras, all designed to be discarded after little use. Today such spartan self-sufficiency as he exhibited at Walden is little more than a dream—or a nightmare—for nearly all Americans; very few choose such self-imposed exile from comfort. But Thoreau has his disciples. One of the most visible is Wendell Berry, a widely known poet and essayist, whose essay "Think Little" is a Thoreauvian exhortation to simplify our lives and take responsibility for changing that which we believe is wrong. The essay is included in this chapter.

Both small and large problems face the inhabitants of earth at the dawn of the twenty-first century. Some problems individuals can go far to solve—we each can cut down on our trash, for instance—but others are founded in millennia of custom and religion. Paul and Anne Ehrlich, authors of *The Population Bomb, The Population Explosion,* and other tracts on the dangers of too many humans and too few resources, believe religious and cultural dogmas encouraging large families must be overcome. To the Ehrlichs the population explosion is the single greatest threat to the Earth—more people mean more pollution, more industry, more trash, more famine, disease, and death—in short, degradation of the whole environment.

Other writers in this section detail ways in which attempts to fix problems by "natural" solutions sometimes go awry. In California, for instance, authorities released supposedly sterile fruit flies to combat an infestation of these insects, but the "sterile" fruit flies multiplied by the millions, and entire communities were then subjected to pesticides sprayed by helicopter. How does an individual respond to such official acts? What can one person do to ensure that those who are supposed to know what to do in emergencies really do know and will act responsibly? .

The majority of writers in this section seem to come back to the dictum of personal responsibility for everything that happens. The Reverend Jesse Jackson, not usually thought of as an environmental activist, urges all of us to stop thinking that an ecology club or nature organization or save-the-Earth network will take care of the problems. He says we are all environmentalists, but the poor and those who ignore what goes on around them are the victims of toxic landfills and debilitative emissions from industries. Don't wait for someone to intervene in the pollution of Earth; stop it yourself, Jackson urges.

Vine Deloria, a Sioux Indian best known for his book *Custer Died for Your Sins,* takes a radical view of perceptions about the Earth by Native Americans and non-Native Americans. He believes many current environmental problems stem from the tendency of Euro-Americans to adapt land to their wants instead of respecting unique and differing lands. In this he joins Aldo Leopold, an early voice in the wilderness crying out for a land ethic, a creed of mutual care for the Earth.

Where do we go from here? The national government has approved a Clean Air Act, which strengthens the law and puts new muscle into efforts to reduce the amount of emissions by cars and industry. But these efforts are largely neutralized by the fact that while individual vehicles or factories are polluting less, we have many more vehicles and factories adding their detritus to the air. Many people drive two blocks to the grocery store to buy prepackaged food and then drive back home to settle down for an evening of television with the thermostat set at 72 degrees. Many of us recycle aluminum cans because it pays in cash, yet few recycle glass or paper because there is little monetary incentive. Perhaps the acts of individuals make all the difference.

Thoreau wrote in *Walden* that most people "lead lives of quiet desperation." Were he alive today, he might find our lives more desperate and urge less quietude. Indeed, this entire chapter calls for personal responsibility and an end to the silence.

PAUL R. EHRLICH
and
ANNE H. EHRLICH

Making the Population Connection

By 1950 the world's population amounted to about 2.5 billion people; it had taken 50,000 years to reach that total. Since 1950, however, the population has more than doubled; the Earth currently supports in excess of 5 billion human beings, and population growth shows no real sign of slowing.

Population growth is often linked to the problems of hunger, illiteracy, and disease. But environmental problems are also connected to population growth, and many believe that the carrying capacity of the Earth has been reached or surpassed, that it can support no more people and, indeed, may not be able to support its current population.

Ideas concerning population control are controversial—they stem from cultural, religious, personal, and political beliefs and practices. Nevertheless, many people are beginning to believe that the first environmental problem the world's nations must solve is the growth of their populations.

In 1968 Paul R. Ehrlich and Anne H. Ehrlich published The Population Bomb, *an early warning of the immediacy and danger of overpopulation. What follows is an excerpt from their last look at population,* The Population Explosion. *One of the points it makes is that processes of nature reduce populations that have grown too large. Before you read, write about some of those processes.*

GLOBAL WARMING, acid rain, depletion of the ozone layer, vulnerability to epidemics, and exhaustion of soils and groundwater are all, as we shall see, related to population size. They are also clear and present dangers to the persistence of civilization. Crop failures due to global warming alone might result in the premature deaths of a billion or more people in the next few decades, and the AIDS epidemic could slaughter hundreds of millions. Together these would constitute a harsh "population control" program provided by nature in the face of humanity's refusal to put into place a gentler program of its own.

We shouldn't delude ourselves: the population explosion will come to an end before very long. The only remaining question is whether it will be halted through the humane method of birth control, or by nature wiping out the surplus. We realize that religious and cultural opposition to birth control exists throughout the world; but we believe that people simply don't understand the choice that such opposition implies. Today, anyone opposing birth control is unknowingly voting to have the human population size controlled by a massive increase in early deaths.

Of course, the environmental crisis isn't caused just by expanding human numbers. Burgeoning consumption among the rich and increasing dependence on ecologically unsound technologies to supply that consumption also play major parts. This allows some environmentalists to dodge the population issue by emphasizing the problem of malign technologies. And social commentators can avoid commenting on the problem of too many people by focusing on the serious maldistribution of affluence.

But scientists studying humanity's deepening predicament recognize that a major factor contributing to it is rapidly worsening overpopulation. The Club of Earth, a group whose members all belong to both the U.S. National Academy of Sciences and the American Academy of Arts and Sciences, released a statement in September 1988 that said in part:

> Arresting global population growth should be second in importance only to avoiding nuclear war on humanity's agenda. Overpopulation and rapid population growth are intimately connected with most aspects of the current human predicament, including rapid depletion of nonrenewable resources, deterioration of the environment (including rapid climate change), and increasing international tensions.[1]

When three prestigious scientific organizations cosponsored an international scientific forum, "Global Change," in Washington in 1989, there was general agreement among the speakers that population growth was a substantial contributor toward prospective catastrophe. Newspaper coverage was limited, and while the population component was mentioned in *The New York Times's* article,[2] the point that population limitation will be essential to resolving the predicament was lost. The coverage of environmental issues in the media has been generally excellent in the last few years, but there is still a long way to go to get adequate coverage of the intimately connected population problem.

Even though the media occasionally give coverage to population issues, some people never get the word. In November 1988, Pope John Paul II reaffirmed the Catholic Church's ban on contraception. The occasion was the twentieth anniversary of Pope Paul's anti-birth-control encyclical, *Humanae Vitae*.

Fortunately, the majority of Catholics in the industrial world pay little attention to the encyclical or the Church's official ban on all practical means of birth control. One need only note that Catholic Italy at present has the smallest average completed family size (1.3 children per couple) of any nation. Until contraception and then abortion were legalized there in the 1970s, the Italian birth rate was kept low by an appalling rate of illegal abortion.

The bishops who assembled to celebrate the anniversary defended the encyclical by announcing that "the world's food resources theoretically could feed 40 billion people."[3] In one sense they were right. It's "theoretically possible" to feed 40 billion people—in the same sense that it's theoretically possible for your favorite major-league baseball team to win every single game for fifty straight seasons, or for you to play Russian roulette ten thousand times in a row with five out of six chambers loaded without blowing your brains out.

One might also ask whether feeding 40 billion people is a worthwhile goal for humanity, even if it could be reached. Is any purpose served in turning Earth, in essence, into a gigantic human feedlot? Putting aside the near-certainty that such a miracle couldn't be sustained, what would happen to the *quality* of life?

We wish to emphasize that the population problem is in no sense a "Catholic problem," as some would claim. Around the world, Catholic reproductive performance is much the same as that of non-Catholics in similar cultures and with similar economic status. Nevertheless, the *political* position of the Vatican, traceable in no small part to the extreme conservatism of Pope John Paul II, is an important barrier to solving the population problem.[4] Non-Catholics should be very careful not to confuse Catholics or Catholicism with the Vatican—most American Catholics don't. Furthermore, the Church's position on contraception is distressing to many millions of Catholics, who feel it morally imperative to follow their own consciences in their personal lives and disregard the Vatican's teachings on this subject.

Nor is unwillingness to face the severity of the population problem limited to the Vatican. It's built into our genes and our culture. That's one reason many otherwise bright and humane people behave like fools when confronted with demographic issues. Thus, an economist specializing in mail-order marketing can sell the thesis that the human population could increase essentially forever because people are the "ultimate resource,"[5] and a journalist can urge more population growth in the United States so that we can have a bigger army![6] Even some environmentalists are taken in by the frequent assertion that "there is no population problem, only a problem of distribution." The statement is usually made in a context of a plan for conquering hunger, as if food shortage were the only consequence of overpopulation.

But even in that narrow context, the assertion is wrong. Suppose food *were* distributed equally. If everyone in the world ate as Americans do, less than half the *present* world population could be fed on the record harvests of 1985 and 1986.[7] Of course, everyone doesn't have to eat like Americans. About a third of the world grain harvest—the staples of the human feeding base—is fed to animals to produce eggs, milk, and meat for American-style diets. Wouldn't feeding that grain directly to people solve the problem? If everyone were willing to eat an essentially vegetarian diet, that additional grain would allow perhaps a billion more people to be fed with 1986 production.

Would such radical changes solve the world food problem? Only in the *very* short term. The additional billion people are slated to be with us by the end of the century. Moreover, by the late 1980s, humanity already seemed to be encountering trouble maintaining the production levels of the mid-1980s, let alone keeping up with population growth. The world grain harvest in 1988 was some 10 percent *below* that of 1986. And there is little sign that the rich are about to give up eating animal products.

So there is no reasonable way that the hunger problem can be called "only" one of distribution, even though redistribution of food resources would greatly alleviate hunger today. Unfortunately, an important truth, that maldistribution is a cause of hunger now, has been used as a way to avoid a more important truth—that overpopulation is critical today and may well make the distribution question moot tomorrow.

The food problem, however, attracts little immediate concern among well-fed Americans, who have no reason to be aware of its severity or extent. But other evidence that could make everyone face up to the seriousness of the population dilemma is now all around us, since problems to which overpopulation and population growth make major contributions are worsening at a rapid rate. They often appear on the evening news, although the population connection is almost never made.

Consider the television pictures of barges loaded with garbage wandering like The Flying Dutchman across the seas, and news stories about "no room at the dump."[8] They are showing the results of the interaction between too many affluent people and the environmentally destructive technologies that support that affluence. Growing opportunities to swim in a mixture of sewage and medical debris off American beaches can be traced to the same source. Starving people in sub-Saharan Africa are victims of drought, defective agricultural policies, and an overpopulation of both people and domestic animals—with warfare often dealing the final blow. All of the above are symptoms of humanity's massive and growing negative impact on Earth's life-support systems.

RECOGNIZING THE POPULATION PROBLEM

The average person, even the average scientist, seldom makes the connection between such seemingly disparate events and the population problem, and thus remains unworried. To a degree, this failure to put the pieces together is due to a taboo against frank discussion of the population crisis in many quarters, a taboo generated partly by pressures from the Catholic hierarchy and partly by other groups who are afraid that dealing with population issues will produce socially damaging results.

Many people on the political left are concerned that focusing on overpopulation will divert attention from crucial problems of social justice (which certainly need to be addressed *in addition* to the population problem). Often those on the political right fear that dealing with overpopulation will encourage abortion (it need not) or that halting growth will severely damage the economy (it could, if not handled properly). And people of varied political persuasions who are unfamiliar with the magnitude of the population problem believe in a variety of farfetched technological fixes—such as colonizing outer space—that they think will allow the need for regulating the size of the human population to be avoided forever.[9]

Even the National Academy of Sciences avoided mentioning controlling human numbers in its advice to President Bush on how to deal with global environmental change. Although Academy members who are familiar with the issue are well aware of the critical population component of that change, it was feared that all of the Academy's advice would be ignored if recommendations were included about a subject taboo in the Bush administration. That strategy might have been correct, considering Bush's expressed views on abortion and considering the administration's weak appointments in many environmentally sensitive positions. After all, the Office of Management and Budget even tried to suppress an expert evaluation of the potential seriousness of global warming by altering the congressional testimony of a top NASA scientist, James Hansen, to conform with the administration's less urgent view of the problem.[10]

All of us naturally lean toward the taboo against dealing with population growth. The roots of our aversion to limiting the size of the human population are as deep and pervasive as the roots of human sexual behavior. Through billions of years of evolution, outreproducing other members of your population was the name of the game. It is the very basis of natural selection, the driving force of the evolutionary process.[11] Nonetheless, the taboo must be uprooted and discarded.

Overcoming the Taboo

There is no more time to waste; in fact, there wasn't in 1968 when *The Population Bomb* was published. Human inaction has already condemned hundreds of millions more people to premature deaths from hunger and disease. The population connection must be made in the public mind. Action to end the population explosion *humanely* and start a gradual population *decline* must become a top item on the human agenda: the human birthrate must be lowered to slightly below the human death rate as soon as possible. There still may be time to limit the scope of the impending catastrophe, but not *much* time. Ending the population explosion by controlling births is necessarily a slow process. Only nature's cruel way of solving the problem is likely to be swift.

Of course, if we do wake up and succeed in controlling our population size, that will still leave us with all the other thorny problems to solve. Limiting human numbers will not alone end warfare, environmental deterioration, poverty, racism, religious prejudice, or sexism; it will just buy us the opportunity to do so. As the old saying goes, whatever your cause, it's a lost cause without population control.[12]

America and other rich nations have a clear choice today. They can continue to ignore the population problem and their own massive contributions to it. Then they will be trapped in a downward spiral that may well lead to the end of civilization in a few decades. More frequent droughts, more damaged crops and famines, more dying forests, more smog, more international conflicts, more epidemics, more gridlock, more drugs, more crime, more sewage swimming, and other extreme unpleasantness will mark our course. It is a route already traveled by too many of our less fortunate fellow human beings.

Or we can change our collective minds and take the measures necessary to lower global birthrates dramatically. People can learn to treat growth as the cancerlike disease it is and move toward a sustainable society. The rich can make helping the poor an urgent goal, instead of seeking more wealth and useless military advantage over one another. Then humanity might have a chance to manage all those other seemingly intractable problems. It is a challenging prospect, but at least it will give our species a shot at creating a decent future for itself. More immediately and concretely, taking action now will give our children and their children the possibility of decent lives.

ENDNOTES

1. Statement released Sept. 3, 1988, at the Pugwash Conference on Global Problems and Common Security, at Dagomys, near Sochi, USSR. The signatories were Jared Diamond, UCLA; Paul Ehrlich, Stanford; Thomas Eisner, Cornell; G. Evelyn Hutchinson, Yale; Gene E. Likens, Institute of Ecosystem Studies; Ernst Mayr, Harvard; Charles D. Michener, University of Kansas; Harold A. Mooney, Stanford; Ruth Patrick, Academy of Natural Sciences, Philadelphia; Peter H. Raven, Missouri Botanical Garden; and Edward O. Wilson, Harvard.

    The National Academy of Sciences and the American Academy of Arts and Sciences are the top honorary organizations for American scientists and scholars, respectively. Hutchinson, Patrick, and Wilson also are laureates of the Tyler Prize, the most distinguished international award in ecology.
2. May 4, 1989, by Philip Shabecoff, a fine environmental reporter. In general, the *Times* coverage of the environment is excellent. But even this best of American newspapers reflects the public's lack of understanding of the urgency of the population situation.
3. *Washington Post,* Nov. 19, 1988, p. C-15.
4. Italy is a not freak case. Catholic France has an average completed family size of 1.8 children, the same as Britain and Norway; Catholic Spain, with less than half the per-capita GNP of Protestant Denmark, has the same completed family size of 1.8 children. We are equating "completed family size" here with the *total fertility rate,* the average number of children a woman would bear in her lifetime, assuming that current age-specific birth and death rates remained unchanged during her childbearing years—roughly 15–49. In the United States, a Catholic woman is more likely to seek abortion than a non-Catholic woman (probably because she is likelier to use less-effective contraception). By 1980, Catholic and non-Catholic women in the U.S. (except Hispanic women, for whom cultural factors are strong) had virtually identical family sizes. (W.D. Mosher, "Fertility and Family Planning in the United States: Insights from the National Survey of Family Growth," *Family Planning Perspectives,* vol. 20, no. 5, pp. 202–17. Sept./Oct. 1988.) On the role of the Vatican, see, for instance, Stephen D. Mumford, "The Vatican and Population Growth Control: Why an American Confrontation?," *The Humanist,* September/October 1983, and Penny Lernoux, "The Papal Spiderweb," *The Nation,* April 10 and 17, 1989.
5. J. Simon, *The Ultimate Resource* (Princeton Univ. Press, Princeton, N.J., 1981).
6. B. Wattenberg, *The Birth Dearth* (Pharos Books, New York, 1987).
7. R. W. Kates, R. S. Chen, T. E. Downing, J. X. Kasperson, E. Messer, S. R. Millman, *The Hunger Report: 1988* (The Alan Shawn Feinstein World Hunger Program, Brown University, Providence, R.I., 1988). The data on distribution in this paragraph are from this source.
8. The name of a series of reports on KRON-TV's news programs, San Francisco, the week of May 8, 1989.

9. For an amusing analysis of the "outer-space" fairy tale, see Garrett Hardin's classic essay "Interstellar Migration and the Population Problem," *Journal of Heredity*, vol. 50, pp. 68–70 (1959), reprinted in G. Hardin, ed., *Stalking the Wild Taboo*, 2nd ed. (William Kaufmann, Los Altos, Calif., 1978). Note that some things have changed; to keep the population of Earth from growing today, we would have to export to space 95 million people annually!
10. This story received broad coverage in both electronic and print media; for instance, *New York Times*, May 8, 1989.
11. For a discussion of natural selection and evolution written for nonspecialists, see P. R. Ehrlich, *The Machinery of Nature* (Simon and Schuster, New York, 1986).
12. As discussed in Chapter 10, "population control" does not require coercion, only attention to the needs of society.

Content Considerations

1. How does nature control populations?
2. What environmental and social problems do the authors attribute to overpopulation?
3. Why is population control such a taboo subject?

The Writers' Strategies

1. What opposing evidence do the authors cite? Why do they do so?
2. What is the tone of this excerpt? How effective is it? Why?

Writing Possibilities

1. In an essay, explain the problems you think can be attributed to overpopulation.
2. Explain your own position on population control.
3. With a classmate, come to some conclusions about what affluence should have to do with population growth. Write these conclusions, making sure you acknowledge and cast doubt upon the opposing points of view.

WENDELL BERRY

Think Little

> Kentuckian Wendell Berry left teaching some years ago in order to spend his life farming and writing. A novelist, poet, and essayist, Berry has published more than a dozen books. Environmental themes run strongly through his work, for Berry finds in the land the essence of life and ways of living.
>
> Ways of living, too, become his theme. In "Think Little," an essay from *A Continuous Harmony,* Berry links the ways we live to how the land fares. He proposes changes in philosophy and practice, attitude and action—and not just in environmental matters. He sees ecology as a matter larger than earth and air and sea; it also concerns "Better minds, better friendships, better marriages, better communities."
>
> To prepare for Berry's essay, review what you have done so far today. Then write about how what you have done has affected the land.

FIRST THERE WAS CIVIL RIGHTS, and then there was the War, and now it is the Environment. The first two of this sequence of causes have already risen to the top of the nation's consciousness and declined somewhat in a remarkably short time. I mention this in order to begin with what I believe to be a justifiable skepticism. For it seems to me that the Civil Rights Movement and the Peace Movement, as popular causes in the electronic age, have partaken far too much of the nature of fads. Not for all, certainly, but for too many they have been the fashionable politics of the moment. As causes they have been undertaken too much in ignorance; they have been too much simplified; they have been powered too much by impatience and guilt of conscience and short-term enthusiasm, and too little by an authentic social vision and long-term conviction and deliberation. For most people those causes have remained almost entirely abstract; there has been too little personal involvement, and too much involvement in organizations that were insisting that *other* organizations should do what was right.

There is considerable danger that the Environment Movement will have the same nature: that it will be a public cause, served by organizations that will self-righteously criticize and condemn other organizations, inflated for a while by a lot of public talk in the media, only to be replaced in its turn by another fashionable crisis. I hope that will not happen, and I believe that there are ways to keep it from happening, but I know that if this effort is carried on solely as a public cause, if millions of people cannot or will not undertake it as a *private* cause as well, then it is *sure* to happen. In five years the energy of our present concern will have petered out in a series of public gestures—and no doubt in a series of empty laws—and a great, and perhaps the last, human opportunity will have been lost.

It need not be that way. A better possibility is that the movement to preserve the environment will be seen to be, as I think it has to be, not a digression from the civil rights and peace movements, but the logical culmination of those movements. For I believe that the separation of these three problems is artificial. They have the same cause, and that is the mentality of greed and exploitation. The mentality that exploits and destroys the natural environment is the same that abuses racial and economic minorities, that imposes on young men the tyranny of the military draft, that makes war against peasants and women and children with the indifference of technology. The mentality that destroys a watershed and then panics at the threat of flood is the same mentality that gives institutionalized insult to black people and then panics at the prospect of race riots. It is the same mentality that can mount deliberate warfare against a civilian population and then express moral shock at the logical consequence of such warfare at My Lai. We would be fools to believe that we could solve any one of these problems without solving the others.

To me, one of the most important aspects of the environmental movement is that it brings us not just to another public crisis, but to a crisis of the protest movement itself. For the environmental crisis should make it dramatically clear, as perhaps it has not always been before, that there is no public crisis that is not also private. To most advocates of civil rights, racism has seemed mostly the fault of someone else. For most advocates of peace the war has been a remote reality, and the burden of the blame has seemed to rest mostly on the government. I am certain that these crises have been more private, and that we have each suffered more from them and been more responsible for them, than has been readily apparent, but the connections have been difficult to see. Racism and militarism have been institutionalized among us for too long for our personal involvement in those evils to be easily apparent to us. Think, for example, of all the Northerners who assumed—until black people attempted to move into *their* neighborhoods—that racism was a Southern phenomenon. And think how quickly—one might almost say how naturally—among some of its members the peace movement has spawned policies of deliberate provocation and violence.

But the environmental crisis rises closer to home. Every time we draw a breath, every time we drink a glass of water, every time we eat a bite of food we are suffering from it. And more important, every time we indulge in, or depend on, the wastefulness of our economy—and our economy's first principle is waste—we are *causing* the crisis. Nearly every one of us, nearly every day of his life, is contributing *directly* to the ruin of this planet. A protest meeting on the issue of environmental abuse is not a convocation of accusers, it is a convocation of the guilty. That realization ought to clear the smog of self-righteousness that has almost conventionally hovered over these occasions, and let us see the work that is to be done.

In this crisis it is certain that every one of us has a public responsibility. We must not cease to bother the government and the other institutions to see that they never become comfortable with easy promises. For myself, I want to say that I hope never again to go to Frankfort to present a petition to the governor on an issue so vital as that of strip mining, only to be dealt with by some ignorant functionary—as several of us were not so long ago, the governor himself being "too busy" to receive us. Next time I will go prepared to wait as long as necessary to see that the petitioners' complaints and their arguments are heard *fully*—and by the governor. And then I will hope to find ways to keep those complaints and arguments from being forgotten until something is done to relieve them. The time is past when it was enough merely to elect our officials. We will have to elect them and then go and *watch* them and keep our hands on them, the way the coal companies do. We have made a tradition in Kentucky of putting self-servers, and worse, in charge of our vital interests. I am sick of it. And I think that one way to change it is to make Frankfort a less comfortable place. I believe in American political principles, and I will not sit idly by and see those principles destroyed by sorry practice. I am ashamed and deeply distressed that American government should have become the chief cause of disillusionment with American principles.

And so when the government in Frankfort again proves too stupid or too blind or too corrupt to see the plain truth and to act with simple decency, I intend to be there, and I trust that I won't be alone. I hope, moreover, to be there, not with a sign or a slogan or a button, but with the facts and the arguments. A crowd whose discontent has risen no higher than the level of slogans is *only* a crowd. But a crowd that understands the reasons for its discontent and knows the remedies is a vital community, and it will have to be reckoned with. I would rather go before the government with two men who have a competent understanding of an issue, and who therefore deserve a hearing, than with two thousand who are vaguely dissatisfied.

But even the most articulate public protest is not enough. We don't live in the government or in institutions or in our public utterances and acts, and the environmental crisis has its roots in our *lives*. By the same token, environmental health will also be rooted in our lives. That is, I take it, simply a fact, and in the light of it we can see how superficial and foolish we would be to think that we could correct what is wrong merely by tinkering with the institutional machinery. The changes that are required are fundamental changes in the way we are living.

What we are up against in this country, in any attempt to invoke private responsibility, is that we have nearly destroyed private life. Our people have given up their independence in return for the cheap seductions and the shoddy merchandise of so-called "affluence." We have delegated all our vital functions and responsibilities to salesmen and agents and bureaus and experts of all sorts. We cannot feed or clothe ourselves, or entertain ourselves, or communicate with each other, or be charitable or neighborly or loving, or even respect ourselves, without recourse to a merchant or a corporation or a public-service organization or an agency of the government or a style-setter or an expert. Most of us cannot think of dissenting from the opinions or the actions of one organization without first forming a new organization. Individualism is going around these days in uniform, handing out the party line on individualism. Dissenters want to publish their personal opinions over a thousand signatures.

The Confucian *Great Digest* says that the "chief way for the production of wealth" (and he is talking about real goods, not money) is "that the producers be many and that the mere consumers be few. . . ." But even in the much-publicized rebellion of the young against the materialism of the affluent society, the consumer mentality is too often still intact: the standards of behavior are still those of kind and quantity, the security sought is still the security of numbers, and the chief motive is still the consumer's anxiety that he is missing out on what is "in." In this state of total consumerism—which is to say a state of helpless dependence on things and services and ideas and motives that we have forgotten how to provide ourselves—all meaningful contact between ourselves and the earth is broken. We do not understand the earth in terms either of what it offers us or of what it requires of us, and I think it is the rule that people inevitably destroy what they do not understand. Most of us are not directly responsible for strip mining and extractive agriculture and other forms of environmental abuse. But we are guilty nevertheless, for we connive in them by our ignorance. We are ignorantly dependent on them. We do not know enough about them; we do not have a particular enough sense of their danger. Most of us, for example, not only do not know how to produce the best food in the best way—we don't know how to produce any kind in any way. Our model citizen is a sophisticate who before puberty understands how to produce a baby, but who at the age of thirty will not know how to produce a potato. And for this condition we have elaborate rationalizations, instructing us that dependence for everything on somebody else is efficient and economical and a scientific miracle. I say, instead, that it is madness, mass produced. A man who understands the weather only in terms of golf is participating in a chronic public insanity that either he or his descendants will be bound to realize as suffering. I believe that the death of the world is breeding in such minds much more certainly and much faster than in any political capital or atomic arsenal.

For an index of our loss of contact with the earth we need only look at the condition of the American farmer—who must in our society, as in every society, enact man's dependence on the land, and his responsibility to it. In an age of unparalleled affluence and leisure, the American farmer is harder pressed and harder worked than ever before; his margin of profit is small, his hours are long; his outlays for land and equipment and the expenses of maintenance and operation are growing rapidly greater; he cannot compete with industry for labor; he is being forced more and more to depend on the use of destructive chemicals and on the wasteful methods of haste and anxiety. As a class, farmers are one of the despised minorities. So far as I can see, farming is considered marginal or incidental to the economy of the country, and farmers, when they are thought of at all, are thought of as hicks and yokels, whose lives do not fit into the modern scene. The average American farmer is now an old man whose sons have moved away to the cities. His knowledge, and his intimate connection with the land, are about to be lost. The small independent farmer is going the way of the small independent craftsmen and storekeepers. He is being forced off the land into the cities, his place taken by absentee owners, corporations, and machines. Some would justify all this in the name of efficiency. As I see it, it is an enormous social and economic and cultural blunder. For the small farmers who lived on their farms *cared* about their land. And given their established connection to their land—which was often hereditary and traditional as well as economic—they could have been encouraged to care for it more competently than they have so far. The corporations and machines that replace them will never be bound to the land by the sense of birthright and continuity, or by the love that enforces care. They will be bound by the rule of efficiency, which takes thought only of the volume of the year's produce, and takes no thought of the slow increment of the life of the land, not measurable in pounds or dollars, which will assure the livelihood and the health of the coming generations.

If we are to hope to correct our abuses of each other and of other races and of our land, and if our effort to correct these abuses is to be more than a political fad that will in the long run be only another form of abuse, then we are going to have to go far beyond public protest and political action. We are going to have to rebuild the substance and the integrity of private life in this country. We are going to have to gather up the fragments of knowledge and responsibility that we have parceled out to the bureaus and the corporations and the specialists, and we are going to have to put those fragments back together again in our own minds and in our families and households and neighborhoods. We need better government, no doubt about it. But we also need better minds, better friendships, better marriages, better communities. We need persons and households that do not have to wait upon organizations, but can make necessary changes in themselves, on their own.

For most of the history of this country our motto, implied or spoken, has been Think Big. I have come to believe that a better motto, and an essential one now, is Think Little. That implies the necessary change of thinking and feeling, and suggests the necessary work. Thinking Big has led us to the two biggest and cheapest political dodges of our time: plan-making and law-making. The lotus-eaters of this era are in Washington, D.C., Thinking Big. Somebody comes up with a problem, and somebody in the government comes up with a plan or a law. The result, mostly, has been the persistence of the problem, and the enlargement and enrichment of the government.

But the discipline of thought is not generalization; it is detail, and it is personal behavior. While the government is "studying" and funding and organizing its Big Thought, nothing is being done. But the citizen who is willing to Think Little, and, accepting the discipline of that, to go ahead on his own, is already solving the problem. A man who is trying to live as a neighbor to his neighbors will have a lively and practical understanding of the work of peace and brotherhood, and let there be no mistake about it—he is *doing* that work. A couple who make a good marriage, and raise healthy, morally competent children, are serving the world's future more directly and surely than any political leader, though they never utter a public word. A good farmer who is dealing with the problem of soil erosion on an acre of ground has a sounder grasp of that problem and *cares* more about it and is probably doing more to solve it than any bureaucrat who is talking about it in general. A man who is willing to undertake the discipline and the difficulty of mending his own ways is worth more to the conservation movement than a hundred who are insisting merely that the government and the industries mend *their* ways.

If you are concerned about the proliferation of trash, then by all means start an organization in your community to do something about it. But before—*and while*—you organize, pick up some cans and bottles yourself. That way, at least, you will assure yourself and others that you mean what you say. If you are concerned about air pollution, help push for government controls, but drive your car less, use less fuel in your home. If you are worried about the damming of wilderness rivers, join the Sierra Club, write to the government, but turn off the lights you're not using, don't install an air conditioner, don't be a sucker for electrical gadgets, don't waste water. In other words, if you are fearful of the destruction of the environment, then learn to quit being an environmental parasite. We all are, in one way or another, and the remedies are not always obvious, though they certainly will always be difficult. They require a new kind of life—harder, more laborious, poorer in luxuries and gadgets, but also, I am certain, richer in meaning and more abundant in real pleasure. To have a healthy environment we will all have to give up things we like; we may even have to give up things we have come to think of as necessities. But to be fearful of the disease and yet unwilling to pay for the cure is not just to be hypocritical; it is to be doomed. If you talk a good line without being changed by what you say, then you are not just hypocritical and doomed; you have become an agent of the disease. Consider, for an example, President Nixon, who advertises his grave concern about the destruction of the environment, and who turns up the air conditioner to make it cool enough to build a fire.

Odd as I am sure it will appear to some, I can think of no better form of personal involvement in the cure of the environment than that of gardening. A person who is growing a garden, if he is growing it organically, is improving a piece of the world. He is producing something to eat, which makes him somewhat independent of the grocery business, but he is also enlarging, for himself, the meaning of food and the pleasure of eating. The food he grows will be fresher, more nutritious, less contaminated by poisons and preservatives and dyes than what he can buy at a store. He is reducing the trash problem; a garden is not a disposable container, and it will digest and re-use its own wastes. If he enjoys working in his garden, then he is less dependent on an automobile or a merchant for his pleasure. He is involving himself directly in the work of feeding people.

If you think I'm wandering off the subject, let me remind you that most of the vegetables necessary for a family of four can be grown on a plot of forty by sixty feet. I think we might see in this an economic potential of considerable importance, since we now appear to be facing the possibility of widespread famine. How much food could be grown in the dooryards of cities and suburbs? How much could be grown along the extravagant right-of-ways of the interstate system? Or how much could be grown, by the intensive practices and economics of the small farm, on so-called marginal lands? Louis Bromfield liked to point out that the people of France survived crisis after crisis because they were a nation of gardeners, who in times of want turned with great skill to their own small plots of ground. And F. H. King, an agriculture professor who traveled extensively in the Orient in 1907, talked to a Chinese farmer who supported a family of twelve, "one donkey, one cow . . . and two pigs on 2.5 acres of cultivated land"—and who did this, moreover, by agricultural methods that were sound enough organically to have maintained his land in prime fertility through several thousand years of such use. These are possibilities that are readily apparent and attractive to minds that are prepared to Think Little. To Big Thinkers—the bureaucrats and businessmen of agriculture—they are quite simply invisible. But intensive, organic agriculture kept the farms of the Orient thriving for thousands of years, whereas extensive—which is to say, exploitive or extractive—agriculture has critically reduced the fertility of American farmlands in a few centuries or even a few decades.

A person who undertakes to grow a garden at home, by practices that will preserve rather than exploit the economy of the soil, has set his mind decisively against what is wrong with us. He is helping himself in a way that dignifies him and that is rich in meaning and pleasure. But he is doing something else that is more important: he is making vital contact with the soil and the weather on which his life depends. He will no longer look upon rain as an impediment of traffic, or upon the sun as a holiday decoration. And his sense of man's dependence on the world will have grown precise enough, one would hope, to be politically clarifying and useful.

What I am saying is that if we apply our minds directly and competently to the needs of the earth, then we will have begun to make fundamental and necessary changes in our minds. We will begin to understand and to mistrust *and to change* our wasteful economy, which markets not just the produce of the earth, but also the earth's ability to produce. We will see that beauty and utility are alike dependent upon the health of the world. But we will also see through the fads and the fashions of protest. We will see that war and oppression and pollution are not separate issues, but are aspects of the same issue. Amid the outcries for the liberation of this group or that, we will know that no person is free except in the freedom of other persons, and that man's only real freedom is to know and faithfully occupy his place—a much humbler place than we have been taught to think—in the order of creation.

But the change of mind I am talking about involves not just a change of knowledge, but also a change of attitude toward our essential ignorance, a change in our bearing in the face of mystery. The principle of ecology, if we will take it to heart, should keep us aware that our lives depend upon other lives and upon processes and energies in an interlocking system that, though we can destroy it, we can neither fully understand nor fully control. And our great dangerousness is that, locked in our selfish and myopic economics, we have been willing to change or destroy far beyond our power to understand. We are not humble enough or reverent enough.

Some time ago, I heard a representative of a paper company refer to conservation as a "no-return investment." This man's thinking was exclusively oriented to the annual profit of his industry. Circumscribed by the demand that the profit be great, he simply could not be answerable to any other demand—not even to the obvious needs of his own children.

Consider, in contrast, the profound ecological intelligence of Black Elk, "a holy man of the Oglala Sioux," who in telling his story said that it was not his own life that was important to him, but what he had shared with all life: "It is the story of all life that is holy and it is good to tell, and of us two-leggeds sharing in it with the four-leggeds and the wings of the air and all green things. . . ." And of the great vision that came to him when he was a child he said: "I saw that the sacred hoop of my people was one of many hoops that made one circle, wide as daylight and as starlight, and in the center grew one mighty flowering tree to shelter all the children of one mother and father. And I saw that it was holy."

Content Considerations

1. In what way is the separation of the civil rights, anti-Vietnam War, and environmental movements artificial?
2. What part does an individual play in the environmental state of the world?
3. What does Berry mean when he says that American society is in a "state of total consumerism"?

The Writer's Strategies

1. How does Berry make it clear that he considers himself one of the guilty?
2. What is Berry arguing for? How does he structure his argument?

Writing Possibilities

1. Explore more fully the notion of thinking little in order to make fewer contributions to the degradation of our environment. What would it mean for you to think little?
2. Berry says that we have given up "private life," the ability to supply our own needs. In a reply, support or refute his argument.
3. Examine again what Berry says about public and private responsibility and action. Expand or modify his comments according to your own position and respond to his statement that "We are going to have to rebuild the substance and the integrity of private life in this country" (paragraph 12).

VINE DELORIA, JR.

The Artificial Universe

Vine Deloria, Jr., a Standing Rock Sioux born in South Dakota, is the author of Custer Died for Your Sins *(1969) and* We Talk, You Listen *(1970) as well as several other books. An Indian spokesman, philosopher, professor, and social activist, Deloria has proposed the notion that the difficulties America has experienced in the latter half of the twentieth century could be remedied if its energies were directed toward the needs of humanity rather than spent sustaining "the god of economics." He advocates a return to the philosophies and practices of "tribalism," or community, as the most practical and beneficial way of living in harmony with one another and the land.*

In "The Artificial Universe," a chapter of We Talk, You Listen, *Deloria compares non-Native American and Native American attitudes and practices toward land, a difference he believes has resulted in supreme artificiality because the non-Native American perspective has dominated. Most of American society is alienated from the land, he says, and that alienation has had disastrous and painful consequences, for in the land are the roots of identity.*

Before you read his essay, examine your attitudes toward land, its uses, and human relationships to it. Write about these attitudes; write also your beliefs about what America has made of itself.

THE JUSTIFICATION FOR TAKING LANDS from Indian people has always been that the needs and requirements of civilized people had to come first. Settlers arriving on these shores saw a virtual paradise untouched by the works of man. They drooled at the prospect of developing the land according to their own dictates. Thus a policy of genocide was advocated that would clear the land of the original inhabitants to make way for towns, cities, farms, factories, and highways. This was progress.

Even today Indian people hold their land at the sufferance of the non-Indian. The typical white attitude is that Indians can have land as long as whites have no use for it. When it becomes useful, then it naturally follows that the land must be taken by whites to put to a better use. I have often heard the remark "what happens to the Indian land base if we decide we need more land?" The fact that Indian rights to land is guaranteed by the Constitution of the United States, over four hundred treaties, and some six thousand statutes seems irrelevant to a people hungry for land and dedicated to law and order.

The major reason why whites have seen fit to steal Indian lands is that they feel that their method of using land is so much better than that of the Indian. It follows that God would want them to develop the land. During the Seneca fight against Kinzua Dam, sympathetic whites would raise the question of Indian legal rights and they would be shouted down by people who said that the Indians had had the land for two hundred years and did *nothing* with it. It would be far better, they argued, to let whites take the land and develop something on it.

From the days of the earliest treaties, Indians were shocked at the white man's attitude toward land. The tribal elders laughed contemptuously at the idea that a man could sell land. "Why not sell the air we breathe, the water we drink, the animals we hunt?" some replied. It was ludicrous to Indians that people would consider land as commodity that could be owned by one man. The land, they would answer, supports all life. It is given to all people. No one has a superior claim to exclusive use of land, much less does anyone have the right to fence off a portion and deny others its use.

In the closing decades of the last century, Indian tribes fought fiercely for their lands. Reservations were agreed upon and tribes held a fragment of the once expansive hunting grounds they had roamed. But no sooner had Indians settled on the reservations, than the government, ably led by the churches, decided that the reservation areas should be divided into tiny plots of land for farming purposes. In many reservation areas it was virtually impossible to farm such lands. The situation in California was so desperate that a report was issued denouncing the government land policy for Indians. The report contained such detrimental material exposing the vast land swindles that it was pigeonholed in the Senate files and *has never been released and cannot be obtained today, nearly a century later!!!*

Tribe after tribe succumbed to the allotment process. After the little plots of land were passed out to individual Indians, the remainder, which should have been held in tribal hands, was declared surplus and opened to settlement. Millions of "excess" acres of lands were thus casually transferred to federal title and given to non-Indian settlers. Churches rushed in and grabbed the choice allotments for their chapels and cemeteries, and in some cases simply for income-producing purposes. They had been the chief advocates of allotment—on the basis that creating greed and selfishness among the Indians was the first step in civilizing them and making them Christians.

For years the development of the land did make it seem as if the whites had been correct in their theory of land use. Cities were built, productive farms were created, the wilderness was made safe, and superhighways were built linking one portion of the nation with the others. In some areas the very landscape was changed as massive earth-moving machines relocated mountains and streams, filled valleys, and created lakes out of wandering streams.

Where Indian people had had a reverence for the productiveness of the land, whites wanted to make the land support their way of life whether it was suited to do so or not. Much of San Francisco Bay was filled in and whole areas of the city were built upon the new land. Swamps were drained in the Chicago area and large portions of the city were built on them. A great portion of Ohio had been swamp and grassland and this was drained and farmed. Land was the great capital asset for speculation. People purchased apparently worthless desert land in Arizona, only to have the cities grow outward to their doorstep, raising land prices hundreds of percents. Land worth pennies an acre in the 1930s became worth thousands of dollars a front foot in the 1960s.

The rapid increase of population, technology, and capital has produced the present situation where the struggle for land will surpass anything that can be conceived. We are now on the verge of incredible development of certain areas into strip cities that will extend hundreds of miles along the coasts, major rivers, and mountain ranges. At the same time, many areas of the country are steadily losing population. Advanced farming techniques allow one man to do the work that several others formerly did, so that the total population needed in agricultural states continues to decline without a corresponding decline in productivity.

The result of rapid industrialization has been the creation of innumerable problems. Farm surpluses have lowered prices on agricultural products so that the federal government has had to enter the marketplace and support prices to ensure an adequate income for farmers. Farm subsidies are no longer a small business. In nine wheat and feed grain-producing counties in eastern Colorado in 1968, $31.4 million was given in farm subsidies. In all of Colorado, $62.8 million was given in 1968 to support farmers. This was a state with a declining farm population. Under the Agricultural Stabilization Conservation Service, some $3.5 billion was paid out in 1968, $675 million paid to 33,395 individual farmers as farm "income maintenance," some receiving amounts in excess of $100,000.

For much of the rural farm areas the economy, the society, and the very structure of life is completely artificial. It depends wholly upon government welfare payments to landowners, a thinly disguised guaranteed annual income for the rich. If the payments were suddenly cut off, millions of acres would become idle because it would not pay to farm them and there would be no way to live on them without income. Our concern for the family farm and the rural areas is thus a desperate effort to maintain the facade of a happy, peace-loving nation of farmers, tillers of the soil who stand as the bastion of rugged individualism.

If rural areas have an artificial economy, the urban areas surpass them in everything. Wilderness transformed into city streets, subways, giant buildings, and factories resulted in the complete substitution of the real world for the artificial world of the urban man. Instead of woods, large buildings rose. Instead of paths, avenues were built. Instead of lakes and streams, sewers and fountains were created. In short, urban man lives in a world of his own making and not in the world that his ancestors first encountered.

Surrounded by an artificial universe where the warning signals are not the shape of the sky, the cry of the animals, the changing of seasons, but simply the flashing of the traffic light and the wail of the ambulance and police car, urban people have no idea what the natural universe is like. They are devoured by the goddess of progress, and progress is defined solely in terms of convenience within the artificial technological universe with which they are familiar. Technological progress totally defines the outlook of most of America, so that as long as newer buildings and fancier roads can be built, additional lighting and electric appliances can be sold, and conveniences for modern living can be created there is not the slightest indication that urban man realizes that his artificial universe is dependent on the real world.

Milk comes in cartons, and cows are so strange an animal that hunters from large cities kill a substantial number of cattle every year on their annual hunting orgies. This despite the fact that in many areas farmers paint the word COW on the side of their animals to identify them. Food comes in plastic containers highly tinged with artificial sweeteners, colors, and preservatives. The very conception of plants, growing seasons, rainfall, and drought is foreign to city people. Artificial criteria of comfort define everything that urban areas need and therefore dominate the producing rural areas as to commercial products.

The total result of this strange social order is that there has been total disregard for the natural world. The earth is considered simply another commodity used to support additional suburbs and superhighways. Plant and animal life are subject to destruction at the whim of industrial development. Rivers are no more than wasted space separating areas of the large cities. In many areas they are open sewers carrying off the millions of tons of refuse discarded by the urban consumer.

The Indian lived with his land. He feared to destroy it by changing its natural shape because he realized that it was more than a useful tool for exploitation. It sustained all life, and without other forms of life, man himself could not survive. People used to laugh at the Indian respect for smaller animals. Indians called them little brother. The Plains Indians appeased the buffalo after they had slain them for food. They well understood that without all life respecting itself and each other no society could indefinitely maintain itself. All of this understanding was ruthlessly wiped out to make room for the white man so that civilization could progress according to God's divine plan.

In recent years we have come to understand what progress is. It is the total replacement of nature by an artificial technology. Progress is the absolute destruction of the real world in favor of a technology that creates a comfortable way of life for a few fortunately situated people. Within our lifetime the difference between the Indian use of land and the white use of land will become crystal clear. The Indian lived with his land. *The white destroyed his land. He destroyed the planet earth.*

Non-Indians have recently come to realize that the natural world supports the artificial world of which they are so fond. Destruction of nature will result in total extinction of the human race. There is a limit beyond which man cannot go in reorganizing the land to suit his own needs. Barry Commoner, Director of the Center for the Biology of Natural Systems at Washington University in St. Louis, has been adamant about the destruction of nature. He told a Senate Subcommittee on Intergovernmental Affairs that the present system of technology would destroy the natural capital, the land, air, water, and other resources within the next fifty years. He further pointed out that the massive use of inorganic fertilizers may increase crop yields for a time but inevitably changes the physical character of the soil and destroys the self-purifying capability of the rivers. Thus the rivers in Illinois have been almost totally destroyed, while the nitrate level of rivers in the Midwest and California has risen above the safe level for use as drinking water.

A conference on pollution in Brussels outlined the same problem and had a much earlier deadline in mind. Scientists there predicted the end of life on the planet within a minimum of thirty-five years. Elimination of the oxygen in the atmosphere was credited to jet engines, destruction of oxygen-producing forests, and fertilizers and pesticides such as DDT that destroy oxygen-producing microorganisms. Combining all of the factors that are eliminating the atmosphere, the scientists could not see any future for mankind. Realization of the situation is devastating.

Even where forests and plant life exist, the situation is critical. In southern California millions of trees are dying from polluted air. A recent aerial survey by the Forest Service in November, 1969, showed 161,000 acres of conifers already dead or dying in southern California. The situation has been critical since 1955, when residents of the area discovered trees turning yellow, but no one even bothered to inquire until 1962. In the San Bernardino forest 46,000 acres of pine are already dead and close to 120,000 acres more are nearly dead.

With strip cities being developed that will belch billions of tons of pollutants skyward every day the pace will rapidly increase so that optimistic projections of fifty to a hundred years more of life must be telescoped to account for the very rapid disappearance of plant life by geometrically increasing pollution. The struggle for use of land has polarized between conservationists, who understand that mankind will shortly become extinct, and developers, who continue to press for immediate short-term financial gains by land exploitation.

The Bureau of Land Management, alleged guardian of public lands, has recently been involved in several controversial incidents with regard to its policies. In one case Bureau officials reversed themselves and acceded to Governor Jack Williams' request to transfer 40,000 acres of federal range to the state "so the land could be leased to ranchers." Stewart Udall, the great conservationist, upheld the original decision of the Bureau of Land Management because he thought that federal lands closer to cities could be obtained for development purposes. The overall effect of government policies on land is to silently give the best lands to state or private development without regard for the conservation issue or the public welfare.

We can be relatively certain that the federal and state governments will not take an objective view of land use. Agencies established to protect the public interest are subject to heavy political pressure to allow land to slip away from their trusteeship for short-sighted gains by interest groups. This much is certain: at the moment there is not the slightest chance that mankind will survive the next half century. The American public is totally unconcerned about the destruction of the land base. It still believes in the infallibility of its science, technology, and government. Sporadic and symbolic efforts will receive great publicity as the future administrations carefully avoid the issue of land destruction. Indian people will find their lands under continual attack and will probably lose most of them because of the strongly held belief that progress is inevitable and good.

With the justification of progress supporting the destruction of Indian tribes and lands, the question of results becomes important. Four hundred years of lies, cheating, and genocide were necessary in order for American society to destroy the whole planet. The United States government is thus left without even the flimsiest excuse for what has happened to Indian people, since the net result of its machinations is to destroy the atmosphere, thus suffocating mankind.

There is a grim humor in the situation. People used to make fun of Indians because of their reverence for the different forms of life. In our lifetime we may very well revert to panicked superstition and piously worship the plankton of the sea, begging it to produce oxygen so that we can breathe. We may well initiate blood sacrifices to trees, searching for a way to make them productive again. In our lifetime, society as a whole will probably curse the day that white men landed on this continent, because it will all ultimately end in nothingness.

Meanwhile, American society could save itself by listening to tribal people. While this would take a radical reorientation of concepts and values, it would be well worth the effort. The land-use philosophy of Indians is so utterly simple that it seems stupid to repeat it: man must live with other forms of life on the land and not destroy it. The implications of this philosophy are very far-reaching for the contemporary political and economic system. Reorientation would mean that public interest, indeed the interest in the survival of humanity as a species, must take precedent over special economic interests. In some areas the present policies would have to be completely overturned, causing great political dislocations in the power structure.

In addition to cleaning up streams and rivers and cutting down on air pollution, a total change in land use should be instituted. Increase in oxygen-producing plants and organisms should be made first priority. In order to do this, vast land areas should be reforested and bays should be returned to their natural state. At present, millions of acres of land lie idle every year under the various farm programs. A great many more acres produce marginal farming communities. Erosion and destruction of topsoil by wind reduces effectiveness of conservation efforts. All of this must change drastically so that the life cycle will be restored.

Because this is a total social problem and the current solutions such as sporadic national and state parks and soil banks are inadequate answers, a land-use plan for the entire nation should be instituted. The government should repurchase all marginal farmlands and a substantial number of farms in remote areas. This land should be planted with its original growth, whether forest or grassland sod. The entire upper midwest plains area of the Dakotas and Montana and upper Wyoming should become open-plains range with title in public hands. Deer, buffalo, and antelope should gradually replace cattle as herd animals. Outside of the larger established towns, smaller towns should be merely residences for people employed to redevelop the area as a wilderness.

Creeks and streams should be cleared of mining wastes and their banks replanted with bushes and trees. The Missouri should be returned to its primitive condition, except where massive dams have already been built. These should remain primarily as power-generating sites without the corresponding increase in industry surrounding them. Mining and tourism should be cut to a minimum and eventually prohibited. The present population could well be employed in a total conservation effort to produce an immense grasslands filled with wildlife.

The concept is not impossible. Already a rancher in Colorado has tried the idea of grazing wild animals and beef cattle on his range with excellent results. Tom Lasater has a 26,000-acre ranch east of Colorado Springs, Colorado. He has pursued a no-shooting, no-poisoning, no-killing program for his land. There has already been a substantial increase in game animals, primarily mule deer and antelope, without any disturbance to his beef animals. Lasater first decided to allow wild animals to remain on his land when his foreman remarked, after the prairie dogs had been exterminated, that the grass always grew better when the prairie dogs had been allowed to live on the land.

The result of Lasater's allowing the land to return to its primitive state has been the notable decrease of weeds. Lasater feels that the smaller animals, such as gophers, ground squirrels, badgers, and prairie dogs, that dig holes all provided a better means of aerating the ground and introducing more oxygen into it than modern farming methods of periodically turning the sod by plowing. All of the wildlife use on the land produced better grazing land and reduced the danger of overgrazing in a remarkable way. The fantastic thing about Lasater's ranch is that it returns almost double the income from beef cattle, because of the improved conditions of the soil and the better grasses, than would the average ranch of comparable acreage using the so-called modern techniques of ranching.

The genius of returning the land to its original animals is that the whole program cuts down on labor costs, maintains fertility far better than modern techniques, increases environmental stability, and protects the soil from water and wind erosion. The net result is that the land supports much more life, wild and domestic, and is in better shape to continue to support life once the program is underway. Returning the major portion of the Great Plains to this type of program would be the first step in creating a livable continental environment. But introduction of this kind of program would mean dropping the political platitudes of the rancher and farmer as America's last rugged individualists, admitting that they are drinking high on the public trough through subsidies, and instituting a new kind of land use for the areas involved.

In the East and Far West, all land that is not immediately productive of agricultural products for the urban areas should be returned to forest. This would mean purchasing substantial acreage in Wisconsin, Ohio, Michigan, New York, New Jersey, and Pennsylvania and planting new forests. With the exception of settled urban areas, the remainder of those states would probably become vast woods as they were originally. Wildlife would be brought in to live on the land since it is an irreplaceable part of the forest ecology. With the exception of highspeed lanes for transportation facilities, the major land areas of the East Coast would become forest and woodlands. The presence of great areas of vegetation would give carbon-dioxide-consuming plants a chance to contribute to the elimination of smog and air pollution.

The social structure of the East would have to change considerably. In New York City the number of taxicabs is limited because unimpeded registration of cabs would produce a city so snarled with traffic that there would be no transportation. In the same manner anyone owning a farm of substantial acreage would have to be licensed by the state. The rest of the land would become wilderness with a wildlife cycle supporting the artificial universe of the cities by producing relatively clean air and water. The countless millions now on welfare in the eastern cities could be resettled outside the cities with conservation jobs and in retirement towns to ensure that the green belt of oxygen-producing plants would be stabilized.

In the coal mining states strip mining would be banned and a substantial number of people could be employed in work to return the land to its natural state. Additional people could harvest the game animals and the food supply would partially depend upon meat from wild animals instead of DDT-bearing beef animals. Mines would be filled in and vegetation planted where only ugly gashes in the earth now exist. People disenchanted with urban society would be allowed to live in the forests with a minimum of interference. Any who might want to live in small communities and exist on hunting and fishing economies would be permitted to do so.

On the seacoasts, pollution should be cut to a minimum. Where there are now gigantic ports for world shipping, these would be limited to a select few large enough to handle the trade. Others would have to become simply ports for pleasure boats and recreation. Some of the large commercial centers on both coasts would have to change their economy to take into account the absence of world trade and shipping. Beaches would have to be cleaned and set aside as wilderness areas or used by carefully selected people as living areas. Lobster beds, oyster beds, and areas that used to produce edible seafood would have to be returned to their original condition.

The pollution crisis presents the ultimate question on tribalism. If mankind is to survive until the end of the century, a substantial portion of America's land area must be returned to its original state of forest and grasslands. This is fundamentally because these plants produce oxygen and support the life cycle at the top of which is man. Without air to breathe it is ridiculous to speak of progress, culture, civilization, or technology. Machines may be able to live in the present environment, but it is becoming certain that people cannot.

By returning the land to its original state, society will have to acknowledge that it can no longer support two hundred million people at an artificial level of existence in an artificial universe of flashing lights and instantaneous communications. To survive, white society must return the land to the Indians in the sense that it restores the land to the condition it was in before the white man came. And then to support the population we now have on the land that will be available, a great number of people will have to return to the life of the hunter, living in the forests and hunting animals for food.

Whenever I broach this subject to whites, they cringe in horror at the mere prospect of such a development. They always seem to ask how anyone could consider returning to such a *savage and unhappy state,* as the government reports always describe Indian life. Yet there is a real question as to which kind of life is really more savage. Does the fact that one lives in a small community hunting and fishing for food really indicate that one has no sensitive feelings for humanity? Exactly how is this kind of life primitive when affluent white hunters pay thousands of dollars every fall merely for the chance to roam the wilderness shooting at one another in the hopes of also bringing down a deer?

In 1967 I served on the Board of Inquiry for Hunger and Malnutrition in the United States. We discovered that a substantial number of Americans of all colors and backgrounds went to bed hungry every night. Many were living on less than starvation diets and were so weakened that the slightest sickness would carry them off. The black children in the Mississippi delta lands were eating red clay every other day to fill their stomachs to prevent hunger pains. Yet the Agricultural Department had millions of tons of food in giant storehouses that went undistributed every year. Is this type of society more savage than living simply as hunters and fishermen? Is it worth being civilized to have millions of people languishing every year for lack of food while the warehouses are filled with food that cannot be distributed?

Last Christmas in California a federal judge, disgusted at the snarls of red tape that prevented distribution of food to hungry people, ordered a warehouse opened and the food distributed in spite of the pleas of bureaucrats that it was against regulations. In the field of hunger alone the government had better act before hungry people take the law into their own hands.

For years Indian people have sat and listened to speeches by non-Indians that gave glowing accounts of how good the country is now that it is developed. We have listened to people piously tell us that we must drop everything Indian as it is impossible for Indians to maintain their life style in a modern civilized world. We have watched as land was stolen so that giant dams and factories could be built. Every time we have objected to the use of land as a commodity, we have been told that progress is necessary to the American way of life.

Now the laugh is ours. After four centuries of gleeful rape, the white man stands a mere generation away from extinguishing life on this planet. Granted that Indians will also be destroyed—it is not because we did not realize what was happening. It is not because we did not fight back. And it is not because we refused to speak. We have carried our responsibilities well. If people do not choose to listen and instead overwhelm us, then they must bear the ultimate responsibility.

What is the ultimate irony is that the white man must drop his dollar-chasing civilization and return to a simple, tribal, game-hunting, berry-picking life if he is to survive. He must quickly adopt not just the contemporary Indian world view but the ancient Indian world view to survive. He must give up the concept of the earth as a divisible land area that he can market. The lands of the United States must be returned to public ownership and turned into wilderness if man is to live. It will soon be apparent that one man cannot fence off certain areas and do with the land what he will. Such activity will be considered too dangerous to society. Small animals and plants will soon have an equal and perhaps a greater value for human life than humans themselves.

Such a program is, of course, impossible under the American economic and political system at the present time. It would interfere with vested economic interests whose motto has always been "the public be damned." Government policy will continue to advocate cultural oppression against Indian tribes, thinking that the white way of life is best. This past year, five powerful government agencies fought the tiny Lummi tribe of Western Washington to prevent it from developing a bay that the tribe owned as a sealife sanctuary. The agencies wanted to build massive projects for commercial use on the bay, the Indians wanted it developed as a conservation area restoring its original food-producing species such as fish, clams, and oysters. Fortunately, the tribe won the fight, much to the chagrin of the Army Corp of Engineers, which makes a specialty of destroying Indian lands.

The white man's conception of nature can be characterized as obscene, but that does not even begin to describe it. It is totally artificial and the very existence of the Astrodome with its artificial grass symbolizes better than words the world visualized by the non-Indian. In any world there is an aspect of violence that cannot be avoided; Nature is arbitrary and men must adjust to her whims. The white man has tried to make Nature adjust to his whims by creating the artificial world of the city. But even here he has failed. Politicians now speak reverently of corridors of safety in the urban areas. They are main lines of transportation where your chances of being robbed or mugged are greatly reduced. Everywhere else there is indiscriminate violence. Urban man has produced even an artificial jungle, where only the fittest or luckiest survive.

With the rising crime rate, even these corridors of safety will disappear. People will only be able to go about in the urban areas in gangs, tribes if you will. Yet the whole situation will be artificial from start to finish. The ultimate conclusion of American society will be that even with respect to personal safety it was much safer and more humane when Indians controlled the whole continent. The only answer will be to adopt Indian ways to survive. For the white man even to exist, he must adopt a total Indian way of life. That is really what he had to do when he came to this land. It is what he will have to do before he leaves it once again.

Content Considerations

1. What is the basic difference between how settlers and Native Americans viewed the land?
2. What specific objections does Deloria make to how the land is used?
3. Summarize the specifics of the land-use policy Deloria proposes. Upon what general principle does he base them?

The Writer's Strategies

1. What is Deloria's tone? How is irony a part of his approach?
2. What does Deloria mean by his title "The Artificial Universe"? What examples of it does he give?

Writing Possibilities

1. Build your own argument for or against Deloria's proposal for land use and its underlying principle.
2. Respond to Deloria's account of how land went from Indian control to white. How does it fit your own perspective? Write a reply to Deloria.
3. In an essay, write your response to the notion of an artificial universe. Do you think you are living in such a universe? What is your relationship with the land?
4. With a classmate, discuss what would happen if Deloria's land-use policy were adopted. Consider economic, social, political, and personal consequences; write your predictions.

— JESSE L. JACKSON, SR. —

Making Lions Lay Down With Lambs

Faced with a multitude of environmental problems, an individual may feel overwhelmed, powerless, and without control. The central problem is a global one—the health of the Earth must be maintained, and in ways improved, if the world and its inhabitants are to thrive or, according to some predictions, survive. In the last few years many books, magazine articles, and television broadcasts have addressed the role of the individual in preserving the planet.

Individuals can recycle papers, aluminum and glass; they can precycle by buying products that do not harm the environment. They can reduce their consumption of products and their use of energy; they can develop the habit of considering the environmental consequences of each of their actions and decisions.

In the following the Reverend Jesse L. Jackson, Sr. speaks to the role of the individual in creating harmony between humans and Earth. Before you read his essay, taken from his comments on Earth Day, 1990, and published in the environmental magazine Trilogy, *write for a few minutes about what it means to be empowered.*

TWENTY-FIVE YEARS AGO, we marched from Selma to Montgomery for the right to vote. We marched for the right to public accommodations, for an end to discrimination, for the right to education and equal access.

Today we must fight for the most fundamental right of all—the right to breathe free. This is the great American promise, the right to breathe. "Give me your poor, your tired, your huddled masses, who yearn to breathe free." Yet when we marched from Selma, we took this right for granted. We did not realize that without it, all other rights become meaningless. Maybe we thought the environment was a luxury, the concern of the rich. Maybe we wrote off the environmental struggle because we were not a part of that movement. But you don't need to join the Sierra Club to be an environmentalist. Everyone who breathes is an environmentalist. Don't wait for an environmental organization to give you back your clean air—go out and fight for it yourselves.

You cannot separate environment from empowerment. Toxic waste dumps are put in communities where people are the poorest, the least organized, the least registered to vote. If you are poor, you are a target for toxic waste. If you are unregistered to vote, you are a target. You must organize and register to fight back. It pains me greatly to think that so much blood was shed for the right to vote, and yet so many are unregistered. Dr. King died so that you might vote. Nelson Mandela sat in jail for 27 years in the fight for freedom. He is out of jail, but he is not free. He cannot vote. You cannot cheer for Mandela if you are not registered to vote, registered to protect the environment.

You cannot separate environment from empowerment. You cannot breathe free if you are choking on pesticides. You cannot breathe free if your water is contaminated. It's not just a matter of science or legislation. It's a matter of morality. We have an ethical obligation to care for the earth. The earth is the Lord's. We did not make the earth, and no one has the right to destroy what the Lord gave to us all. Corporations do not make air or water. If corporations made some air, then they could negotiate it. If they made some soil, then they could negotiate it. But they do not make it and they are destroying what God has made. Destroying the earth is a sin. The wages of sin are death. We need a zero-based pollution policy—no toleration for toxics, carcinogens, and poison.

Some time ago, they had a nuclear accident at Chernobyl in the Soviet Union. The media and the politicians here got up on their podiums and said, "The Russians have a problem, but they aren't being open about it. They are not telling the world about it, and therefore they are not as democratic and excellent as we are." Have we forgotten Three Mile Island? So the pundits crowed that the Russians had a problem. But on the third day, the wind blew. And then Europe had a problem. And on the fourth day, the wind kept blowing and cows in Oregon had a problem.

We live on one spaceship earth. We cannot allow these false divisions to separate us, for toxic waste does not know the boundaries of nation states. It makes no sense to ban pesticides in this country, export them to poor third world countries, and then import food back into this country with pesticides on it. These divisions are an illusion. It is as ridiculous as when they tell you that rows one, two, and three are smoking, and rows four, five and six are non-smoking. It's all the same air. If you are in that airplane, you are either smoking or getting smoked. We live in one spaceship earth.

Yes, they put toxic waste in poor communities. But those who live in the rich neighborhoods can take no consolation from this. For as surely as the wind blows, that waste ends up on the other side of town. As surely as water flows, groundwater contamination hurts rich and poor, black and white alike. You cannot hide from pollution.

They say that pollution and waste ends up in minority communities. And it does. But in the world, the minority is the majority. What do I mean? I mean that half the world is Asian, and one half of them are Chinese. One eighth of the world is African, and a quarter of them are Nigerian. Most of the world is yellow, black and brown, poor, female, young, and does not speak English. When Bush and Gorbachev meet, they represent one eighth of the human race. Seven-eighths of the world is not at that meeting, and it won't be long before the rest of the world stops waiting to hear what that one eighth has planned for them, and for their environment. We must stop letting them play these color games and fight for respect for our earth.

I now understand something in the Bible which I never understood before. The Bible tells us that we will have peace when lions and lambs lay down together. This always seemed like a contradiction to me. After all, lions have a tendency to eat lambs, and lambs, even the most timid, have enough sense to run away from lions. It sounds like a way of saying that peace is a long way off. But the Bible tells us that there will be peace in the valley. What is it, what force, could make lions and lambs form a coalition? How can the lion and the lamb find common ground? Now I understand what the Bible meant, for neither the lion nor the lamb want acid rain on their backs. When the lion and the lamb know their common survival is threatened, then they will form a coalition and we will have peace in the valley.

You know they say honeybees do not have brains. They have a buzz, or a stinger, or an instinct, but they do not have sense. Yet honeybees keep on making other honeybees, and there are no slum honeybee hives. Honeybees don't have Ph.D. or master degrees but there are no homeless honeybees. How do they manage? They follow natural law. And let me remind you that you cannot break natural law. You only prove it. If you jump off of a 50-story building, you do not break natural law. You break your neck. When honeybees go into a flower and take its nectar, they leave pollen behind. That is natural law. They know if they don't, the flower will die and other honeybees will die. When corporations suck the nectar from communities and leave behind pollution instead of pollen, they are killing the community. They are going against natural law. Corporations cannot be allowed to suck the pollen out of the community and deposit their profits elsewhere. Let those who make the mess clean it up. Let them invest in where they take from, create local jobs for local people and better health for the community. If they will not do it voluntarily, then levy a tax to put pollen back into the community. We can make saving the environment job and health intensive.

For too long, people have been told that they had to keep quiet about the environment in order to have a job. They have been handed a deal by corporations and governments to sacrifice their health for employment. But what is the use of a good job and a house, if they build a toxic waste dump behind your house and your children get asthma and blood poisoning? What is the use of working 20 years, and then spending your retirement money on chemotherapy? The tradeoff between jobs and health is wrong. It is a bad deal. We have the right to both a job and good health. We deserve environmental as well as economic security for our children. When a man beats a woman and treats her badly, and then tells her that at least she has a place to stay, that is morally wrong and an unacceptable tradeoff. Women have the right to respect and a place to rest their head. We all have the right to respect, to employment, and environment.

We need to act to save the environment. We have the tools at hand. If you want a clean environment, vote about it. If you don't want a toxic dump in your backyard, vote about it. We've done the research. We are doing the education. We will sit down with the corporations and do the negotiation. If we are successful, we will all participate in the celebration but if not, we do not draw back from confrontation. It is time to end the tradeoff between jobs and health. It is time to give respect to the Lord's earth and all that is on it. It is time to fight for the right to breathe free.

Content Considerations

1. In what sense do all people "live in one spaceship earth"?
2. How does empowerment relate to the environment?
3. What environmental responsibility does Jackson assign individuals? Corporations? Government?

The Writer's Strategies

1. What is Jackson trying to persuade the reader to believe and do? How does he try to persuade?
2. What metaphors, analogies, and allusions does Jackson use? What relationships do they clarify?

Writing Possibilities

1. In an essay, form and explain your own definition of an environmentalist.
2. Write a reply to Jackson supporting or challenging his assertions and conclusions.

3. With a classmate, discuss what individual actions would help provide a healthy and safe environment in the workplace and in the home. Then explain your conclusions, addressing the issue of how much individuals can actually achieve.

ELLEN GOODMAN

The Killer Bee Syndrome

Awarded a Pulitzer Prize for commentary in 1980, Ellen Goodman writes a column that is carried by more than 400 newspapers across America. She also has published one work of nonfiction and several collections of her syndicated columns. These columns range among topics, but most of them dwell upon modern life, its obstacles and confusions, its dangers and joys.

In the following piece, originally published in July 1981, Goodman ponders a situation she calls the "killer bee syndrome." She refers to the accidental release in the 1970s of intensely aggressive bees, which have been traveling steadily northward from South America.

Before you read her column, work with your classmates to compile a list of mistakes that have had frightening consequences. Who bears responsibility for such mistakes?

BOSTON—The tree outside my front door is safe for now. It stands there, half its leaves brown and nibbled, looking like a banquet table deserted by guests in the middle of the salad course.

That's pretty much what happened. The guests—rude, greedy creatures—were gypsy moth caterpillars. They had just eaten their way in from the suburbs and begun on my tree when nature called them to their cocoon.

This time I was lucky. But looking at the leaves I remember the original gypsy moth immigrants, whose descendants have decimated the Northeast. They were brought to this country from France by a scientist who thought they would produce silk.

Anybody could make a mistake.

While my tree stands in Boston, the people in California are dealing with another imported pest, the Mediterranean fruit fly. The planes there sprayed people as well as land. The public's outrage there was palpable.

How did this new outbreak occur? In part because the 200,000 sterile fruit flies released to mate turned out to be fertile. "We got burned on a shipment from Peru," said a state official.

Sterile. Fertile. Anybody could make a mistake.

On both sides of the country, then, we have examples of that fun couple, scientific method and human error. Call it the Killer Bee Syndrome if you will. Someone sets out to breed a bee that will produce lots of honey; some technician unlatches the cage with the killer bees.

Anybody who could make a mistake usually does.

The most scientific system in the world with fail-safes and triple checks, and computer backups is devised by people, run by people, used and misused and screwed up by people. To put it as simply as possible, the more dangerous the science, the more terrifying our fallibility.

This is something I would like to see cross-stitched on the walls when the leaders of all the major industrial countries get together in Ottowa. One of the chief subjects on the agenda is that ultimate killer bee, nuclear know-how.

For decades, America has been the chief exporter of the most dangerous scientific species. Not only have we built bombs and used them, we have passed out most of the nuclear information for what we used to call Atoms for Peace. We've exported uranium, exported sixty small research reactors, and lent $5 billion for seventy commercial reactors for energy production.

While the construction of nuclear plants has slowed in this country—by the sheer tug of public protest over safety issues—we continue to sell overseas, the same way we sell banned chemicals. We worry about our own technicians, our own safety standards, our own hazardous waste. But we regard "foreign" problems as if they had some private stock of air and water.

More to the point, each "peaceful" nuclear reactor produces the raw material for nuclear weapons. One 1000-megawatt reactor produces enough plutonium for more than twenty bombs a year. That is why Israel bombed Iraq's French-made plant.

The Reagan attitude toward countries that want to develop the bomb was expressed best in his campaign quote, "I just don't think it's any of our business."

That's changing now, but slowly. In preparation for Ottawa, the administration has come out with a stronger statement against extending the nuclear "family." But it also promised to remain a "clearly reliable and credible supplier" of nuclear technology for peaceful purposes.

The problem is that no one knows how to control the spread of nuclear weapons while expanding the market for nuclear energy. Even the Reagan government offers only some vague idea about monitoring or retrieving the plutonium from foreign countries.

In short, we still seem to be dealing with nuclear bombs the way we've dealt with handguns. I can almost see the bumper sticker: A-BOMBS DON'T KILL PEOPLE. PEOPLE KILL PEOPLE.

But the more bombs we build and store, and the greater the number of people involved, the greater the risk. The risk is multiplied by each country—with its own enemies and instabilities, its own leaders and technicians—that gets nuclear knowledge.

Surely, nobody wants to blow up the world. But, as I can tell from my tree, anybody can make a mistake. 20

Content Considerations

1. What relation exists between "scientific method and human error"?
2. Why is it unwise to treat other countries as if "they had some private stock of air and water"?
3. At what point might all precautions and backups fail?

The Writer's Strategies

1. How do the beginning and ending paragraphs of this essay connect?
2. How does using several examples strengthen Goodman's argument?

Writing Possibilities

1. Focus on and then expand the tenth paragraph of this piece (beginning "The most scientific . . . "). To what extent do you agree with Goodman?
2. Explain your own position on the export of nuclear technology and materials to other countries as well as their use in America.
3. With classmates, discuss ways humans have introduced species into new environments to solve a problem, only to find a more serious problem has been created as a result. Write your discoveries and conclusions.

ALBERT L. HUEBNER

The Medfly Wars

In the early 1980s California experienced a proliferation of Mediterranean fruit flies (Medflies), pests that damaged millions of dollars worth of fruit crops. Biological control was attempted by the release of more than 100,000 supposedly sterile Medflies, but a mistake was made—the Medflies were fertile and more crops were lost. The federal government then threatened to impose a quarantine on all fruit produced in California in order to confine the epidemic.

In response the state of California implemented a spraying program, using the pesticide malathion to eradicate the Medfly. The safety of malathion to humans, domestic animals, and forms of wildlife is and has been in dispute.

Spraying recurred during the 1980s, the last bout beginning after one Medfly was seen in July 1989. The following article by California writer Albert L. Huebner focuses on that last bout of spraying—how it was conducted and how people responded. Before you read the article, published in The Progressive *in June 1990, write down the concerns you would have if helicopters were going to fly over your neighborhood once a week, spraying gallons of pesticide on everything below.*

IT'S USUALLY QUITE LATE IN THE EVENING when the jarringly noisy helicopters, flying in military formation, appear over the rooftops of the target area. They begin dropping their poisons on the people huddled inside their homes below, the attack often lasting into the early hours of the morning.

The scene isn't a flashback to Vietnam in the 1960s. It's been happening for months over dozens of cities, covering 400 square miles, in Los Angeles and Orange counties. The onslaught began with discovery of a single Mediterranean fruit fly, or Medfly, in a park near Dodger Stadium last July 20. Since then, more than sixty cities, large and small, have been sprayed with the pesticide malathion, some as many as five times by now.

And plans are to increase the spraying rate. Infested areas were sprayed every twenty-one days during the winter months. As the weather warms, people in the area face the prospect of being sprayed every fifteen days, then every seven days.

Not surprisingly, the spraying has caused widespread concern. The fear and outrage have produced a network of grassroots organizations, some new, some veterans of earlier spraying programs, joined together under the banner of the Coalition Against Urban Spraying (CAUSE).

"Public outrage is growing and growing and growing," says Daniel Bender, president of Safe Alternatives for Fruit Fly Eradication. He observes that "new groups keep popping up out of nowhere" to demand that their local governments and their state representatives take action against the spraying. According to Bender, "I don't even have to come up with ideas for demonstrations any more. They just call me up and say, 'Have you heard about this demonstration?' "

"I have never seen anything like it in my whole life," comments Jacqui Dole, organizer of Orange County Citizens Against Urban Spraying, referring to a unanimous vote by one city council to reverse a previous position and oppose further spraying. "The council is all superconservative Orange County—if they can change, anything can happen."

The reasons for intense feeling about the spraying are not difficult to understand. "The one thing you have control over is your home, your yard, your family," says Dole. "The spraying has been an attack against the sanctity of the home. It has struck a deep psychological chord."

Patty Prickett, of Residents Against Spraying of Pesticides (RASP), adds, "It's being locked in your house with all the windows shut and hearing the helicopters flying over and knowing you can't go out. It's horrible, this feeling of being trapped in your house for no reason." For many members of Prickett's group, this isn't the first time they've encountered the problem: RASP was originally formed to oppose malathion spraying during the 1985 Mexican fruit-fly eradication program.

Some residents can't remain "locked in" because they work at night. And some people are, in effect, locked out. Those especially sensitive to the chemicals, or whose children develop a range of nasty symptoms after each spraying, are forced to leave their homes completely, for as long as possible, just before the helicopters come. While the psychological effects, and the sheer noise, are oppressive, this direct physical impact on health is by far the greatest concern of people in the target areas.

Repeated assurances by state officials about the safety of the sprayed mixture, consisting of malathion and a "bait" of protein and corn syrup, give no comfort to people like Robina Suwol. After the first aerial spraying of her community, her son developed swollen glands, a sore throat, and a fever. When the symptoms recurred after the second spraying, she and her husband took him out of the area the evening of the third. The next day, they took the boy to a park to play—a park in the sprayed area—and that night the flu-like symptoms returned.

Julie Thornton left a window slightly open on a warm evening. When the helicopters suddenly appeared and began spraying—it isn't always clear at what time, or even on which evenings, they will spray a given community—she remembered the window and quickly shut it. Later that evening she experienced "a sickeningly sweet taste" in her mouth. At 5:30 A.M. the following morning, she was awakened by her eighteen-month-old son thrashing about in his crib. When she lifted him he immediately vomited, and so began a five-day siege of a violent illness. The child's physician suggested malathion as the possible culprit.

While acute allergic reactions and severe flu-like symptoms are cause enough for concern, there are deeper problems. Malathion is an organophosphate pesticide that can be absorbed by the body through the lungs, mouth, and skin. Organophosphates are nerve- and immune-system poisons. Malathion can produce optic nerve damage and blindness, paralysis, coma, and death.

In addition, some highly reputable researchers worry that it causes cancer and genetic damage.

California's Governor George Deukmejian, abetted by state and county agriculture and health departments, justifies the aerial bombardment on grounds that the Medfly is a major threat to the state's economy, while the malathion concentration is so low that it is harmless. The experience of Suwol, Thornton, and many others is dismissed as "anecdotal" rather than scientific evidence. Official after official has echoed the words of Jack Parnell of the U.S. Department of Agriculture: Rather than give in to "emotional appeals," the "solution must be left to science."

Even if the state's "science" were sound, periodically dropping a gooey mess from helicopters on homes, cars, and pets would violate some basic rights. But the coalition of organizations opposed to spraying, supported by strong allies among environmental groups and independent scientists, has carefully investigated the scientific case for safety.

By any reasonable standard, that case hasn't held up. The chief author of a 1980 study used by the state to "demonstrate" the safety of malathion charges that California officials altered the risk assessment. According to Marc Lappé, now professor of health policy and ethics at the University of Illinois, "the change made was consistent with the wholesale exclusion of the major portion of our conclusions regarding the uncertainty which existed (and still exists) over the long-term safety of such a spray operation."

More specifically, Lappé points out that his estimate of the cancer risk was cut significantly; that his estimates were based on use of pure malathion, while the commercial grade being sprayed contains impurities that increase the toxicity; and that his risk estimates were based on six sprayings, not the much greater number possible if current plans are followed. He concludes that "there is significant danger of chronic toxicity and possible genetic damage" to vulnerable populations, such as infants, children, the elderly, the sick, the malnourished, and the homeless.

State health officials, dismissing him as a "fearmonger," chose to ignore Lappé's comments about general toxicity, instead emphasizing that studies done after his 1980 analysis found malathion noncarcinogenic. But here too, claims of safety were less scientific than they purported to be. Brian Dementi is an EPA toxicologist who was charged with reviewing studies of the health effects of malathion. In mid-1989 he wrote an internal memorandum that sharply dissented from the agency's official finding that the "EPA does not have concerns that malathion may be a carcinogen." Dementi called for new studies to clarify the carcinogenicity of malathion, and the EPA, while insisting that Dementi's memo "overstated" the risks, is doing these studies.

Commenting on this incident, Mary Nichols of the Natural Resources Defense Council concluded, "The jury is out even more than we thought."

A letter to physicians from county health officials points up the sharp dichotomy between assurances to the public of complete safety and an abundance of disturbing uncertainties. The letter surfaced in connection with a suit brought by advocates for the steadily increasing number of homeless. The county officials acknowledged that "persons directly sprayed with the pesticides, or breathing fairly concentrated vapors from sprayed surfaces, might experience eye and throat irritation or chest tightness. If exposure is prolonged, headache or nausea might develop."

Robert Cohen, of the Legal Aid Society of Orange County, says of the letter, "This is outrageous. They're telling the public one thing and physicians the other." He's referring to a malathion fact sheet in which officials tell the public, "If convenient, stay indoors at the moment of application in your area to avoid spotting personal apparel." There is no mention of health effects. "If you're a homeless person, getting your clothes spotted may not be a top priority, but getting sick would be," says Cohen.

The state's predictions about the cost of Medfly infestation don't hold up much better than its assertions about malathion. From Southern California, goes the argument, it could migrate to the fertile Central Valley. There it would threaten such fruit crops as citrus, peaches, and plums, which account for $5 billion of the state's $16 billion agriculture industry. Agribusiness and the California Department of Food and

Agriculture, which speak with one voice, insist that the result would be higher food costs, embargoes on export, loss of jobs, and more use of pesticides.

The prospect of Medflies in the Central Valley isn't attractive to anyone, but the consequences have been vastly overstated. State officials, running interference for agribusiness, imply that nearly one-third of California's agriculture would be wiped out. According to a University of California study they often mention, however, if the fly got a toehold in the state, it would eat up only about 4 per cent of the threatened crops and a much smaller fraction of the entire agriculture industry.

There would be additional expenses to fumigate and chill export crops, to meet Japanese quarantine requirements. David Bunn of Pesticide Watch, a citizens' antipesticide group, doesn't believe those costs to agribusiness justify the air war, paid for by the public. "I don't think we should spray one million people fourteen times to sell expensive oranges to rich people in Japan," says Bunn.

James Carey, who has analyzed Medfly outbreaks during the past fifteen years, believes the Medfly is already entrenched in Southern California and can't be eradicated. "It's like a disease," he says. "Remission and cure are two very different things. We can spray this into remission, but that can't go on forever." Several of his fellow entomologists at the University of California agree.

Eradication, perhaps as futile as it is unsafe, has an alternative—control of the Medfly. But despite recurring infestations, the state has chosen to count on spraying, and has grossly underfunded a number of promising, safe, environmentally benign methods of biological control.

As the spraying continued last winter, President Bush came to Los Angeles to join Governor Deukmejian for a big Republican fundraiser. Hundreds of demonstrators across the street, appealing to the "environmentalist" President to intervene, were ignored. In fact, USDA's Parnell later said the Federal Government would take over the spraying if the state stopped.

In contrast, local governments haven't been able to ignore the rising anger. As a result of organized grass-roots opposition, city councils in Los Angeles, Burbank, and dozens of small communities are bringing suit to stop the spraying. Although polls now show overwhelming public sentiment against spraying, Charles Getz, the deputy attorney general who will defend the state's position in court, says he expects to win.

Pasadena took a more innovative approach by banning the flight of low-flying helicopters over that city, but the ordinance was brushed aside by state officials who insist that their emergency powers to combat the Medfly override everything in their flight path.

Resentment at this arrogance led one person in a target area to make a comparison with the struggle for democracy in Eastern Europe. In a letter to the *Los Angeles Times,* he went on to suggest that "Californians should be taking to the streets." Virginia Johannessen of CAUSE recalls that in the early 1980s, some helicopters spraying in Northern California eventually were met by ground fire. While her coalition is committed to ending the spraying by "legal and nonviolent means," Johannessen says, "It doesn't surprise me that people are becoming increasingly frustrated with what's going on."

The current conflict is still far from resolution. But as *Times* columnist Robert A. Jones commented, "If a new Medfly emergency is announced in 1991 and the spraying starts again, you will have to roll tanks in the streets of Los Angeles to keep the populace under control."

Content Considerations

1. Why did some people object to spraying for the fruit fly?
2. What was the government's position on the necessity and safety of spraying begun in the summer of 1989?
3. Summarize the method used to eradicate the fruit fly in Los Angeles and Orange counties.

The Writer's Strategies

1. What tone do the first two paragraphs set for the article?
2. What is the author's attitude toward the situation he writes about? How is that attitude expressed?

Writing Possibilities

1. According to your position, support or criticize the steps people took to halt the spraying.
2. Write about the precautions that should be taken or the consequences to be considered before such a controversial eradication system is implemented.
3. Locating additional information as it is needed, discuss the ways pesticides affect ecosystems, including the human ones.

Aldo Leopold

Conservation

Forester and first professor of game management at the University of Wisconsin, Aldo Leopold was born in Iowa in 1886 and died in 1948 fighting a fire on land near his farm in Wisconsin. He was the author of hundreds of articles and two volumes on game management; his two best-known works, A Sand County Almanac *and* Round River *were published after his death.*

Leopold spoke early and long about the need to protect wilderness areas and worked to preserve forest lands and wildlife. He believed strongly that all parts of the land work together in ways of which humans may never be aware; thus, people must become stewards of the land and its beauty, for they are themselves part of an intricate interdependence. Beyond the practical reasons for taking care of the land, however, Leopold also invoked the ethical and moral responsibility of doing so.

Some of these issues arise in the following essay from Round River. *Before you read it, write a page or so about the ways in which people may indeed be part of an interdependent system of life.*

---

1   CONSERVATION IS A BIRD that flies faster than the shot we aim at it.

2   I can remember the day when I was sure that reforming the Game Commission would give us conservation. A group of us worked like Trojans cleaning house at the Capitol. When we got through we found we had just started. We learned that you can't conserve game by itself; to rebuild the game resource you must first rebuild the game range, and this means rebuilding the people who use it, and all of the things they use it for. The job we aspired to perform with a dozen volunteers is now baffling a hundred professionals. The job we thought would take five years will barely be started in fifty.

3   Our target, then, is a receding one. The task grows greater year by year, but so does its importance. We begin by seeking a few trees or birds; to get them we must build a new relationship between men and land.

4   Conservation is a state of harmony between men and land. By land is meant all of the things on, over, or in the earth. Harmony with land is like harmony with a friend; you cannot cherish his right hand and chop off his left. That is to say, you cannot love game and hate

predators; you cannot conserve the waters and waste the ranges; you cannot build the forest and mine the farm. The land is one organism. Its parts, like our own parts, compete with each other and co-operate with each other. The competitions are as much a part of the inner workings as the co-operations. You can regulate them—cautiously—but not abolish them.

The outstanding scientific discovery of the twentieth century is not television, or radio, but rather the complexity of the land organism. Only those who know the most about it can appreciate how little we know about it. The last word in ignorance is the man who says of an animal or plant: 'What good is it?' If the land mechanism as a whole is good, then every part is good, whether we understand it or not. If the biota, in the course of aeons, has built something we like but do not understand, then who but a fool would discard seemingly useless parts? To keep every cog and wheel is the first precaution of intelligent tinkering.

Have we learned this first principle of conservation: to preserve all the parts of the land mechanism? No, because even the scientist does not recognize all of them.

In Germany there is a mountain called the Spessart. Its south slope bears the most magnificent oaks in the world. American cabinetmakers, when they want the last word in quality, use Spessart oak. The north slope, which should be the better, bears an indifferent stand of Scotch pine. Why? Both slopes are part of the same state forest; both have been managed with equally scrupulous care for two centuries. Why the difference?

Kick up the litter under the oaks and you will see that the leaves rot almost as fast as they fall. Under the pines, though, the needles pile up as a thick duff; decay is much slower. Why? Because in the Middle Ages the south slope was preserved as a deer forest by a hunting bishop; the north slope was pastured, plowed, and cut by settlers, just as we do with our woodlots in Wisconsin and Iowa today. Only after this period of abuse was the north slope replanted to pines. During this period of abuse something happened to the microscopic flora and fauna of the soil. The number of species was greatly reduced, i.e. the digestive apparatus of the soil lost some of its parts. Two centuries of conservation have not sufficed to restore these losses. It required the modern microscope, and a century of research in soil science, to discover the existence of these 'small cogs and wheels' which determine harmony or disharmony between men and land in the Spessart.

American conservation is, I fear, still concerned for the most part with show pieces. We have not yet learned to think in terms of small cogs and wheels. Look at our own back yard: at the prairies of Iowa and southern Wisconsin. What is the most valuable part of the prairie? The fat black soil, the chernozem. Who built the chernozem? The black prairie was built by the prairie plants, a hundred distinctive species of grasses, herbs, and shrubs; by the prairie fungi, insects, and bacteria; by the prairie mammals and birds, all interlocked in one humming community of co-operations and competitions, one biota. This biota, through ten thousand years of living and dying, burning and growing, preying and fleeing, freezing and thawing, built that dark and bloody ground we call prairie.

Our grandfathers did not, could not, know the origin of their prairie empire. They killed off the prairie fauna and they drove the flora to a last refuge on railroad embankments and roadsides. To our engineers this flora is merely weeds and brush; they ply it with grader and mower. Through processes of plant succession predictable by any botanist, the prairie garden becomes a refuge for quack grass. After the garden is gone, the highway department employs landscapers to dot the quack with elms, and with artistic clumps of Scotch pine, Japanese barberry, and Spiraea. Conservation Committees, en route to some important convention, whiz by and applaud this zeal for roadside beauty.

Some day we may need this prairie flora not only to look at but to rebuild the wasting soil of prairie farms. Many species may then be missing. We have our hearts in the right place, but we do not yet recognize the small cogs and wheels.

In our attempts to save the bigger cogs and wheels, we are still pretty naïve. A little repentance just before a species goes over the brink is enough to make us feel virtuous. When the species is gone we have a good cry and repeat the performance.

The recent extermination of the grizzly from most of the western stock-raising states is a case in point. Yes, we still have grizzlies in the Yellowstone. But the species is ridden by imported parasites; the rifles wait on every refuge boundary; new dude ranches and new roads constantly shrink the remaining range; every year sees fewer grizzlies on fewer ranges in fewer states. We console ourselves with the comfortable fallacy that a single museum-piece will do, ignoring the clear dictum of history that a species must be saved *in many places* if it is to be saved at all.

The ivory-billed woodpecker, the California condor, and the desert sheep are the next candidates for rescue. The rescues will not be effective until we discard the idea that one sample will do; until we insist on living with our flora and fauna in as many places as possible.

We need knowledge—public awareness—of the small cogs and

wheels, but sometimes I think there is something we need even more. It is the thing that *Forest and Stream,* on its editorial masthead, once called 'a refined taste in natural objects.' Have we made any headway in developing 'a refined taste in natural objects'?

In the northern parts of the lake states we have a few wolves left. Each state offers a bounty on wolves. In addition, it may invoke the expert services of the U.S. Fish and Wildlife Service in wolf-control. Yet both this agency and the several conservation commissions complain of an increasing number of localities where there are too many deer for the available feed. Foresters complain of periodic damage from too many rabbits. Why, then, continue the public policy of wolf-extermination? We debate such questions in terms of economics and biology. The mammalogists assert the wolf is the natural check on too many deer. The sportsmen reply they will take care of excess deer. Another decade of argument and there will be no wolves to argue about. One conservation inkpot cancels another until the resource is gone. Why? Because the basic question has not been debated at all. The basic question hinges on 'a refined taste in natural objects.' Is a wolfless north woods any north woods at all?

The hawk and owl question seems to me a parallel one. When you band a hundred hawks in fall, twenty are shot and the bands returned during the subsequent year. No four-egged bird on earth can withstand such a kill. Our raptors are on the toboggan.

Science has been trying for a generation to classify hawks and owls into 'good' and 'bad' species, the 'good' being those that do more economic good than harm. It seems to me a mistake to call the issue on economic grounds, even sound ones. The basic issue transcends economics. The basic question is whether a hawkless, owl-less countryside is a livable countryside for Americans with eyes to see and ears to hear. Hawks and owls are a part of the land mechanism. Shall we discard them because they compete with game and poultry? Can we assume that these competitions which we perceive are more important than the co-operations which we do not perceive?

The fish-predator question is likewise parallel. I worked one summer for a club that owns (and cherishes) a delectable trout stream, set in a matrix of virgin forest. There are 30,000 acres of the stuff that dreams are made on. But look more closely and you fail to see what 'a refined taste in natural objects' demands of such a setting. Only once in a great while does a kingfisher rattle his praise of rushing water. Only here and there does an otter-slide on the bank tell the story of pups rollicking in the night. At sunset you may or may not see a heron; the rookery has been shot out. This club is in the throes of a genuine educational process. One faction wants simply more trout; another wants trout plus all the trimmings, and has employed a fish ecologist to find ways and means. Superficially the issue again is 'good' and 'bad' predators, but basically the issue is deeper. Any club privileged to own such a piece of land is morally obligated to keep all its parts, even though it means a few less trout in the creel.

In the lake states we are proud of our forest nurseries, and of the progress we are making in replanting what was once the north woods. But look in these nurseries and you will find no white cedar, no tamarack. Why no cedar? It grows too slowly, the deer eat it, the alders choke it. The prospect of a cedarless north woods does not depress our foresters; cedar has, in effect, been purged on grounds of economic inefficiency. For the same reason beech has been purged from the future forests of the Southeast. To these voluntary expungements of species from our future flora, we must add the involuntary ones arising from the importation of diseases: chestnut, persimmon, white pine. Is it sound economics to regard any plant as a separate entity, to proscribe or encourage it on the grounds of its individual performance? What will be the effect on animal life, on the soil, and on the health of the forest as an organism? Is there not an aesthetic as well as an economic issue? Is there, at bottom, any real distinction between aesthetics and economics? I do not know the answers, but I can see in each of these questions another receding target for conservation.

I had a bird dog named Gus. When Gus couldn't find pheasants he worked up an enthusiasm for Sora rails and meadowlarks. This whipped-up zeal for unsatisfactory substitutes masked his failure to find the real thing. It assuaged his inner frustration.

We conservationists are like that. We set out a generation ago to convince the American landowner to control fire, to grow forests, to manage wildlife. He did not respond very well. We have virtually no forestry, and mighty little range management, game management, wildflower management, pollution control, or erosion control being practiced voluntarily by private landowners. In many instances the abuse of private land is worse than it was before we started. If you don't believe that, watch the strawstacks burn on the Canadian prairies; watch the fertile mud flowing down the Rio Grande; watch the gullies climb the hillsides in the Palouse, in the Ozarks, in the riverbreaks of southern Iowa and western Wisconsin.

To assuage our inner frustration over this failure, we have found us a meadowlark. I don't know which dog first caught the scent; I do know that every dog on the field whipped into an enthusiastic backing-

point. I did myself. The meadowlark was the idea that if the private landowner won't practice conservation, let's build a bureau to do it for him.

Like the meadowlark, this substitute has its good points. It smells like success. It is satisfactory on poor land which bureaus can buy. The trouble is that it contains no device for preventing good private land from becoming poor public land. There is danger in the assuagement of honest frustration; it helps us forget we have not yet found a pheasant.

I'm afraid the meadowlark is not going to remind us. He is flattered by his sudden importance.

Why is it that conservation is so rarely practiced by those who must extract a living from the land? It is said to boil down, in the last analysis, to economic obstacles. Take forestry as an example: the lumberman says he will crop his timber when stumpage values rise high enough, and when wood substitutes quit underselling him. He said this decades ago. In the interim, stumpage values have gone down, not up; substitutes have increased, not decreased. Forest devastation goes on as before. I admit the reality of this predicament. I suspect that the forces inherent in unguided economic evolution are not all beneficent. Like the forces inside our own bodies, they may become malignant, pathogenic. I believe that many of the economic forces inside the modern bodypolitic are pathogenic in respect to harmony with land.

What to do? Right now there is a revival of the old idea of legislative compulsion. I fear it's another meadowlark. I think we should seek some organic remedy—something that works from the inside of the economic structure.

We have learned to use our votes and our dollars for conservation. Must we perhaps use our purchasing power also? If exploitation-lumber and forestry-lumber were each labeled as such, would we prefer to buy the conservation product? If the wheat threshed from burning strawstacks could be labeled as such, would we have the courage to ask for conservation-wheat, and pay for it? If pollution-paper could be distinguished from clean paper, would we pay the extra penny? Over-grazing beef vs. range-management beef? Corn from chernozem, not subsoil? Butter from pasture slopes under 20 percent? Celery from ditchless marshes? Broiled whitefish from five-inch nets? Oranges from unpoisoned groves? A trip to Europe on liners that do not dump their bilgewater? Gasoline from capped wells?

The trouble is that we have developed, along with our skill in the exploitation of land, a prodigious skill in false advertising. I do not want to be told by advertisers what is a conservation product. The only alternative is a consumer-discrimination unthinkably perfect, or else a new batch of bureaus to certify 'this product is clean.' The one we can't hope for, the other we don't want. Thus does conservation in a democracy grow ever bigger, ever farther.

Not all the straws that denote the wind are cause for sadness. There are several that hearten me. In a single decade conservation has become a profession and a career for hundreds of young 'technicians.' Ill-trained, many of them; intellectually tethered by bureaucratic superiors, most of them; but in dead earnest, nearly all of them. I look at these youngsters and believe they are hungry to learn new cogs and wheels, eager to build a better taste in natural objects. They are the first generation of leaders in conservation who ever learned to say, 'I don't know.' After all, one can't be too discouraged about an idea which hundreds of young men believe in and live for.

Another hopeful sign: Conservation research, in a single decade, has blown its seeds across three continents. Nearly every university from Oxford to Oregon State has established new research or new teaching in some field of conservation. Barriers of language do not prevent the confluence of ideas.

Once poor as a church mouse, American conservation research now dispenses 'federal aid' of several kinds in many ciphers.

These new foci of cerebration are developing not only new facts, which I hope is important, but also a new land philosophy, which I know is important. Our first crop of conservation prophets followed the evangelical pattern; their teachings generated much heat but little light. An entirely new group of thinkers is now emerging. It consists of men who first made a reputation in science, and now seek to interpret the land mechanism in terms any scientist can approve and any layman understand, men like Robert Cushman Murphy, Charles Elton, Fraser Darling. Is it possible that science, once seeking only easier ways to live off the land, is now to seek better ways to live with it?

We shall never achieve harmony with land, any more than we shall achieve justice or liberty for people. In these higher aspirations the important thing is not to achieve, but to strive. It is only in mechanical enterprises that we can expect that early or complete fruition of effort which we call 'success.'

The problem, then, is how to bring about a striving for harmony with land among a people many of whom have forgotten there is any such thing as land, among whom education and culture have become almost synonymous with landlessness. This is the problem of 'conservation education.'

When we say 'striving,' we admit at the outset that the thing we need must grow from within. No striving for an idea was ever injected wholly from without.

When we say 'striving,' I think we imply an effort of the mind as well as a disturbance of the emotions. It is inconceivable to me that we can adjust ourselves to the complexities of the land mechanism without an intense curiosity to understand its workings and an habitual personal study of those workings. The urge to comprehend must precede the urge to reform.

When we say 'striving,' we likewise disqualify at least in part the two vehicles which conservation propagandists have most often used: fear and indignation. He who by a lifetime of observation and reflection has learned much about our maladjustments with land is entitled to fear, and would be something less than honest if he were not indignant. But for teaching the fresh mind, these are outmoded tools. They belong to history.

My own gropings come to a dead end when I try to appraise the profit motive. For a full generation the American conservation movement has been substituting the profit motive for the fear motive, yet it has failed to motivate. We can all see profit in conservation practice, but the profit accrues to society rather than to the individual. This, of course, explains the trend, at this moment, to wish the whole job on the government.

When one considers the prodigious achievements of the profit motive in wrecking land, one hesitates to reject it as a vehicle for restoring land. I incline to believe we have overestimated the scope of the profit motive. Is it profitable for the individual to build a beautiful home? To give his children a higher education? No, it is seldom profitable, yet we do both. These are, in fact, ethical and aesthetic premises which underlie the economic system. Once accepted, economic forces tend to align the smaller details of social organization into harmony with them.

No such ethical and aesthetic premise yet exists for the condition of the land these children must live in. Our children are our signature to the roster of history; our land is merely the place our money was made. There is as yet no social stigma in the possession of a gullied farm, a wrecked forest, or a polluted stream, provided the dividends suffice to send the youngsters to college. Whatever ails the land, the government will fix it.

I think we have here the root of the problem. What conservation education must build is an ethical underpinning for land economics and a universal curiosity to understand the land mechanism. Conservation may then follow.

Content Considerations

1. Why must the "cogs and wheels" be saved?
2. What does Leopold mean by the "land mechanism"?
3. What part does economics play in conservation?

The Writer's Strategies

1. Leopold begins this essay with a metaphor. How does he return to the metaphor in other parts of the essay? What does the metaphor mean?
2. What is Leopold trying to persuade the reader of?

Writing Possibilities

1. What does it mean to possess "a refined taste in natural objects"? Write an essay exploring the phrase and discussing the extent to which you possess such a taste.
2. Respond to Leopold's contention that the "first principle of conservation" is "to preserve all the parts of the land mechanism." To what extent have we learned to do this in the almost forty years since this essay was published?
3. With a classmate, work on a proposal for conservation education. When would it start? How would it be handled? What results would it work toward?

DAN GROSSMAN
and
SETH SHULMAN

Down in the Dumps

Of the 3.5 pounds of trash produced each day by every person in America, about 10 percent is recycled. Another 10 percent is incinerated, and the rest of it gets hauled to the dump.

What happens to it there is the topic of the next article, an article about landfill excavation undertaken to discover what we discard and what happens to our trash once it has landed in the community refuse pile.

The problem of refuse is a growing one—dumps are filling rapidly, and communities are increasingly reluctant to allow new ones to open within their borders. Landfills are not pretty sights, and many communities fear that contaminants will leak from them into groundwater, streams, air, and land.

Some experts suggest that the first defense against such dangers is source reduction—producing less trash by buying more wisely and economically. That defense is aided by recycling—by finding ways to use material over and over again. Others suggest changing our practices—rather than bagging fallen leaves and other yard wastes, for example, we should leave them on lawns or construct our own compost piles in which they can decompose. Others suggest we refuse to buy products that are "overpackaged," since about half of what we throw away each day is packaging materials.

The following article by Dan Grossman and Seth Shulman appeared in Discover *in April 1990. In it they tell of their trip to the dump, of what a community discarded decades ago. Before you read, write a paragraph or two on what you think they find.*

IN AN OPEN TRACT OF DESERT an archeological crew toils under a punishing sun. Researchers in hats and visors huddle in separate work groups: some of them crouch along the lip of a deep trench, recording its size and depth in a small notebook; others sift through the desert soil, bagging and labeling any treasures they've pulled from the ground; still others load the finds into a white pickup to be carried back to the lab.

The hellish conditions at the site challenge even the heartiest diggers. After only a few minutes at the pit the dry desert heat almost seems to have weight. The only refuge to be found is under a small nylon canopy strung from a van to provide shade for a rest area and a cooler of soda.

In most aspects, then, the scene is that of an archetypal, romantic search for the past. But some things are amiss: The horizon is broken up not by looming pyramids or African hills but by high-tension wires. The nearest big city is not exotic Cairo or Nairobi but familiar, friendly Phoenix. And the artifacts being dug from the ground are not the treasures of a long-lost pharaoh or a mysterious warrior-king but detritus from the age of Eisenhower and Uncle Miltie. This archeological team is unearthing the contents of a landfill in Arizona. And the culture they are excavating is our own.

As the workers sift through the fetid soil, the finds they extract seem hardly worth the trouble: a cosmetics jar from the 1950s, a Schlitz beer can, an official Girl Scout shoe, a surprisingly fresh soupbone, a container of Shinola shoe cream, a 1952 edition of the *Phoenix Gazette* in near-mint condition. "American pilots shoot down 12 Communist jets in a renewed outbreak of major air combat over North Korea," begins one front-page story.

And the workers—all trained archeologists or university students—are outfitted appropriately for their task: most toil in old denims they could discard without missing and T-shirts that have seen far fresher days. Many of the shirts are emblazoned with a trashcan logo and the project's lighthearted title: LE PROJET DU GARBAGE.

Supervising the group's activities is archeologist William Rathje of the University of Arizona, dressed in his customary safari shirt and aviator sunglasses. Although trained in the more traditional practices of archeology, Rathje decided some years ago that what really caught his fancy was trash. After all, most digs, no matter how dramatic they may seem, are essentially acts of organized trash-picking. The broken pottery or encrusted beads that are prized so highly today were once nothing more than the throwaways of somebody else's society. What makes such finds valuable is what they tell us about the people who discarded them—or, in the present case, what they tell us about ourselves.

This afternoon the heat is high in the Arizona desert and the Tempe landfill is ripe. Dozens of yards away the pungent stench of methane already spices the air. At closer range the site takes on the bouquet of a multitude of overflowing trash cans on a stifling day in the city; the scent quickly sticks to hair and clothing. Some workers wear respirators to protect themselves from the acrid stink, but the thick plastic masks offer only a limited defense and add to the workers' discomfort.

A young researcher at the far side of the trench sifts through the rubble, looking for artifacts that can help date the site. Reading the bottom of a glass bottle, he announces, "We've got a bottle here from 1952—the Owens-Illinois Glass Company." The find is added to a small pile that includes dented cans, newspaper, pieces of cardboard packaging, even an ear of corn with its tassel intact.

Helping to catalog the new artifacts is Rathje's bearded, good-natured codirector, Wilson Hughes. An on-site traffic cop and troubleshooter, Hughes helps to organize the diggers and—more important—to jolly them through their often disagreeable work. "I dream up these schemes," Rathje says as he surveys the site, "but Wilson has to figure out how to do it all."

Odious—and odorous—as the Tempe dig can be, it is a job that must be done. By now everyone in the United States has heard the grim story of the nation's trash: we generate 160 million tons of rubbish a year—three and a half pounds a day for every man, woman, and child in the nation. A convoy of ten-ton garbage trucks carrying the nation's annual waste would reach halfway to the moon. A hundred millennia from now the human engineering feats that will most dramatically mark our era may not be our skyscrapers or our highways but our landfills. The nation's largest, Fresh Kills landfill on Staten Island, New York, contains enough accumulated refuse to fill the Panama Canal twice over.

Every year we discard 220 million tires, 1.6 billion pens, and 2 billion razors. Half of the average trash can is filled just with the containers and packages that hold our purchases. Even newborns are chronic trash-makers; a typical baby uses nearly 8,000 disposable diapers before being toilet trained, contributing to an astronomical 18 billion diapers thrown away every year—not to mention 700 million pounds of accompanying feces.

Roughly 10 percent of the mountain of trash we produce is recycled. Another 10 percent is incinerated. The remaining 80 percent, however, is sent to old-fashioned dumps and landfills. But landfills around the country are rapidly filling up, and new ones aren't being opened fast enough. The Environmental Protection Agency estimates that within two years, one-third of our existing landfills will be full; in 20 years, four-fifths will be closed. The most obvious solution is to reduce the amount of trash generated in the first place. But how?

Rathje believes any answer must start with accurate data about the makeup of our refuse. Since 1973, when he began studying the contents of household trash bags (in Phoenix more than $6 million worth of recyclable aluminum and nearly $2 million of newsprint was going to waste each year), he has been studying what we throw away and what happens to it. In 1987 he began his first excavation, at the Vincent H. Mullins landfill in Tucson. To date he has studied eight landfills in five states. All the digs have yielded some surprises, and the current Arizona project has been no exception.

On the second day of the Arizona dig the team rises with the sun, appearing at the landfill at 6:30 in the morning. Part of the motivation for the early start comes from the hope of finishing the day's work before the afternoon heat becomes too oppressive. Just as important, though, is the anticipation of what they might find today. The increasingly older trash they have been uncovering suggests that with just a little more digging they might strike their first sample of honest-to-goodness, vintage 1940s junk. Hughes is ecstatic. "Trash of the forties," he exclaims, "that's what I dream of."

Almost immediately things get pungent as the backhoe digs into the narrow pit. The heat of the waste is considerable—11 feet down it is a sticky 84 degrees—but Hughes remembers hotter trash: at a previous dig in San Francisco the refuse came out steaming.

The day's first load is deposited on a thick piece of plywood. From here it is shoveled by hand into trash barrels to be weighed, measured, and then sifted through a half-inch mesh. The biggest samples stay on top of the mesh, the smallest ones fall into a container below. Analysts will later determine if the fine-grain finds are plastic, metal, Styrofoam, or some other material.

Unearthed in today's garbage are a 1948 bottle of Clorox, the top of a preserves jar, a ceramic mug, and a sock. Also uncovered is a broom, intact down to its decades-old bristles. To the workers of the Garbage Project the excellent condition of the trash they find has thrown into question the entire notion of biodegradability.

"The perception is that a landfill is a large compost pile," says project archeologist Doug Wilson. "But in reality, it seems to be the best place to preserve things." Wilson speculates that landfills preserve garbage so well because the garbage is tightly packed and completely covered and is exposed to little light or moisture.

Rathje agrees. "Nothing has as popular an image as biodegradability in landfills. It's 'natural.' It's the 'right thing.' Unfortunately, though, it simply doesn't happen."

Rathje acknowledges that some food debris and yard waste does degrade, but at a very slow rate—about 25 percent in the first 15 years, with little or no additional change for at least another 40 years. Most of the remaining trash in landfills seems to retain its original weight, volume, and form for at least four decades. And, Rathje stresses, the findings of intact trash are not exclusive to Arizona's arid climate. Similar results have been found in digs from Chicago to California.

All of which makes garbologists like Wilson and Rathje especially 21
skeptical of the so-called biodegradable plastics that are appearing increasingly in household products. Many of these products are "photodegradable," meaning they have to remain on the surface and be exposed to steady sunlight in order to break down. Other plastics use an additive such as cornstarch to allow them to decompose in the soil. But their decomposition depends on there being enough cornstarch-eating microorganisms in the soil.

And even if a sufficient population of microorganisms exists, some 22
environmentalists warn, this is not necessarily a good thing: they eat only the starch matrix, not the little pieces of plastic left behind, which are free to migrate away from the landfill and find their way into groundwater. Carl Hultberg, a staffer at the Environmental Action Coalition in New York, claims that by "pandering to the public's real concern about the environment," the manufacturers of biodegradable plastics are actually "foiling legitimate plastic recycling efforts."

Perhaps the clearest example of how little biodegrading actually 23
takes place in landfills comes up in nearly every shovelful: newspapers. Rathje's group finds that newspapers are the largest single commodity in landfills—taking up as much as 16 percent of the volume of waste dumped at the average site. To Rathje the figures clearly demonstrate the significant impact recycling programs could have on decreasing the total amount of garbage dumped. The smallest proportion of excavated newspapers comes from digs in the San Francisco Bay area where the recycling rate is higher than in other areas Rathje has studied. Yet even there newspapers account for at least 10.5 percent of the fill.

Most of the newspapers Rathje has found, no matter the age, come 24
out of the ground in remarkably readable shape. In fact, the team's researchers routinely use them to pinpoint the age of the day's trash sample. CONTROL OF UNION LINKED TO KREMLIN screams one headline. REDS BLASTED RELENTLESSLY reads another. They were published within a few days of each other in August 1952, during the bleak days of the Korean War.

The papers may date back less than 40 years, but the city they 25
reveal seems centuries past. Although the *Phoenix Gazette* reported the international tensions of the Cold War and the excitement of the upcoming Eisenhower-Stevenson election, the city still had very much the insular flavor of a frontier town. Outlaws, tough lawmen, and gambling were major preoccupations; news that the year's cotton crop would for the first time top one million bales ranked higher than reports of congressional maneuverings in Washington.

But change was coming to Arizona and the rest of the nation in 26
the form of massive postwar product innovations. A backhoe's load of

Tempe trash from 1952 reveals the town's then-burgeoning appetite for new, nationally distributed consumer goods. JUST PRESS, NO FUSS promises the label of an early aerosol can that looks more like a knobby piece of military hardware than a modern spray can. SPARE YOUR KNEES promises a can of Glo-Coat liquid wax. Next to a can of A-1 beer, manufactured by Phoenix's Arizona Brewing Company, is a rusting Schlitz can—manufactured by the Joseph Schlitz Brewing Company in distant Milwaukee. Hidden beneath a set of home canning supplies is an empty can of Del Monte green beans.

The garbage team does not spend every day digging in the landfill. Many mornings they gather in a University of Arizona shack that serves as the group's laboratory. Fittingly, the structure is located in the corner of a yard used to store the large receptacles into which campus trash is dumped. The sorting room of the makeshift lab is filled with tables made from large plywood sheets laid across upright oil drums. On one wall is the crushed head of a plastic doll, hanging by its hair—a prize find from a previous excavation.

Outside, on the far side of the yard, Hughes opens a large steel shipping container, pulls out two yellow trash bags, and carries them into the sorting room. "Are you awake?" he asks one student garbed in heavy rubber gloves and a white apron. "Open up that bag, take a deep breath, and that'll wake you right up," he says, grinning.

As the bag is emptied out, a sour stench blankets the room, renewing the unpalatable memory of the dig. Heaped before the students is a small mound of finds, shaken clean of the sand, soil, and small stones from the landfill. The researchers begin to pull out pale-colored objects and place them in more than a dozen different white plastic bins designated for each class of material, such as packaging paper, plastic, yard waste, wood, and rubber. The work goes quickly, except when the researchers pause to puzzle over an unusual item or to read a story from a scrap of newspaper. Hughes comes across an intriguing old postage stamp, and discreetly plucks it away to store among his treasures. "It's not going to affect the weight or volume too much," he says sheepishly.

As the work progresses, the bin that holds the plastic waste draws special attention. Plastics have been blamed for a third of all trash in the United States; but the Garbage Project's findings suggest that the volume of plastic products in landfills has risen only slightly as a percentage of the total in recent decades—from 10 percent in the 1960s to 13 percent today. The finding both confounds common sense and contradicts studies concluding that the volume of plastics dumped every year has risen dramatically. Rathje's explanation is that individual plastic products were made thinner and lighter as more of them were produced, allowing them to be squashed flat and take up less space.

Along one wall of the Garbage Project's lab a smaller group of students sits before a long folding table, meticulously sorting a pile of the tiny finds that the mesh separated out. Nearby are several muffin tins lined with colored paper cups for each category of material. A look into the tins shows that the homogeneous-looking sample pile actually contains little shards of colored glass, pieces of plastic, and twisted scraps of metal. Joe Banchy, a University of Arizona graduate who is working at the table, joined the Garbage Project after seeing an episode of the *Donahue* show about the New York garbage barge that traveled 6,000 miles through the Atlantic, the Caribbean, and the Gulf of Mexico looking for a place to dump its load. Banchy holds a small beige object the size of a button in his tweezers. It is smooth on one side and rough on the other, and seems pliable. "Could be part of a car seat," he muses, and places it in the "textile" cup.

Despite questions of accuracy in the sorting process, there is little doubt that the Garbage Project's research is producing data more reliable than any that can be obtained through other, perhaps cleaner, approaches: such as by studying products and materials at the other end of their lives—the manufacturing end. That technique is known as materials flow calculation and was first tried out in the early 1970s by the Environmental Protection Agency.

Today materials flow estimates are conducted for the EPA by waste-disposal expert Marjorie Franklin. Franklin begins with figures representing the quantities of all products made in the United States that could reasonably be expected to find their way into a dump; the products are grouped into broad categories; durable goods such as major appliances, nondurable goods such as clothing, and containers and packaging. She also estimates such undocumented rubbish as food and yard waste by commissioning samples from landfills and garbage trucks. To these she adds all products imported into the country and subtracts all products exported. She reasons that unless a product is stored indefinitely—the way, say, a book is—it must either be recycled, incinerated, or discarded. "Our principle," she says, "is that if a product is created, it has to go somewhere."

Materials flow calculations, however, include much trash, such as roadside litter, that does not in fact end up in landfills. Conversely, as Rathje notes, yard clippings are known to take up a large share of the nation's dumps, yet obviously there are no production figures for them as there are for industrial products. Similarly, reliable records are not kept of concrete, bricks, and other construction debris, yet Rathje's findings indicate that this kind of heavy rubbish represents fully 11 percent of the contents of landfills.

For these reasons, Rathje believes, landfill excavations will gain increasing acceptance as the best way to determine just what's happening on the nation's garbage front. Other investigations conducted by the Garbage Project before the landfill excavations began also showed the value of discarding paper-and-pencil research technique in favor of down-and-dirty trash-picking.

In 1984, for example, the Garbage Project began measuring the amount of hazardous waste in municipal trash. Household cleaners, automotive products, flashlight batteries, paints, pesticides, cosmetics, even prescription drugs contain a dizzying assortment of highly toxic ingredients. But no one can calculate exactly how much hazardous waste makes it into the trash can, without actually peering into one. Donning their protective gear, Garbage Project staffers burrowed into trash bags and found that household trash cans contain up to one percent hazardous waste.

While this may not seem like much, Rathje's calculations showed that the accumulated hazardous material discarded by an entire city could be alarming. The residents of New Orleans, according to his estimates, throw away over 600 tons of hazardous material a year. In this study the largest category of items containing toxic chemicals is batteries and other electrical supplies. In fact, Rathje estimates that nearly 2 million such throwaways—containing sulfuric acid, mercuric oxide, and lead—are dumped in New Orleans every year.

Rathje's 17 years of rubbish work have yielded reams of these kinds of data. Other digs following the Tempe dig—including a recently completed excavation on Staten Island and one in progress near Tampa—promise to unearth even more. Yet for all that Rathje has accomplished since he first took a shine to trash, he, more than anyone, is uncertain how the data should be used.

"Archeologists don't know much about anything except digging," he says, taking a break near the end of a day of sorting. "The only thing we're trying to do is identify what's in the ground and what's happening to it. These data can then be used by the people who do have to make the policy decisions. If we were to become advocates for one waste control plan or the other, we would lose the credibility our data demand."

Some garbage activists suggest that lawmakers could take advantage of Rathje's research to force recycling or regulation of the items that make up the largest percentage of our trash. One system now in place in Seattle and several other localities requires residents to pay for the privilege to dump; the more you discard, the higher your bill. This, plus regulation limiting the amount of household hazardous waste permitted in landfills, and incentives to corporations to discard less, could help citizens figure out their own ways to limit their trash output.

"Ultimately," Rathje says, "we probably won't be buried in our own garbage. We'll do what every other civilization has done: simply live on top of however much trash we do produce."

For the time being, Rathje will continue to spend his days living not on top of the nation's trash but within it. While legislators will no doubt turn the data this way and that to help arrive at workable waste disposal policies, Rathje will be taking a somewhat longer view.

"The history of garbage," he says, "parallels nothing less than the history of civilization. The closer we get to looking at our trash, the closer we get to looking at ourselves."

Content Considerations

1. What do Rathje and other archeologists hope to discover by sifting through trash?
2. Summarize the processes of landfill excavation and materials flow estimation. What information does each type of research reveal?
3. What conclusions about trash do the landfill digs suggest?

The Writers' Strategies

1. How do the authors emphasize the idea that most refuse dumped in a landfill is preserved?
2. What do the authors do to help the reader assimilate the numbers and facts in this article?

Writing Possibilities

1. Expand upon the article's closing quotation that "the closer we get to looking at our trash, the closer we get to looking at ourselves."
2. Find out what happens to the trash from your community; write a report of your findings.
3. With a classmate, save everything you would have discarded over two or three days—paper, empty pens, disposable razors, food containers—and bring the whole collection to class. Examine these artifacts, then write about what they say about you—and whether you like the picture.

Additional Writing and Research

Connections

1. In what ways to the essays by Deloria and Leopold carry much the same message? How do they differ? Is it possible for Native Americans and non-Native Americans to ever view the land exactly the same way?
2. Huebner and Goodman both write about the Medfly and the disaster in California. Describe the theme of each writer.
3. The theme of personal responsibility runs through the essays by Berry and Jackson. Who is the audience for each? How does that affect the tone of the essays?
4. The essays by Grossman/Shulman and Berry all deal with American lifestyles and their resulting effects on the environment. Summarize these effects.
5. How would population growth, the subject of the essay by the Ehrlichs, affect any of the issues dealt with by other writers in this chapter?
6. Summarize each of the essays in this chapter in a paragraph or two, then write about how attitudes, practices, and population size are interrelated.

Researches

1. Study another culture, possibly Native American culture, to learn how its view of the earth affects its practices regarding the environment. In what ways has this culture been affected by the attitudes of industrial nations?
2. The campaign to clean up auto emissions is based primarily on the continued use of oil. What efforts are under way to produce cars that use other, less polluting forms of energy? Why have none been successful thus far?
3. What are some ways in which the glut of garbage in America today is being tackled? Which holds the best promise of success?
4. Learn about the demographics of toxic landfills or even standard city dumps or landfills. Who lives near them? What is the neighborhood's socioeconomic status? Did the residents want the landfill? Why was the decision made to place it where it is?
5. Aldo Leopold was once little known outside professional game management and academia; today his books are widely regarded as prescient. Discover the people in your community who are quietly doing work for the environment. Write about them and their views.
6. Research efforts to curtail population growth in other cultures. Where have these been successful? Where not? What factors determine the success of a program to limit population growth?

APPENDIX

Environmental Organizations

STUDENTS MAY WISH to contact one or more of the following organizations for research material:

Environmental Defense Fund
257 Park Avenue, South
New York, NY 10010

GreenPeace U.S.A.
1436 U Street, N.W.
Washington, D.C. 20009

EarthFirst!
P.O. Box 5871
Tucson, AZ 85703

Americans for the Environment
1400 16th Street, N.W.
Washington, D.C. 20036

The Wilderness Society
1400 I Street, N.W.
Washington, D.C. 20005

Sierra Club
730 Polk Street
San Francisco, CA 94109

National Wildlife Federation
1412 16th Street, N.W.
Washington, D.C. 20036

Environmental Action
1525 New Hampshire, N.W.
Washington, D.C. 20036

Greenhouse Crisis Foundation
1130 17th Street, N.W.
Suite 630
Washington, D.C. 20036

National Audubon Society
645 Pennsylvania Avenue, S.E.
Washington, D.C. 20003

National Resources Defense Council
40 West 20th Street
New York, NY 10011

Oceanic Society
218 D Street, S.E.
Washington, D.C. 20003

Rainforest Action Network
300 Broadway, Suite 28
San Francisco, CA 94133

World Resources Institute
1735 New York Avenue, N.W.
Washington, D.C. 20006

Worldwatch Institute
1776 Massachusetts Avenue, N.W.
Washington, D.C. 20036

Acknowledgments (Continued)
This page constitutes a continuation of the copyright page.
Page 22. "Halting Land Degradation," by Sandra Postel, from "*Focus,* State of the World, 1990," Worldwatch Institute, Washington, DC. Reprinted by permission.
Page 29. "A Wilderness of Light." From *Stories From Under the Sky* by John Madson. Iowa Heritage Collection Edition, 1988; © 1961 Iowa State University Press, Ames, Iowa, 50010. Reprinted by permission.
Page 33. "The Thin Edge." Reprinted by permission of Sterling Lord Literistic, Inc. Copyright © 1978 by Anne W. Simon.
Page 41. "The Frontier Dream We Call Alaska," by Thomas A. Lewis. Copyright 1990 by the National Wildlife Federation. Reprinted from the June/July 1990 issue of *National Wildlife.*
Page 48. "The Solace of Open Spaces." From *The Solace of Open Spaces* by Gretel Ehrlich. Copyright © 1985 by Gretel Ehrlich. Reprinted by permission of Viking Penguin, a division of Penguin Books USA, Inc.

Chapter 2
Page 63. "Playing Dice with Megadeath," by Jared Diamond. From *Discover* magazine, April 1990. © 1990 Discover Publications.
Page 73. "The Owls of Night." From: *Pastorale: A Natural History of Sorts.* Reprinted by permission of Jake Page.
Page 82. "Sharks Under Attack," by Rudy Abramson. Reprinted by permission of Los Angeles Times Syndicate.
Page 86. "Unlikely Harbingers," by Michael Milstein. Reprinted by permission from *National Parks* magazine, July/August, 1990. Copyright © 1990 by National Parks and Conservation Association.
Page 94. "Africa Daze Montana Knights" by Margaret Knox. Copyright © 1990, *Buzzworm: The Environmental Journal,* July/August, Vol II, No. 4. May not be reprinted or photocopied without prior written permission.
Page 105. "Natural Man." From *The Lives of a Cell* by Lewis Thomas. Copyright © 1971 by the Massachusetts Medical Society. Reprinted by permission of Viking Penguin, a division of Penguin Books USA Inc.

Chapter 3
Page 114. "The Flow of the River." From *The Immense Journey* by Loren Eiseley. Copyright © 1957 by Loren Eiseley. Reprinted by permission of Random House Inc.
Page 122. "The Return of Beaver to the Missouri River." From *Sundancers and River Demons* by Conger Beasley, Jr. Copyright © 1990 by Conger Beasley, Jr. Reprinted with permission.
Page 144. "Los Angeles Against the Mountains." Excerpt from *The Control of Nature* by John McPhee. Copyright © 1989 by John McPhee. Reprinted by permission of Farrar, Straus and Giroux, Inc.
Page 155. "The Present." Excerpt from *Pilgrim at Tinker Creek* by Annie Dillard. Copyright © 1974 by Annie Dillard. Reprinted by permission of HarperCollins Publishers.
Page 161. "Creating an Endangered Ecosystems Act." Reprinted from *Down by the River: The Impact of Federal Water Projects and Policies on Biological Diversity* by permission of Island Press. Copyright © 1988 Constance E. Hunt.

Chapter 4
Page 178. "A Treasure of Complexities," by John Alcock. From *Wilderness* magazine, Summer 1990. Reprinted by permission of *Wilderness.*
Page 184. "An Overture." Excerpted from *The Desert Smells Like Rain,* Copyright © 1982 by Gary Paul Nabhan. Published by North Point Press and reprinted by permission.

Acknowledgments

Page 189. "Desert Sojourn," by Charlie Haas. From *Esquire* magazine, October 1987. Reprinted by permission of Charlie Haas.
Page 196. "Meeting on the Mesa." From *On the Mesa* by John Nichols. Reprinted by permission of John Nichols.
Page 201. "The Water Profiteers." Used by permission of Peter Steinhart. Reprinted from *Audubon*, the magazine of the National Audubon Society, March 1990.
Page 214. "A Poet, a Painter, and the Lonesome Triangle." Used by permission of Kim Heacox. Reprinted from *Audubon*, the magazine of the National Audubon Society, May 1990.

Chapter 5
Page 231. "The Ocean System." Paul Colinvaux, *Why Big Fierce Animals Are Rare: An Ecologist's Perspective*. Copyright © 1978 by Paul A. Colinvaux. Excerpt pp. 89–96 reprinted by permission of Princeton University Press.
Page 237. "Wealth from the Salt Seas." From *The Sea Around Us*, Revised Edition, by Rachel L. Carson. Copyright © 1950, 1951, 1961 by Rachel L. Carson; renewed 1979 by Roger Christie. Reprinted by permission of Oxford University Press, Inc.
Page 249. "Oil Spill." From *The Atlantic*, February 1990. Reprinted by permission of Peter Sears.
Page 250. "In the Wake of the *Exxon Valdez:* Marine Birds." From *In the Wake of the Exxon Valdez*, by Art Davidson. Copyright © 1990 by Art Davidson. Reprinted with permission of Sierra Club Books.
Page 257. "What We Should Do to Stop Spills," by Peter Nulty. From *Fortune*, July 16, 1990. Copyright © 1990 The Time Inc. Magazine Company. All rights reserved.
Page 263. "Holding Back the Sea," by Jodi L. Jacobson. From *The Futurist*, September/October 1990, published by the World Future Society, 4916 Saint Elmo Avenue, Bethesda, Maryland 20814. Reprinted with permission.
Page 274. "Sludge." From *The Late Great Lakes* by William Ashworth. Copyright © 1986 by William Ashworth. Reprinted by permission of Alfred A. Knopf, Inc.
Page 286. "The Tainted Cup." From *Undersea Life* by Joseph S. Levine. Published by Stewart, Tabori & Chang, New York. Text copyright © 1985 Joseph S. Levine.

Chapter 6
Page 299. "Our Animal Rites," by Anna Quindlen. From *The New York Times*, August 5, 1990. Copyright 1990 by The New York Times Company. Reprinted by permission.
Page 302. "Yukon-Charley: The Shape of Wilderness" is reprinted with permission of Charles Scribner's Sons, an imprint of Macmillan Publishing Company, from *Crossing Open Ground* by Barry Lopez. Copyright © 1983, 1988, Barry Holstun Lopez. First appeared under a different title and in a different form in Wilderness, Fall 1982.
Page 314. "The Great American Desert." From *The Journey Home: Some Words in Defense of the American West* by Edward Abbey. Copyright © 1977 by Edward Abbey. Reprinted by permission of the publisher, Dutton, an imprint of New American Library, a division of Penguin Books USA Inc.
Page 323. "Summer." From *A Country Year: Living the Questions* by Sue Hubbell. Copyright © 1983, 1984, 1985, 1986 by Sue Hubbell. Reprinted by permission of Random House, Inc.
Page 327. "Eco-tourism: The Light at the End of the Terminal," by Christian Kallen. Reprinted with permission from *E—The Environmental Magazine,* July/August 1990, subscriptions $20/year; P.O. Box 6667, Syracuse, NY 13217/(800) 825-0061.
Page 335. "The Dolphins of Monkey Mia," by Lori Nelson. From *SeaFrontiers* magazine, Volume 36, No. 4. Reprinted by permission from *SeaFrontiers* © 1990 by the International Oceanographic Foundation, 4600 Rickenbacker Causeway, Virginia Key, Miami, Florida 33149.

Page 342. "The Great Swamp." Excerpt from *This Bright Land* by Brooks Atkinson, copyright © 1970, 1971, 1972, by Brooks Atkinson. Used by permission of Doubleday, a division of Bantam Doubleday Dell Publishing Group, Inc.

Chapter 7
Page 352. "Cove and Forest." From *In Defense of Nature*. Reprinted by permission of John Hay.
Page 358. "The Last Stand?" by Laura Tangley. From *Earthwatch*, June 1990. Reprinted by Permission of *Earthwatch*.
Page 363. "Green Giants," by Doug Stewart. From *Discover* magazine, April 1990. Copyright © 1990 Discover Publications.
Page 369. "Forestry: Only God Can Make a Tree, But . . ." Reprinted by permission of The University of Tennessee Press. From Michael Frome's *Conscience of a Conservationist*. Copyright © 1989 by The University of Tennessee Press.
Page 378. "Expanses of Trees Fall Sick and Die," by John Nielsen. Copyright 1990 by the National Wildlife Federation. Reprinted from the March/April 1990 issue of *International Wildlife*.
Page 383. "Acid Rain and Forest Decline," by Don Hinrichsen. Reproduced from *The Earth Report*, published by HP Books, a division of Price Stern Sloan, Inc., Los Angeles, California. Copyright © 1988 by Mitchell Beazley Publishers.
Page 395. "The Tropical Rain-Forest Setting," by Julie S. Denslow; Christine Paddoch. *People of the Tropical Rain Forest*, pages 25–36. Copyright © 1988 Smithsonian Institution. Reprinted by permission of the publisher, The University of California Press.

Chapter 8
Page 409. "Making the Population Connection." From *The Population Explosion*. Copyright © 1990 by Paul R. Ehrlich and Anne H. Ehrlich. Reprinted by permission of Simon & Schuster, Inc.
Page 417. "Think Little," by Wendell Berry. From *A Continuous Harmony: Essays Cultural and Agricultural*, Copyright © 1972 by Wendell Berry, reprinted by permission of Harcourt Brace Jovanovich, Inc.
Page 427. "The Artificial Universe." Reprinted with permission of Macmillan Publishing Company from *We Talk, You Listen* by Vine Deloria, Jr. Copyright © 1970 by Vine Deloria, Jr.
Page 439. "Making Lions Lay Down With Lambs." Copyright 1991: Rev. Jesse Jackson, Sr. Reprinted by permission.
Page 444. "The Killer Bee Syndrome," by Ellen Goodman. from *Keeping in Touch*. Copyright © 1985 by The Washington Post Company. Reprinted by permission of Summit Books, a division of Simon & Schuster, Inc.
Page 447. "The Medfly Wars," by Albert L. Huebner. From *The Progressive*, June 1990. Reprinted by permission from *The Progressive*, 409 East Main Street, Madison, WI 53703.
Page 453. "Conservation," by Aldo Leopold. From *Round River: From the Journals of Aldo Leopold*. Copyright 1953 by Oxford University Press, Inc.; renewed 1981 by Luna B. Leopold and Oxford University Press, Inc. Reprinted by permission of the publisher.
Page 462. "Down in the Dumps," by Dan Grossman and Seth Shulman. From *Discover* magazine, April 1990. Copyright © 1990 Discover Publications.

Index

Abbey, Edward *(The Great American Desert)*, 314–321
Abramson, Rudy *(Sharks Under Attack)*, 82–85
Acid Rain and Forest Decline (Hinrichsen), 383–393
acid rain and snow, 90–92, 383–393, 409
Africa Daze Montana Knights (Knox), 94–104
agriculture, 201–212, 228–236, 263, 267–269, 358–362, 378–381, 395–403, 409–416, 423–425, 447–452, 453–460
Alcock, John *(A Treasure of Complexities)*, 178–183
The Artificial Universe (Deloria), 427–438
Ashworth, William *(Sludge)*, 274–284
Atkinson, Brooks *(The Great Swamp)*, 342–347
Austin, Mary *(Land of Little Rain)*, 171–176

Beasley, Conger, Jr., *(The Return of Beaver to the Missouri River)*, 122–128
Berry, Wendell *(Think Little)*, 417–425

Carson, Rachel *(Wealth from the Salt Seas)*, 237–247
clearcutting, 90, 98–100, 369–377, 378–381, 395–403
coastal regions, 33–40, 232–233, 250–256, 263–272, 274–284, 286–293, 335–341, 435
Colinvaux, Paul *(The Ocean System)*, 231–236
Colorado River 162, 192, 201–212
Concord River (Thoreau), 130–134
Conservation (Leopold), 453–460
conservation ethics, 8–16, 63–71, 73–80, 105–108, 94–104, 122–128, 178–183, 302–312, 342–347,
417–425, 427–438, 439–442, 453–460
Cove and Forest (Hay), 352–356
Creating an Endangered Ecosystems Act (Hunt), 161–166

Davidson, Art *(In the Wake of the Exxon Valdez: Marine Birds)*, 250–256
deforestation, 45, 352–356, 358–362, 363–367, 369–377, 378–381, 383–393, 395–403
Deloria, Vine, Jr. *(The Artificial Universe)*, 427–438
Denslow, Julie Sloan *(The Tropical Rain-Forest Setting)*, 395–403
"Desert Sojourn" (Haas), 189–195
desertification, 17–20, 22–27, 352–356, 383–393, 395–403
deserts
  development, 171–176, 178–183, 189–195, 196–199, 201–212, 214–226, 314–321
  life, 184–188, 189–195, 196–199, 214–226, 314–321
  soil, 13, 173
development
  Alaska, 41–46
  coasts, 33–40, 263–272, 274–284, 288, 435
  deserts, 171–176, 178–183, 189–195, 196–199, 201–212, 214–226, 314–321
  housing and industry, 299–301, 342–347, 378–381
  rivers, 170–173, 201–212
  Wyoming, 48–57
Diamond, Jared *(Playing Dice with Megadeath)*, 63–71
Dillard, Annie *(The Present)*, 155–159
The Dolphins of Monkey Mia (Nelson), 335–341
Down in the Dumps (Grossman and Shulman), 462–470

ecosystems, 86–93, 161–166, 171–

176, 178–183, 222–223, 228–236, 237–247, 250–256, 263–272, 274–284, 286–293, 314–321, 342–347, 352–356, 358–362, 363–367, 369–377, 383–393, 395–403, 427–438, 453–460, 462–470
Eco-tourism: The Light at the End of the Terminal (Kallen), 327–333
Ehrlich, Gretel *(The Solace of Open Spaces)*, 48–57
Ehrlich, Paul R. and Anne H. *(Making the Population Connection)*, 409–416
Eiseley, Loren, *(The Flow of the River)*, 114–120
Expanses of Trees Fall Sick and Die (Nielsen), 378–381
extinction, 2, 61, 63–71, 73–80, 86–93, 94–104, 161–166, 178–183, 286–293, 363–367

flooding, 122–128, 130–134, 136–143, 144–153, 201–212, 263–272, 358–362
The Flow of the River (Eiseley), 114–120
folklore, 122–128, 323–326, 378–381
forest
  Alaska, 45
  dry, 358–362
  floods, 137
  pollutants, 383–393
  soil, 10
  temperate, 378–381
  tropical rain 70, 76, 395–403
  United States, 45, 98–100, 352–356, 363–367, 369–377
Forestry: Only God Can Make a Tree, But . . . (Frome), 369–377
Frome, Michael *(Forestry: Only God Can Make A Tree, But . . .)*, 369–377
The Frontier Dream We Call Alaska (Lewis), 41–46
fruit flies, 444, 447–452
garbage, 462–470
global warming, 2, 171, 263–272, 395–403, 409

Goodman, Ellen *(The Killer Bee Syndrome)*, 444–446
grasslands, 17–20, 29–31, 130–134, 180, 196–199, 358–362
The Great American Desert (Abbey), 314–321
Great Lakes, 274–284, 289–290
The Great Swamp (Atkinson), 342–347
Green Giants (Stewart), 363–367
Grossman, Dan (and Seth Shulman, *Down in the Dumps*), 462–470

Haas, Charlie *(Desert Sojourn)*, 189–195
habitat destruction, 65, 69–70, 76, 78, 86–93, 123–124, 161–166, 178–183, 196–199, 201–212, 214–226, 231–236, 250–256, 263–272, 274–284, 286–293, 299–301, 314–321, 327–333, 335–341, 352–356, 358–362, 363–367, 369–377, 378–381, 383–393, 395–403, 417–425, 427–438, 439–442, 453–460
Halting Land Degradation: Abuse It and Lose It (Postel), 22–27
Hay, John *(Cove and Forest)*, 352–356
Heacox, Kim *(A Poet, a Painter, and the Lonesome Triangle)*, 214–226
Hinrichsen, Don *(Acid Rain and Forest Decline)*, 383–393
Holding Back the Sea (Jacobson), 263–272
Hubbell, Sue *(Summer)*, 323–326
Huebner, Albert L. *(The Medfly Wars)*, 447–452
Hunt, Constance Elizabeth *(Creating an Endangered Ecosystems Act)*, 161–166
hunting, 94–104, 235, 359, 436–437

In the Wake of the Exxon Valdez: *Marine Birds* (Davidson), 250–256

Jackson, Jesse L., 1, *(Making Lions Lay Down With Lambs)*, 439–442
Jacobson, Jodi L. *(Holding Back the Sea)*, 263–272

Index

Kallen, Christian *(Eco-tourism: The Light at the End of the Terminal)*, 327–333
The Killer Bee Syndrome (Goodman), 444–446
Knox, Margaret *(Africa Daze Montana Knights)*, 94–104

The Land of Little Rain (Austin), 171–176
landscape, 48–57, 116, 122–128, 178–183, 302–312, 314–321, 342–347, 352–356
The Last Stand? (Tangley), 358–362
Leopold, Aldo *(Conservation)*, 453–460
Levin, Joseph S. *(The Tainted Cup)*, 286–293
Lewis, Thomas A. *(The Frontier Dream We Call Alaska)*, 41–46
Lopez, Barry *(Yukon-Charley: The Shape of Wilderness)*, 302–312
Los Angeles Against the Mountains (McPhee), 144–153
Lovelock, John, 6

Madson, John *(A Wilderness of Light)*, 29–31
Making Lions Lay Down With Lambs (Jackson), 439–442
Making the Population Connection (Ehrlich and Ehrlich), 409–416
McPhee, John *(Los Angeles Against the Mountains)*, 144–153
The Medfly Wars (Huebner), 447–452
Meeting on the Mesa (Nichols), 196–199
Milstein, Michael *(Unlikely Harbingers)*, 86–93
Missouri River, 122–128, 433
Muir, John *(The River Floods)*, 136–143

Nabhan, Gary Paul *(An Overture)*, 184–188
Natural Man (Thomas), 105–108
Nelson, Lori *(The Dolphins of Monkey Mia)*, 335–341

Nichols, John *(Meeting on the Mesa)*, 196–199
Nielsen, John *(Expanses of Trees Fall Sick and Die)*, 378–381
nuclear power, 444–446
Nulty, Peter *(What We Should Do to Stop Spills)*, 257–261

The Ocean System (Colinvaux), 231–236
oceans, 33–40, 231–236, 237–247, 250–256, 257–261, 263–272, 286–293, 335–341
oil, 36, 39, 43–44, 244–247, 250–256, 257–261, 286–293
Oil Spill (Sears), 249
organization of life, 73–80, 107–108, 114–120, 130–134, 136–143, 155–159, 228–236, 237–247, 290–293, 342–347, 352–356, 417–425, 427–438, 439–442, 453–460
Our Animal Rites (Quindlen), 299–301
An Overture (Nabhan), 184–188
The Owls of Night (Page), 73–80

Page, Jake *(The Owls of Night)*, 73–80
Platte River, 114–120
Playing Dice with Megadeath (Diamond), 63–71
A Poet, a Painter, and the Lonesome Triangle (Heacox), 214–226
pollution, 2, 86–93, 233, 249, 250–256, 257–261, 263–272, 274–284, 286–293, 335–341, 363–367, 378–381, 383–393, 395–403, 427–438, 439–442, 447–452, 453–461
population growth, 2, 19, 23, 36–37, 48–57, 69, 172, 204–205, 268, 270–272, 292, 327–333, 347, 409–416, 427–438
Postel, Sandra *(Halting Land Degradation: Abuse it and Lose It)*, 22–27
prairies, 11–12, 29–31
The Present (Dillard), 155–159

Quindlen, Anna *(Our Animal Rites)*, 299–301

The Return of Beaver to the Missouri River (Beasley), 122–128
The River Floods (Muir), 136–143
rivers, 114–120, 122–128, 130–134, 136–143, 144–153, 155–159, 161–166, 201–212, 233–234, 238, 266–268, 274, 283–284, 302–312

San Gabriel Mountains, 144–153
sea levels, 2, 37–38, 171–172, 238 263–272, 395–403
Sears, Paul B. *(Soil)*, 8–16
Sears, Peter *(Oil Spill)*, 249
Sharks Under Attack (Abramson), 82–85
Shulman, Seth (and Dan Grossman, *Down in the Dumps)*, 462–470
Sierras, 136–143, 173
Simon, Anne W. *(The Thin Edge)*, 33–40
Sludge (Ashwroth), 274–284
The Solace of Open Spaces (Ehrlich), 48–57
Soil (Sears), 8–16
species endangerment, 63–71, 73–80, 82–85, 86–93, 94–104, 161–166, 178–183, 214–226, 250–256, 274–284, 286–293, 327–333, 335–341, 342–347, 358–362, 363–367, 369–377, 378–381, 383–393, 395–403, 453–460
Steinhart, Peter *(The Water Profiteers)*, 201–212
Stewart, Doug *(Green Giants)*, 363–367
Summer (Hubbell), 323–326
Sunquist, Fiona *(Vast Green Seas Shrivel to Desert)*, 17–20

The Tainted Cup (Levine), 286–293
Tangley, Laura *(The Last Stand?)*, 358–362
The Thin Edge (Simon), 33–40
Think Little (Berry), 417–425

Thomas, Lewis *(Natural Man)*, 105–108
Thoreau, Henry David, 117, *(Concord River)*, 130–134, 406
A Treasure of Complexities (Alcock), 178–183
The Tropical Rain-Forest Setting (Denslow), 395–403
tropics
  forests, 358–362, 395–403
  soil, 14–15
  species, 65–67, 71

Unlikely Harbingers (Milstein), 86–93

Vast Green Seas Shrivel to Desert (Sunquist), 17–20

water, 114–120, 122–128, 130–134, 136–143, 144–153, 155–159, 184–188, 201–212, 231–236, 237–247, 263–272, 274–284, 286–293, 342–347, 383–393

The Water Profiteers (Steinhart), 201–212
Wealth from the Salt Seas (Carson), 237–247
What We Should Do to Stop Spills (Nulty), 257–261
wilderness, 29–31, 41–46, 48–57, 73–80, 86–93, 94–104, 171–176, 178–183, 214–226, 250–256, 286–293, 299–301, 302–312, 314–321, 327–333, 335–341, 342–347, 352–356, 358–362, 363–367, 378–381, 427–438, 453–460
A Wilderness of Light (Madson), 29–31
wildlife, 48–57, 73–80, 82–85, 86–93, 94–104, 161–166, 178–183, 214–226, 250–256, 274–284, 286–293, 299–301, 302–312, 314–321, 323–326, 327–333, 335–341, 342–347, 352–356, 358–362, 453–460